Global Power Shift

Comparative Analysis and Perspectives

Ample empirical evidence points to recent power shifts in multiple areas of international relations taking place between industrialized countries and emerging powers, and between states and non-state actors. Yet there is a dearth of theoretical interpretation and synthesis of these findings, and a growing need for coherent approaches to understand and measure the transformation. This edited series aims to bring together scholars from all major world regions as well as different disciplines in order to discuss and possibly blend their different approaches and provide new frameworks for the understanding of global affairs and the governance of global power shifts.

More information about this series at
http://www.springer.com/series/10201

Maximilian Mayer • Mariana Carpes •
Ruth Knoblich
Editors

The Global Politics of Science and Technology - Vol. 2

Perspectives, Cases and Methods

Springer

Editors
Maximilian Mayer
Center for Global Studies
University of Bonn
Bonn
Germany

Mariana Carpes
GIGA German Institute of Global and Area Studies
Hamburg
Germany

Ruth Knoblich
Institute of Development Research and Development Policy (IEE)
Ruhr-University Bochum
Bochum
Germany

and

Institute of Political Science and Sociology
University of Bonn
Bonn
Germany

ISSN 2198-7343 ISSN 2198-7351 (electronic)
ISBN 978-3-642-55009-6 ISBN 978-3-642-55010-2 (eBook)
DOI 10.1007/978-3-642-55010-2
Springer Heidelberg New York Dordrecht London

Library of Congress Control Number: 2014948207

Printed on acid-free paper

Springer is part of Springer Science+Business Media (www.springer.com)

Acknowledgements

Since we have started working on this project about 2 years ago we have been receiving an enormous amount of support from institutions, colleagues, and friends. Since the initial phase, the support and encouragement especially of JP Singh, Sheila Jasanoff, Johan Eriksson, and Stefan Fritsch were crucial. The discussions with many colleagues at the ISA annual conventions (2011–2013) and elsewhere were a warm welcome for us to pursue this project: gather the diverse research on the global politics of science and technology as a subfield of International Relations. Many more authors than we initially had expected were willing to contribute with their ideas, insights, and comments. Thus, the original plan of editing a book rapidly turned into the two volumes collection we have now the pleasure to present. We hope that this handbook will be not only helpful as an empirical and conceptual overview, a collection of theories, and a toolbox of research frameworks. It also aims at encouraging students and experts in IR and other disciplines to deepen their collaborations and jointly further this important area of study.

We want to thank all contributing authors and interview partners for their persistent engagement and dedication throughout all steps of this project. As there was no opportunity to invite all authors to a conference, the constant exchange and development of ideas between authors, interviewees, reviewers, and the editors—mostly via e-mail conversation—became very fruitful for improving the coherence, comprehensiveness, and structure of this compendium. In this journey, we are especially grateful to those researchers coming from "outside" IR. Their patience, willingness to an extended engagement, and inspiring perspectives were responsible for some of the highlights within this project. Merging further the knowledge of their fields of expertise with the discipline of IR is a worthwhile task. Special thanks go, at this point, also to Peer Schouten who has contributed one interview to the double-volume and offered to cooperate with his website Theory Talks.

Unwavering support of our home institutions was instrumental for realising the intellectual and practical parts of this project. We want to thank the Center for Global Studies, University of Bonn, and especially its student assistants who did a brilliant and precise job on formatting and checking chapters as well as transcribing interviews. Similarly important was the support of the Institute of Development Research and Development Policy, Ruhr-University Bochum, the Institute of Political Science and Sociology, University of Bonn, and GIGA German Institute of Global and Area Studies Hamburg. We are especially thankful to Douglas Howland, Peer Schouten, Anna Agathangelou, and Christian Bueger for their critical engagement with our two introductory texts. We furthermore want to thank Barbara Fess, Marion Kreisel, G. Ramya Prakash and the entire team of the section Economics and Political Science at Springer publishing for their guidance and kind assistance throughout the project. Finally, we are very grateful to our families and close friends for all their support and understanding.

Contents

Volume II: Perspectives, Cases and Methods

Anna M. Agathangelou teaches at York University and is co-editor with Nevzat Soguk of *Arab Revolutions and World Transformations* (Routledge, 2013), co-author with L.H.M. Ling of *Transforming World Politics: From Empire to Multiple Worlds* (Routledge, 2009), and author of *Global Political Economy of Sex: Desire, Violence, and Insecurity in Mediterranean Nation-States* (MacMillan, 2004).

Katharina C. Below M.A. works at the research group 'Global Processes' at the University of Konstanz. She studied political science in Bonn and Istanbul and holds a master's degree from University of Bonn. Her main research interests lie in the field of International Relations Theories and Science and Technology Studies with special emphasis on large-scale research as well as knowledge and innovation.

Mariana Carpes is a research fellow and Ph.D. candidate at the GIGA German Institute of Global and Area Studies where she is working on a historical comparative analysis of the nuclear strategies of Brazil, India, and South Africa. Her main areas of interest are history and theory of IR, nuclear politics, and foreign policy analysis. Recently Mariana has also started to work on strategic partnership between EU and Brazil. In 2012 and 2013 she held presentations on this topic at the European External Action Service (EEAS). She holds a bachelor's degree in History from the Universidade Federal Fluminense (Brazil) and a master's degree in International Relations from the Pontificia Universidade Catolica do Rio de Janeiro.

Sean S. Costigan is an educator and consultant in technology and security. His current work includes an educational initiative in environmental policy and an international project on emerging security challenges. From 2010–2012 he led MIT CogNet at the Massachusetts Institute of Technology prior to which he was Visiting Fellow at the University of Calcutta's Institute of Foreign Policy Studies. He currently teaches information technology and international affairs at the New School University in New York and lectures at The NATO School in Oberammergau, Germany. From 2008 to 2010 he led a US Department of Energy initiative examining energy, economic, and environmental security, prior to which he was Director for Strategic Initiatives, Center for Security Studies, ETH Zurich. Mr. Costigan is Senior

Adviser to the NATO/GCSP/PfPC Emerging Security Challenges study group, co-chair of the Economics of Security Study Group, and Chair of the Editorial Board, Partnership for Peace Consortium of Defense Academies and Security Studies Institutes. His most recent book is *Cyberspaces and Global Affairs*.

Erica Dingman is an Associate Fellow at the World Policy Institute whose work focuses on a broad spectrum of issues facing the Arctic. From the geopolitical theatre of Arctic stakeholders to interested non-Arctic parties, Erica's research turns to the interconnection between climate change, hydrocarbon and mineral exploration, and the political participation of Inuit, one of the Arctic's indigenous groups. She holds a Master's in International Affairs from The New School.

Matthias Englert is a researcher at the Interdisciplinary Research Group Science Technology and Security (IANUS) at Darmstadt University of Technology in Germany where he also earned his Ph.D. in physics. Before joining IANUS he was a postdoctoral fellow at Stanford's Center for International Security and Cooperation (CISAC) from 2009 to 2011. His research interests include nonproliferation, disarmament, arms control, nuclear postures, and warheads; fissile material and production technologies; the civil use of nuclear power and its role in future energy scenarios; and the historical, social, and political aspects of the use of nuclear technologies.

Johan Eriksson is Head of Research at the Swedish Institute of International Affairs and Professor of Political Science at Södertörn University, Stockholm. His research interests are international relations, foreign and security policy, ethnopolitics and regionalism, the politics of expertise, emerging technologies, and territoriality. Eriksson has published more than 70 academic texts, including seven books, and numerous peer-reviewed articles. His most recent articles are "On the Policy Relevance of Grand Theory" (*International Studies Perspectives*, 2013) and "Governance Beyond the Global: Who Controls the Extraterrestrial?" (*Globalizations*, 2013, with Lindy M. Newlove-Eriksson).

Giampiero Giacomello is Assistant Professor of strategic studies with Department of Social and Political Sciences, University of Bologna, Italy. He has held research and teaching positions at Cornell University, Columbia University, the Dickinson College, and the James Madison University. His research interests include strategic theory, cybersecurity, simulation techniques, and foreign policy analysis. He has authored and co-edited five volumes and published in several peer-reviewed journals such as *European Political Science*, *International Studies Review*, *European Security*, *Contemporary Politics*, and *International Political Science Review*.

Benoît Godin is professor at Institut National de la Recherche Scientifique (Montreal, Canada). He holds a DPhil in science policy from Sussex University (UK). He has written extensively on science policy and statistics. Since 2000, he has conducted a project on the history of science and technology statistics from which two books have been published: *Measurement and Statistics on S&T: 1920 to the Present*, London: Routledge, 2005, and *La science sous observation: cent ans de measures sur les scientifiques, 1906–2006*, Québec: Presses de l'Université Laval, 2005. More recently, he has launched a large-scale project on the intellectual history of innovation as a category, from the Renaissance to the present day.

Olof Hallonsten holds a Ph.D. in Research Policy and is currently postdoctoral researcher at the University of Gothenburg, Sweden, and Wuppertal University, Germany. His main research interest is in the history, politics, and sociology of Big Science, and he has published extensively on several aspects of this broad topic.

Anne I. Harrington is an American Political Science Association Congressional Fellow in Washington, DC. Since earning her doctorate from the Department of Political Science at the University of Chicago, she has held postdoctoral fellowships at Stanford University's Center for International Security and Cooperation and at the James Martin Center for Nonproliferation Studies. Her award-winning articles have appeared in the *Nonproliferation Review* and *Millennium: Journal of International Studies*.

Sarah Herweg M.A. holds a master's degree from Free University Berlin. She studied political science in Bonn, Paris, and Berlin. Her research interests include the transnational regulation of knowledge and information as well as the interrelation of Intellectual Property Rights and innovation.

Alejandra S. Kern. Ph.D. in Social Science at FLACSO Argentina (2009). She studied International Relations in Universidad del Salvador, Argentina, and specialised in international cooperation. She is Director of the Research Center on International Cooperation and Development at the National University of San Martín in Argentina.

Ruth Knoblich is a research fellow at the Institute of Development Research and Development Policy (IEE), Ruhr-University Bochum and at the Institute of Political Science and Sociology, University of Bonn; she teaches courses in both institutes and at the South African – German Centre for Development Research, University of the Western Cape. She studied in Bonn and Madrid and holds a master's degree in Political Science, History, and Philosophy from the University of Bonn as well as two basic degrees in Medicine. Her research interests include innovation policies and innovation systems, emerging economies and South-South science and technology cooperation, global politics of intellectual property rights, science and technology in the international political economy, and national policy making in the context of globalisation and global governance.

Joseph P. Lane engages government, academic, and industry sectors in the deliberate and systematic generation of technology-based innovations with beneficial socio-economic impacts, through the related processes of knowledge translation (KT) of scientific discoveries, the technology transfer (TT) of engineering inventions, and commercial transactions (CT) for industrial innovations. The combined strategies of supply push, demand pull, and corporate collaboration have brokered the market launch of over 50 new products, while advancing understanding of the underlying mechanisms through several dozen peer-reviewed articles and hundreds of conference presentations. Mr. Lane directs a national centre funded by the US Department

of Education (http://www.kt4tt.buffalo.edu), as part of an organised research unit (http://www.cat.buffalo.edu) within State University of New York, at Buffalo.
Henriette Langstrup is associate professor at Department of Public Health, Center for Medical Science and Technology at the University of Copenhagen. Henriette's research is particularly concerned with the role of Information and Communication Technologies in shaping identities, relations, and institutions in healthcare.

Maximilian Mayer is program co-chair of STAIR (Science, Technology, Arts and International Relations) a newly established section of the International Studies Association. He studied in Bochum and Shanghai and holds a master's degree in East Asian politics from the Ruhr-University Bochum. He received scholarships from the German National Academic Foundation twice from 2004 to 2007 and from 2009 to 2010. He joined the CGS in October 2009 and has worked there as managing assistant (2009–2013). He leads the CGS research group on "rising knowledge powers". His research interests include the global politics/ policy of science, innovation, and technology; Chinese foreign and energy policy and International Relation theories.

Ruth Müller is a postdoctoral fellow at the Research Policy Group, Lund University, Sweden. Her (inter)disciplinary background is Science & Technology Studies, Sociology, Gender Studies and Molecular Biology. Besides her work on the internalisation of STI, her research focuses on the relations between research policy and scientific work practices. This paper is based on research conducted during her time as a postdoctoral fellow at the Austrian Institute for International Affairs, Vienna.

Annegrete Juul Nielsen is assistant professor at Department of Public Health, Center for Medical Science and Technology Studies at the University of Copenhagen. Currently, her research is concerned with ethnographic explorations of the sociopolitical and organisational implications of transferring health technologies from one health-care system to another.

Susanne Peters is currently teaching international relations for Kent State and Webster University in Geneva. Her research areas cover energy politics, geopolitics of energy, and European security policy. She holds a Doctorate in Political Science from the European University Institute in Florence.

Jayita Sarkar is a Ph.D. candidate in the Department of International History at the Graduate Institute Geneva, where she is writing a dissertation on Franco-Indian nuclear relations and US counter-proliferation efforts during the Cold War. Jay was a Gallatin pre-doctoral fellow at Yale University's MacMillan Center in 2013 and has held visiting positions at the Woodrow Wilson International Center for Scholars in Washington, DC, the Institute for Defence Studies and Analyses in New Delhi, and the Norwegian Institute for Defence Studies in Oslo. An earlier version of this chapter was presented at the 2013 annual meeting of the History of Science Society in Boston, MA.

Niklas Schörnig is a Senior Research Fellow at the Peace Research Institute Frankfurt (PRIF) and received his Ph.D. from the Goethe University Frankfurt, where he was appointed substitute Professor in Winter 2011/2012. In 2012 he received the ZIB-Award from the *Zeitschrift für Internationale Beziehungen* (German Journal of International Relations) for best article 2006–2011. His most recent publications include 'Killer Drones – The Silver Bullet of Democratic Warfare?' (*Security Dialogue* 43 (4), 2012; with Frank Sauer) and '*The Militant Face of Democracy – Liberal Forces for Good*' (Cambridge University Press, 2013, edited with Anna Geis and Harald Müller).

Peer Schouten holds a PhD from the School of Global Studies, University of Gothenburg. He is also editor-in-chief of Theory Talks. Finally, Peer regularly does consultancy work on issues related to the nexus of transnational corporations, security governance, and resource extraction in Sub-Sahara Africa.

Denis Fred Simon serves as Vice-Provost for International Strategic Initiatives at Arizona State University. He also is Foundation Professor of Contemporary Chinese Affairs. Dr. Simon was awarded the China National Friendship Award by former Premier Wen Jiabao in the Great Hall of the People in 2006. He also was made an Honorary Citizen of Dalian by former Mayor Xia Deren in 2008. Currently, he is a member of the American Experts Group as part of the US–China Innovation Dialogue. Dr. Simon has written and spoken extensively about science and technology policy, innovation strategy, and the supply and demand of high-end talent in China. He holds an M.A. in Asian Studies (1975) and a Ph.D. in Political Science (1980) from UC Berkeley.

JP Singh is Professor of Global Affairs and Cultural Studies at George Mason University. Singh has authored four monographs, edited two books, and published dozens of scholarly articles. His book *Globalized Arts: The Entertainment Economy and Cultural Identity* (Columbia, 2011) won the American Political Science Association's award for best book in information technology and politics in 2012. He has advised international organisations such as UNESCO, the World Bank, and the World Trade Organization, played a leadership role in several professional organisations, and served as Editor from 2006 to 2009 and dramatically increased the impact of *Review of Policy Research*, the journal specialising in the politics and policy of science and technology. He holds a Ph.D. in Political Economy and Public Policy from the University of Southern California.

Richard P. Suttmeier is a Professor of Political Science, Emeritus, at the University of Oregon. He has written widely on science and technology development issues in China. Dr. Suttmeier has served as Senior Analyst, Office of Technology Assessment, and Director of the Beijing Office, National Academy of Sciences/Committee on Scholarly Communication with the People's Republic of China, and as a consultant to the World Bank, the UNDP, and the US government. He was appointed a Senior Visiting Scholar at the Institute of Policy and Management of the Chinese Academy of Sciences in 2010. He holds the A.B. from Dartmouth College and the Ph.D. from Indiana University.

Hernán Thomas. Ph.D. in Scientific and Technological Policy at Universidad de Campinas (Brazil). He is currently Director of the Institute of Studies on Science and Technology at the Universidad Nacional de Quilmes. He is also Principal Researcher at the CONICET. His recent research includes the analysis of Strategic Technologies, knowledge-intensive technologies, as well as the development and implementation of Technologies for Social Inclusion.

Krystin Unverzagt B.A. holds a Bachelor's degree from the University of Bonn. She has studied political science in Bonn and Copenhagen. Her main research interest centres on the politics of information and communications technology.

Nina Witjes is a scientific researcher at the Austrian Institute for International Affairs in Vienna and a lecturer at Vienna University, Department for Development Studies. Her studies include history, European cultural studies, international development, and sociology. Nina is currently conducting a research project on the internationalisation of STI with a particular focus on knowledge bases and research landscapes. Her research interests include the (new) geographies of science and the circulation of knowledge.

Werner Zittel studied Physics at the Ludwig-Maximilians-Universität in Munich and received his Ph.D. from Technical University Darmstadt. In 1989 he joined the Ludwig-Bölkow-Systemtechnik GmbH (LBST) in Ottobrunn, a strategy and technology consulting firm for sustainable energy and transport concepts where he works as Senior Scientist. Fields of interest are studies on resource availability, sustainable energy systems, and material flow analysis. He is a founding member of ASPO international in 2001. Since 2011 he is head of the Ludwig-Boelkow-Foundation.

A Toolbox for Studying the Global Politics of Science and Technology

Maximilian Mayer, Mariana Carpes, and Ruth Knoblich

Abstract The growing preeminence of science and technology in today's world no longer fits into most existing analytical frameworks. Material elements, technical instruments, and scientific practices are intertwined with basically every aspect of global politics. Nevertheless, the discipline of International Relations (IR) as a whole tends to conceptualize this topic as an exogenous phenomenon. By adopting the notion of techno-politics, we argue that it is neither sufficient to treat sciences and technologies as external to "social" relations, nor as dominating human behavior and determining political outcomes. We propose rather to open up a middle zone in order to study the intersection of science and technology with international and global affairs. Conceptually, the notion of techno-politics involves two broad sets of approaches: interaction and co-production. This introductory chapter presents the chapters of the volume as examples of how the global politics of science and technology might be studied. As a toolbox of methodological insights, the contributions also point towards pathways for future research that enhances the global politics of science and technology as subfield of IR.

Keywords Technology • Science • IR theory • Methods • Techno-politics • Interdisciplinarity

---

M. Mayer (✉)
CGS, University of Bonn, Bonn, Germany
e-mail: maximilian.mayer@uni-bonn.de

M. Carpes
GIGA, Hamburg, Germany
e-mail: mariana.carpes@giga-hamburg.de

R. Knoblich
IPWS, University of Bonn, Bonn, Germany

IEE, Ruhr-University Bochum, Bochum, Germany
e-mail: ruth.knoblich@rub.de

M. Mayer et al. (eds.), *The Global Politics of Science and Technology - Vol. 2*,
Global Power Shift, DOI 10.1007/978-3-642-55010-2_1,

The growing preeminence of science and technology in today's world no longer fits into most existing analytical frameworks. International security, statehood, warfare, as well as diplomacy, power, and global governance are strongly interwoven with and embedded in material elements, technical instruments, and scientific practices. Yet the discipline of International Relations (IR) as a whole tends to conceptualize the subject matter as exogenous to global politics. Although various studies exist and increasing numbers of IR scholars focus their research on science and technologies, the respective subfield as such has remained unrecognized. The authors of this double volume are taking seriously how sciences and technologies lead to transformations or changes that affect every subject matter of IR in a non-trivial way. Against this backdrop, our introduction to companion volume *The Global Politics of Science and Technology Vol. I* proposes the notion of "techno-politics" as an umbrella term (see Mayer et al. 2014). "Techno-politics" implies an understanding of science and technologies beyond the framework of social constructivism on the one hand, and technological determinism on the other. Neither are technologies simply byproducts or external to "social" relations, nor do sterile technologies merely determine social behavior and political outcomes. Instead, the global politics of science and technology should be placed in a conceptual zone that enhances awareness of the complexity, contingency, hybridity, and dynamism present in issue areas, research subjects, and empirical puzzles that are embedded in it (see Mitchell 2002; Hecht 1998; Fritsch 2011).

Construing a conceptual space around the notion of techno-politics supports the development of the subfield of global science and technology in four crucial ways: It connects existing IR scholarship within a specific academic context; it encourages the development of innovative research methods, analytical concepts and theoretical frameworks; it allows for different metatheoretical commitments[1]; and it functions as a boundary concept that brings IR into a productive conversation with neighboring disciplines engaged with science and technology. Widening this conceptual middle zone by means of integrating existing IR approaches and non-IR research might represent the best way to advance theoretical discourse and innovative empirical research. The study of the global politics of science and technology also sheds fresh light on traditional puzzles within IR. The subfield, in sum, contributes to broader reflections of ontology and methodology, which structure current debates in IR.

This volume advances two ontologically distinct sets of approaches based on the contributions of each chapter. The first set emphasizes *interactions* and keeps a clear distinction between the sphere of "international/global politics" and the sphere of "technology/science". It is structured around the question: *how are preexisting entities, processes, practices, and actors affected and tranformed by sciences and technologies? And how do they respond and adapt?* The second set of approaches focuses on the *productive capacity* of sciences and technologies. In this

[1] For instance, different studies of the global techno-politics may well employ Patrick Jackson's four methodological positions (2011).

case, elements and processes of "international/global politics" are seen as inevitably intermingled and jointly produced with "technology/science". The main research question arising is: in which ways are sciences, technologies, and global affairs co-constitutive and emerge together? This simplified twofold structure reflects recent developments of the subfield. It reinforces links and systematizes comparisons and helps to further the debate about the general theoretical and empirical issues involved. In sum, the contributions of this volume exemplify ways in which the notion of techno-politics can function as an umbrella for various IR approaches.

This introduction briefly elaborates two forms of techno-politics that structure the entire book: *interaction* and *co-production.* It goes on to summarize the individual contributions of the volume, yet clusters them along an alternative logic, including *constructivist studies of technology, assemblage theories, critical and subaltern approaches, approaches to technology-based power*, and conceptualizations of techno-politics that dialogue with *International Political Economy*. In addition to the distinction between interaction and co-production, we thereby return to a conceptual debate about techno-political approaches that has been sketched out in the Volume I. Thus, this introduction not only offers specific perspectives and detailed empirical cases, but also indicates an open-ended conceptual and methodical toolbox that can be applied elsewhere to support future research work. In doing so, this introduction encourages methodological reflections and points out potential future research directions within this subfield.

1 Techno-Politics as Interaction or Co-production

Treating the international or global context of relations between politics, sciences and technology in their own right requires novel theoretical frameworks. As IR has not granted much conceptual space to scientific practices, technical systems, and technological infrastructures and artifacts, the challenge is formidable. It immediately connects to defining debates within the discipline. The studied cases of global science and technology thus also inform theoretical, normative, and methodical developments of IR theory. However, the challenge of such a conceptual zone is that it does "too easily become vague and insubstantial", as Geoffrey L Herrera states. "Placing oneself there should be done with care." (2003: 576). The main difficulty for the study of science and technology is that, at first glance, no "middle ground" exists between technological determinism and social constructivism. The conceptual place of techno-politics in IR does not simply resemble a convergence or complementarities of both, because blending two ontologically exclusive domains of reality is tantamount to conceptual self-contradiction (see Law 2004; Latour 2005). Two ideal-types or social/material dichotomies cannot overlap. Meshing together two similarly inadequate notions—a purely social and a purely technical world—does not produce viable analytical positions.

Table 1 Situating the conceptual middle ground of techno-politics

Conceptual frameworks	Main ontological domain	Mode of explanation	Carriers of agency
Social constructivism	Ideational factors	Intersubjective meaning invested into materials/technology is a matter of interpretation	Social actors, speech, texts, intersubjective practices
Technological determinism	Material factors	Material determination of social formations	Material/ technical systems and structures
Institutional externalism	Rational interactions	Institutional constraints/ conditions and patterns of calculation	Social actors, strategic practices
Techno-politics	Symmetrical treatment of ideational and material factors, hybridization, and post-dualism	Interaction/co-production creates/changes political order/ collectives	Collectives, assemblages, networks, mediating coalitions

However, if one understands techno-politics instead in pragmatic terms–that is, as different real-types of research designs and analytical frameworks–then there is a viable "middle zone". Table 1 illustrates the underlying ideas by comparing the framework techno-politics with technological determinism, social constructivism and institutional externalism (see Mayer et al. 2014). Crucially, a balanced or symmetrical treatment of material and social factors is assumed, while the modes of explanation and location of agency include substantial conceptualizations of technologies and sciences. It is in this sense that the notion of techno-politics directly speaks to the issue of rebalancing ideational and material factors in ontological frameworks (see Wight 2006; Deudney 2007). It implies, however, a fusion or an intimate interconnection and entanglement of realms that IR scholars usually have kept at great distance: the "material" vs. the "political" or the "technical" vs. the "social".

Furthermore, we suggest distinguishing between interactional and co-productive approaches. The elaboration of these varieties, by drawing on Sheila Jasanoff's work (2004: 19–22), helps to carve out a conceptual zone conducive to the study of global techno-politics. The interactional approach explores the interactions, boundary conflicts and entanglements among prior fixed entities, groups or processes such as social practices and technological artifacts. The co-productive approach sheds light on the emergence, co-production and stabilization of new things, groups, or practices such as scientific fields, objects, or technological systems. Accordingly, IR approaches can be divided into two groups.

One group, the interactional, deals with the question of how established practices or principles such as sovereignty, state authority or foreign policy are challenged by technological changes or scientific knowledge (Skolnikoff 1994; Litfin 1999). This literature comprises the examination of the consequences of the digital revolution and information technologies for the regulatory capacity, sovereignty, and legitimacy of nation states.[2] A second major research focus is the impact of the evolving data networks, cyber weapons, and cyber space in general on "national security".[3] A related set of questions concerns the changing conduct of war affected by digital technologies in particular.[4] And also, for instance, how the use of drones impacts the ideas of "just war" (Brunstetter and Braun 2011). In addition, various authors study the responses of international institutions, norms and networks of global governance to the information age.[5] In general, the 2013 issue of the *International Studies Review* (Simmons 2013) documents increasing scholarly interest in information technologies.

Other scholars aim at reframing the concept of "power" to accommodate to the information age (Keohane and Nye 1998; Nye 2004; Mayer 2012; Singh 2013). Duvall and Havercroft (2008), for example, show how imperial sovereignty is newly framed with reference to space-based weapons. At the same time, the internet is seen as instrumental for empowering political transformation and, possibly, democratization processes. Social media have become the crystallization point for both hopes about the influence of "technologies of liberty" and fears of tools of suppression, control, and censorship. Many authoritarian states, meanwhile, have adopted highly sophisticated approaches to cyberspace that employ social networks for selective censoring and surveillance at the same time.[6] Analyzing the interplay between the "technical" and "social", the authors focus on interactions or mutual influence of pre-given concepts such as "the state", "power capacities", "society" or "national security" on the one hand, and certain technologies on the other. Their interactive puzzles put diverging emphasis on the respective sides of the equation.

The second group, co-production, starts from the idea that "the realities of human experience emerge as the joint achievements of scientific, technical and social enterprise" (Jasanoff 2004: 17). Accounting for the emergence of new structures, actors, practices, and identities, it also zooms in at the politics and sites of contestation, resistance and negotiation (Luke 1994; Whatmore 2009; Flyverbom 2011). Sheila Jasanoff's exploration of the "biotechnological empire", Ruth Oldenziel's (2011) study of the critical role of small Islands for the global extension of US military technological networks, and Karen T. Litfin's analysis of space technologies are suggestive: Litfin shows how satellites, as a technology of surveillance, have socialized a global gaze that made possible arms control both

[2] Castells (2011), Mowlana (1997), Drezner and Farrell (2004), Eriksson and Giacomello (2009), Mueller (2010).

[3] For example, Eriksson and Giacomello (2006), Kremer and Müller (2013) see Deibert and Rohozinski (2010).

[4] See Bousquet (2009), Halpin et al. (2006), Edwards (1996), Manjikian (2010), Deibert et al. (2012).

[5] Warkentin and Mingst (2000), Deibert and Crete-Nishihata (2013), Mueller et al. (2013), Costigan and Perry (2012).

[6] See e.g. Boas (2004), Mueller and Chango (2008), Diamond (2010), Anduiza et al. (2012), Hussain and Howard (2013), Deibert et al. (2010).

technically and psychologically. This military-dominated technology, at the same time, has enabled a plethora of non-state actors who utilize images or real-time footage in order to reinforce environmental protection or monitor human rights (Litfin 1999, 2002). Litfin retains a state focus, whereas Jasanoff is unearthing a vast structure that reaches across national boundaries that tie together human bodies, metropolitan lifestyles, peripheral agricultural practices, national security policies, high-tech science, and profit strategies of large multinationals (Jasanoff 2005a). Similarly, Barry (2001) demonstrates that a technological fundament, although getting almost no attention from analysts of the European Union, is particularly critical for the success and trajectory of European integration (see also Schipper and Schot 2011). As exemplified by Laura DeNardis' (2014) analysis of the Internet, these approaches render visible previously hidden infrastructures, networks, and interwoven power relationships.

The connection between nuclear weapons and weather monitoring, in the way it illustrates the contingency of processes of techno-political emergence, figures as a major example of co-production. "The United States' foreign policy of containment conceptualized the cold war as a global struggle, reading all conflicts everywhere in the world as part of the contest for military and ideological advantage (. . .) Military technological change also increased the superpowers' appetites for global weather data and forecasts. (. . .) Tactical nuclear strategy depended on knowing the likely path of fallout clouds and the distances they might travel on the wind." (Edwards 2006: 242–243) Paradoxically, rather than deepening international conflict lines, "geostrategy and technological change", as Paul N. Edwards lucidly details, "aligned military interests with the informational globalism of scientists" (2006: 243). The exchange of monitoring weather data was nearly uninterrupted during the Cold War and collaborative efforts to simulate weather and climate eventually transformed the scientific understanding of the earth (Miller 2004).

The enormous influence of the scientific ensembles—consisting of data collection technologies, computer models and shared expertise, international research bodies—is probably best exemplified by the IPCC. Its ability to assert global policy shifts is not only indicative for the removal of power from states to international scientific organizations (cf. Edwards 2006: 250), but also necessitates a sophisticated articulation of the agency-structure problem in IR. The co-productive set of approaches ties into the growing concern with complexity and hybridity of agential forces within IR and global studies (see Glaser 2003; Youngs 1999; Townes 2012). The interweaving of technology and social practices creates a hybrid world in which the separation in human and non-human no longer makes sense. This is most discernible in the military sphere. Technological innovations in warfare have produced a cyborg-world: semi-autonomous machines and human-machine combinations are mutually connected on the battlefields and with command centers back in military headquarters. As such, the rise of network technologies and virtual realities has brought about profound and unforeseen change and feedback processes in surveillance, warfare, and power projection. In this context, war and peace, the domestic and the foreign, humans and non-humans are becoming increasingly merged and indistinguishable, thereby challenging the common dichotomist assumption of the domestic-foreign divide constitutive to many IR theories.[7]

2 A Toolbox of Perspectives, Cases, and Methods

While the notions of interaction and co-production structure the main sections of the book, individual contributions illustrate ways in which sciences and technologies can be studied from an IR viewpoint. So, as the following chapter-summaries provide an overview about cases and perspectives, they can also serve as an alternative reader's guide. In other words, the chapters are reviewed in accordance with theory-driven groupings that relate to the open-ended set of approaches in Mayer, Carpes, and Knoblich (2014: Vol I). Constituting a methodical collection that exemplifies how empirical materials can be connected to theories, conceptual frameworks, and research puzzles, the approaches include (1) constructivist studies technology, (2) assemblage theories, (3) critical and subaltern approaches, (4) technological power, and (5) international techno-political economy. It is worth to notice that these groupings do not entail mutually exclusive analytical frameworks. As some of the chapters might be placed in more than one group, the toolbox emerging from this exercise consists of different, yet complimentary, sets of approaches that display general applicability beyond the presented empirical cases and puzzles.

While ***constructivist studies of technology*** challenge the instrumentalist idea that social actors can simply attach "meaning" to technological artifacts, they do not see scientific practices, epistemic communities and technical designs as objective and neutral phenomena. Instead, the latter evolve always densely interwoven with the societal and political fabric (Litfin 1994; Adler and Berstein 2005). The task is thus to explore ways in which "objective knowledge" and technical standards are enlaced in cultural traditions, ideological views and partial products of political struggles (see MacKenzie 1993; Bijker 1997; for a microsociological perspective see Knorr Cetina and Bruegger 2002). Looking at how Austrian science, technology and innovation (STI) policy-makers and related stakeholders envision and enact a close relationship between China and Austria in the field of green technologies, Ruth Müller and Nina Witjes address intersections of Science and Technology Studies (STS) and International Relations Theories. Analytically, they draw on the concept of 'sociotechnical imaginaries' (Jasanoff and Kim 2009) that stress the entanglement of visions of social order and technological development with the practical making of science and technology policy. Müller and Witjes combine the notion of sociotechnical imaginaries with Maarten Hajer's (1995) approach to argumentative discourse analysis and, particularly, with his work on

[7] Der Derian (2009), Singer (2009), Stroeken (2013), Wall (2011).

discourse coalitions and the institutionalization of discourse. Based on interviews, participant observation and document analysis, they show how Austrian STI actors connect their national positioning activities to the recent focus on sustainability. Their chapter illustrates the interplay of meaning, representation, and emerging technical practices enmeshed in cultural traditions. They trace how these actors attempt to find common ground for relating to China by drawing on a narrative about Austrian technopolitical history and culture that casts the country as a pioneer of environmental awareness and green technologies. They understand and analyze this process as part of a broader Austrian sociotechnical imaginary in the making that constitutes a situated response to an increasingly globalized STI system.

A constructivist perspective on technopolitics can also provide a critical lens to understand bilateral great powers relations, particularly in turbulent environments and under asymmetric conditions. Richard P. Suttmeier and Denis Fred Simon examine the role of science and technology in the development of US-China Relations as a useful tool of diplomacy. They highlight how changing asymmetries of scientific and technological capabilities alter the ways in which the two countries interact. The authors point to the significant role of ethnic Chinese scientists and engineers in the United States for the establishment of a basis for collaboration. But China's growing capabilities in science and technology as well as tectonic change in international political, economic and security environments have been complicating this tool of diplomacy. While the imperatives for sustainable cooperative science and technology relationships are stronger than ever, both sides exhibit notable institutional deficiencies in governance mechanisms for realizing mutual interests. What Suttmeier and Simon thus show is how much science and technology collaboration is influenced by non-scientific factors, including ethnical networks, national identity and the general public atmosphere.

Similarly, Olof Hallonsten shows how far politics affect the genesis and development of European big science collaboration. The accounts of six cases of European big science projects illustrate that political incoherence, discontinuity and a certain degree of arbitrariness delicately impacted each of them in a different way. European collaboration in big science is built on ad hoc solutions and misses a coherent political framework and common regulatory standards. In comparison to the United States, where big science facilities are almost exclusively operating within the National Laboratories system, the lack of coherence in European collaboration made these projects prone to reflecting general trends in European politics at the times of their realization. They were therefore vulnerable to uncertainties indicated by repeated ruptures and break-offs in international science collaboration.

Johan Eriksson and Giampiero Giacomello analyze the threats embedded in the rapid expansion and increasing accessibility to the Internet. As they argue, the Internet has long become a topic of political (and military) interest, shifting from the technical managerial realm to the attention of foreign and security policy (cf. Dunn Cavelty 2013). The analysis of the cyberspace is challenging to the extent that not only the state, but also non-state actors play an important role in securing this realm. Furthermore, it questions the traditional threat perception that would place the enemy in the "real" world. The study of cyberspace and cybersecurity subvert these logics, therefore calling for new analytical tools. To address the topic, the authors combine constructivist theory with content analysis—that is quantitative method focused on key words searching. They argue, on the one hand, that content analysis is a pertinent instrument for the study and practices of cybersecurity, considering that a significant amount of information on the Web is text. On the other hand, constructivism is particularly useful for uncovering the text-based meanings, identities, symbols and ideas that largely constitute the "virtual" world. Together, theory and a set of methods allow for an assessment of how reality is perceived and discourses (of cybersecurity) are built within cyberspace.

Assemblage theories emphasize complexity and make global actor-networks a prime object of study (Acuto and Curtis 2014). The starting point here is the idea that "politics" themselves emerge from an empirical density in which "social" and "material" factors cannot easily and a priori be distinguished (Latour 2005). This insight is exemplified by the extent to which inter-state cooperation, scientific networks, or the spread of technologies is dependent upon and embedded in large-scale infrastructures. The latter facilitate production, trade, finance, communication, surveillance, and defense systems which are far more complex, multi-sited, and interconnected than any state-centric framework of social collective action can cope with.

In this vein, Alejandra Kern and Hernán Thomas analyze international technological cooperation processes between developing and developed countries. Similar to Suttmeier's and Simon's study, an asymmetric relationship is explored. They present an analytical framework based on two notions: First, they refer to Robert Cox's (1987) reflexive approach claiming that production and dissemination of technology is a result of a social process involving power relations and constituting core-periphery divides. Second, they define cooperation processes as socio-technical trajectories drawing on Michel Callon's (1992) definition of "techno-economic networks" (TEN). Adding a political pole (P) at the center of this network, TPEN frames their analysis of Argentine-German cooperation in Informationa and communication technologies. Thereby, they connect approaches from IPE and Social Studies of Science and Technology in order to build up an analytical framework that allows for the analysis of varied TPEN configurations and the path of cooperation processes. The case of German-Argentine cooperation in ICTs illuminates how the structure of international order influenced the way ICTs were politicized in the 1990s—differently in both countries. The case furthermore illustrates that the evolution and the results of scientific and technological cooperation is largely explained by the ways research teams interact.

Drawing on Actor-network theory (ANT) and STS, Annegrete Juul Nielsen and Henriette Langstrup focus on the role and effects of technology in global encounters. The chapter evolves in a case study of the Danish-Indonesian partnership to implement Danish software for diabetes management in Indonesian primary care. It is structured around fieldwork done in Denmark and Jakarta, Indonesia, comprising interviews and analysis of primary documents. Methodologically, they propose an ethnographic study of health technologies—more specifically, global diabetes encounters as an empirical phenomenon. By "encounter" they mean concrete and potentially frictional contacts between specific human and non-human or technological entities meeting and creating relations that transgress national boundaries. The authors claim to identify a process of co-constitution between global encounters and technology. Such a process, they argue, also produces frictions that have to be observed for their insights into the conditions for the intersection of Global Health and International Relations.

Peer Schouten explores technological infrastructures in postcolonial Africa. Combining detailed fieldwork, interviews, and multi-site case studies, the *problematique of absences*—such as roads, statistical tools, and government offices—is addressed against the background of the historical state formation of the Democratic Republic of Congo. Theoretically, this chapter points to an analytical gap emerging at the overlap of political science and science and technology studies. Where political science has been negligent at dealing with the importance of technological infrastructures for the constitution of governmental power, science and technology studies have hitherto been less attentive to situations in which such 'infrastructures of rule' are *absent*. Using the "chukudu"—an improvised wooden transport utility that allows Congolese to go about despite decrepit roads—Schouten's study shows the significance of missing infrastructures, suggesting ways in which IR could draw on insights from Michael Mann's (2008) work about technological infrastructures and statehood. Assemblage theory hence enables novel insights in the importance of technological asymmetries in global power-relations in postcolonial contexts. In terms of method, this chapter combines "thick description" with critical reflections about axiomatic notions underpinning both political theory and IR.

Critical and subaltern approaches stress the technological dimensions of core IR themes such as security, warfare, order, and anarchy that are otherwise hidden and remain unacknowledged.

Anna Agathangelou offers a triangular assessment of global politics that encompasses the IR debate on order/anarchy, and recently emerging biomedical technologies. Building upon a integrative framework of postcolonial studies and STS, she outlines the fact that the scientific and the social/political are always already inside one another and extends this premise to the analysis of the international. The article draws on life science and bio-capital, calling attention to the fact that the moral expectation for capital is shifting our understanding of healthiness and value in their relation to corporeality. In terms of tools for further analyses, one challenging lesson to be taken from Agathangelou's article is the link that she proposes between bioscience and international relations; or as she phrases it, the corporeal and the architecture of the international that can be studied through popular literature and imaginative theatre, complimentary to empirical analysis of transnational techno-economical interactions. The chapter thus offers new avenues to critically analyze and bridge the sides of an apparent divide, that is, the global commercialization of medical science, genetechnology, and the international politics (of bodies).

The chapter by Niklas Schörnig asks why military robots and drones are specifically appealing to Western democracies and where this appeal might come from vis-à-vis alternative weapon systems. When it comes to military robotics in general, the "technological imperative" seems to be a valid and powerful explanation for the worldwide proliferation of drone technology. What is missing from a realist explanation, however, is why drones have become so popular in the first place and especially why Western democracies have been implementing and fostering their development from the very beginning. Referring to liberal theory, the text argues that unmanned military systems offer Western democracies a unique opportunity to implement what has been termed the "New Western Way of War". Instead of viewing "technology push" and "mission pull" as two contradicting explanations of technology development, Schörnig suggests that a combination of both offers a comprehensive picture. While warfare with minimized risks for soldiers and reduced civilian casualties seems to be a worthwhile goal, it demands further attention whether there are less pleasant backlashes of these technologies on democratic norms and institutions. Methodically, this implies a focus on the links between technical details, the evolution of norms, public discussion and actual practices of warfare.

A critical view might study technologies and scientific knowledge production as a *form of ideology*. For example, Susanne Peters and Peter Zittel stress the consequences of the illusive narrative of optimism—the belief in the limitless power of technology that has dominated the Western world since modernity. They use the "tight oil revolution" as a case study showing, on the one hand, that natural factors such as steep depletion rates are alarming symptoms of its lack of sustainability. On the other, they demonstrate that the emergence of this revolution is only partially linked to the "power of technology"; it is also linked to deregulation and Wall Street economics. The authors claim that there are insurmountable limits to technology's "domination of nature" and reflect how far this recognition has to be internalized into IR. They employ their example in order to demonstrate that social scientists and IR scholars in particular need to overcome the discipline's traditional inherent dichotomy between nature and society, and develop theories and concepts explaining the dynamics between energy use and political outcomes. For a peaceful transition from the fossil to the post-fossil world, they argue, we urgently need to develop models, new ideas and new comprehensive approaches integrating nature and society—and also insights from economics, geography, physics, and geology—in order to foster the feasibility of a post-carbon life style.

Jayita Sarkar's analysis of the trajectory of the Indian atomic energy program emphasizes the *irrationality* hidden behind technological ideology. Drawing on primary sources to build a historical narrative of the curiosity-indigeneity paradox, as she calls it, Sarkar analyses the 1951 Franco-Indian nuclear cooperation agreement and India's thorium research and development enterprise. She argues that such paradox is the discursive justification of expensive scientific and technological projects without any tangible impact upon national development. On the one hand, the task of national development in resource-scarce India necessitated scientific projects that made immediate contributions to the socio-economic improvement of the lives of its citizens. On the other hand, the discourse of indigeneity, by underlining the need to establish a domestic scientific knowledge base, justified curiosity-driven research only for the sake of scientific progress itself. The idea that modern technology becomes a mere fetish speaks to earlier insights from IR how weapons are often irrationally turned into instruments national prestige.

Approaches to ***technology-based power*** raise the diversity, unintended consequences, and paradoxes of techno-political power shifts, influence and control.

With the help of proxy indicators, Katharina Below, Sarah Herweg, Ruth Knoblich, and Krystin Unverzagt show that asymmetries persist in cross-country knowledge distribution. Reframing an instrumental understanding of power with the notion of "knowledge power," they make reference to structural approaches such as the concept of structural power by Susan Strange (1988). Structural power impacts the range of choice and the opportunity to develop through network functionalities. These networks, which make up for the global knowledge structure, are assumed to provide an inert environment in which states face restraints on development and their way up to leading positions. As the asymmetric distribution of knowledge capacities derives from past differences in foundational knowledge indicators, power shifts cannot appear out of a sudden, and established powers are likely to prevail. Methodically, this chapter provides a set of indicators, elaborating on options to quantitatively measure global structures.

According to JP Singh, large-scale and dense communicative interactions among multiple actors are creating new and shared understandings in global politics that, in turn, require efforts to re-conceptualize power. The existing "instrumental" conceptions of power tend to underestimate the role of communication in these politics. Singh argues that the increasing intensity of debates on cultural identity exemplify the new meanings that arise in global politics as a result of *large-scale communicative interactions*. Thereby, national identity is not replaced but supplemented with other forms of cultural identity in a networked world. The chapter develops the concept of meta-power in order to capture the emergence of these new. Singh suggests several locations to study the involved techno-politics, ranging from national debates about identity, to the activity of international organizations, and governmental negotiations about intellectual property and the trade of art and cultural products. In addition, a sort of new cosmopolitanism that embodies meaning beyond territorially bounded social formations might emerge due to the transnational network communication practices.

From a different angle, Anna Harrington and Matthias Englert employ concepts from STS to assess the nuclear choices of non-nuclear weapon states. Drawing on the literature about nuclear (non)proliferation and on technical knowledge about uranium enrichment centrifuges, they ask how much nuclear power is enough for a country to perceive itself as safe. Their core argument is that states do proliferate when the benefits outrange the costs of going nuclear. Subsequently, they question a well-accepted discourse in IR about absolute values attributed to nuclear weapons. They argue instead that nuclear technologies are "inherently ambivalent" because different actors can attribute different meanings to them. According to the authors, the deterrence aspect of nuclear proliferation can be achieved without proliferation in itself. In this sense, they offer new lenses to assess deterrence in the case of non-nuclear weapon states: instead of real arsenals, latency—the capability to assemble nuclear weapons in a relative short period of time—can suffice to produce the desired deterrence. The bonus of a virtual arsenal lies in not having to pay the financial and political costs of proliferation.

Infusing conceptualizations of interaction or co-production into theories of ***International Techno-political Economy*** leads to approaches that zoom in at the technological micro-practices and macro-processes that perform and stabilize "the world economy". This conceptual move does not only shed light on how shifting technological monopolies, such as over the collection of big data, produce new landscapes of power and influence (see Nye 2004; Andrejevic 2014). It also leads to a critical inquiry of the evolution of measurments and standards that inevitably underpin our knowledge of economic growth and development.

For instance, stressing the significance of statistics and measurement of science and technology for policymakers and vice versa, Benoît Godin and Joseph Lane illustrate how official statistics of a national scope were transferred into the international arena after World War II when science and technology indicators particularly became crucial for the comparison of European economies and the US. Organizations such as the OECD started measuring cross-country innovation performances and established a standardized set of science and technology indicators that allowed for international comparison, focusing on competitiveness. While the OECD reoriented its whole statistical program in science and technology in the 1990s in order to adapt to globalization processes, the established indicators did not undergo a further restructuring, but were just re-interpreted. The authors expose how far indicators of science, technology and innovation are still driven by competition categories and do not measure what policymakers claim technological and scientific development should serve. Accordingly, methodologies and outputs are necessary to create new nets of social and economic wealth. One example could be the indicators measuring innovation in terms of commercial exploitation and industrial production. Taking into account that statistics today remain primarily national in scope, the authors question whether one could really talk about an international dimension to science and technology for consideration by governmental policymaking bodies—as there is for climate change.

Sean Costigan and ERICA Dingman ask how the immense amount of scientific data that is produced worldwide could be used to improve economic and environmental policy-making on a regional level. They raise the questions of how to make sense of it, and also how to have access to it, considering that an enormous amount of valuable data is not open for free access. Offering a policy-oriented study, the authors draw on the case of the Arctic, an emerging region, to debate the role of technology, environment, energy and access to knowledge in the context of the regionalization of policy-making. They call for improving "access to and share of raw data as well as access to method used to produce the data". They outline that the field of climate policy evidences a ineffective fragmentation of data and the existence of a divide between production of knowledge and access to this knowledge for the sake of policy-making and spreading of information. They suggest that the lack of a "neutral space" to collect, share and communicate needs to be overcome. Given the growing importance of big data, this chapter stresses the technopolitical imperative of enabling efficient policy-making and, thus, to improve the results for involved participants in a far-flung region such as the High North that is especially characterized by a volatile environmental situation.

3 Conclusion

The subfield of global politics of science and technology entails a great variety of approaches. This introduction has focused on two different "techno-political" perspectives, interaction and co-production, indicating that they offer a productive theoretical starting point for research. These perspectives open up novel analytical understandings of classical problems of international politics. In particular, the summarized examples illustrate how scientific practices and technologies of all kinds are both interacting with and constitutive to various levels, units, actors, structures, and spatial dimensions. The "state", the "region", the "international", and the "transnational" are maintained through and changed by technologies. Hence, detailed empirical accounts add a diversity of actors, relationships, and collectives alongside classically tailored case studies and framings of analysis. At the moment, the criteria for knowledge production in this subfield remain open-ended. On the most fundamental level, explorative research in this field is driven by the idea that sciences and technologies cannot be treated merely as by-products or external to "actual" politics (see Table 1). Their limits are not easily definable by means of deduction or a priori categorical delimitation. Instead the characteristic quality of sciences and technologies lies in exploring and reconstructing globally stretching interlinkages and connectivities (see Adger et al. 2008; Linklater 2009; Urry 2003). In doing so, authors emphasize techno-political processes of interaction and co-production, bringing into the limelight the multidimensional and extensive, yet often highly fragile, role of material agency to which societal patterns and meaning are tied (Jasanoff 2004, 2005b; Barry 2013; Connolly 2013). The latter constitutes a deeply political issue that remains usually hidden by conceptual and academic demarcations and compartmentalization which unfortunately are still constitutive for most social theories and IR approaches. Finally, it is not by coincidence that the chapters construct solid bridges to several neighboring disciplines, as serious interdisciplinary engagement and mutual learning are crucial for the methodical and theoretical innovations driving this research field.

Acknowledgements The authors are thankful to Douglas Howland and Anna Agathangelou for their insightful comments and helpful suggestions that were instrumental for writing this chapter.

References

Acuto, M., & Curtis, S. (2014). *Reassembling international theory: Assemblage thinking and international relations*. London: Palgrave Pivot.

Adger, W. N., Eakin, H., & Winkels, A. (2008). Nested and teleconnected vulnerabilities to environmental change. *Frontiers in Ecology and the Environment, 7*(3), 150–57.

Adler, E., & Steven, B. (2005). Knowledge in power: The epistemic construction of global governance. In M. Barnett & R. Duvall (Eds.), *Power and global governance* (pp. 294–318). Cambridge, MA: Cambridge University Press.

Andrejevic, M. (2014). Big data, big questions "The Big Data Divide". *International Journal of Communication, 8*. http://ijoc.org/index.php/ijoc/article/view/2161/1163.

Anduiza, E., Jensen, M. J., & Jorba, L. (Eds.). (2012). *Digital media and political engagement worldwide: A comparative study*. Cambridge: Cambridge University Press.

Barry, A. (2001). *Political machines: Governing a technological society*. London: Athlone Press.

Barry, A. (2013). *Material politics: Disputes along the pipeline*. Oxford: Wiley-Blackwell.

Bijker, W. E. (1997). *Of bicycles, bakelites, and bulbs: Toward a theory of sociotechnical change*. Cambridge: MIT Press.

Boas, T. C. (2004). Weaving the authoritarian web. *Current History, 103*, 438–443.

Bousquet, A. (2009). *The scientific way of war: Order and chaos on the battlefields of modernity*. London: Hurst & Company.

Brunstetter, D., & Braun, M. (2011). The implications of drones on the just war tradition. *Ethics and International Affairs, 25*(3), 337–58.

Callon, M. (1992). The dynamics of techno-economic networks. In P. Saviotti, R. Coombs, & V. Walsh (Eds.), *Technological change and company strategies* (pp. 72–102). London: Academic Press.

Castells, M. (2011). *The rise of the network society: The information age: Economy, society, and culture* (Vol. 1). New York: Wiley.

Connolly, W. E. (2013). The 'new materialism' and the fragility of things. *Millennium-Journal of International Studies, 41*(3), 399–412.

Costigan, S. J., & Perry, J. (2012). *Cyberspaces and global affairs*. Burlington: Ashgate.

Cox, R. W. (1987). *Production, power and world order: Social forces in the making of history*. New York: Columbia University Press.

Deibert, R. J., & Crete-Nishihata, M. (2012). Global governance and the spread of cyberspace controls. *Global Governance: A Review of Multilateralism and International Organizations, 18*(3), 339–61.

Deibert, R. J., & Rohozinski, R. (2010). Risking security: Policies and paradoxes of cyberspace security. *International Political Sociology, 4*(1), 15–32.

Deibert, R. J., Rohozinski, R., & Crete-Nishihata, M. (2012). Cyclones in cyberspace: Information shaping and denial in the 2008 Russia–Georgia war. *Security Dialogue, 43*(1), 3–24.

Deibert, R., Palfrey, J., Rohozinski, R., & Zittrain, J. (Eds.). (2010). *Access controlled: The shaping of power, rights, and rule in cyberspace*. Cambridge, MA: MIT Press.

DeNardis, L. (2014). *The global war for Internet governance*. New Haven, CT: Yale University Press.

Der Derian, J. (2009). *Virtuous war: Mapping the military-industrial-media-entertainment-network*. New York: Routledge.

Deudney, D. (2007). *Bounding power: Republican security theory from the polis to the global village*. Princeton, NJ: Princeton University Press.

Diamond, L. (2010). Liberation technology. *Journal of Democracy, 21*(3), 69–83.

Drezner, D. W., & Farrell, H. (2004). *The power and politics of blogs*. Paper presented at the Annual of the American Political Science Association.

Dunn Cavelty, M. (2013). From cyber-bombs to political fallout: Threat representation with an impact in the cyber-security discourse. *International Studies Review, 15*(1), 105–122.

Duvall, R., & Havercroft, J. (2008). Taking sovereignty out of this world: Space weapons and empire of the future. *Review of International Studies, 34*(4), 755–75.

Edwards, P. N. (1996). *The closed world: Computers and the politics of discourse in Cold War*. Cambridge, MA: MIT Press.

Edwards, P. N. (2006). Meteorology as infrastructural globalism. *Osiris, 21*(1), 229–50.

Eriksson, J., & Giacomello, G. (2006). The information revolution, security, and international relations: (IR) relevant theory? *International Political Science Review, 27*(3), 221–44.

Eriksson, J., & Giacomello, G. (2009). Who controls the internet? Beyond the obstinacy or obsolescence of the State. *International Studies Review, 11*(1), 205–30.

Flyverbom, M. (2011). *The power of networks: Organizing the global politics of the internet*. Cheltenham: Edward Elgar.

Fritsch, S. (2011). Technology and global affairs. *International Studies Perspectives, 12*(1), 27–45.

Glaser, C. L. (2003). Structural realism in a more complex world. *Review of International Studies, 29*(3), 403–14.

Hajer, M. A. (1995). Discourse coalitions and the institutionalization of practice: The case of acid rain in Great Britain. In F. Fischer & J. Forester (Eds.), *The argumentative turn in policy analysis and planning* (pp. 43–67). Durham: Duke University Press.

Halpin, E. F., Trevorrow, P., Webb, D., & Wright, S. (2006). *Cyberwar, netwar and the revolution in military affairs*. New York: Palgrave Macmillan.

Hecht, G. (1998). *The radiance of France: Nuclear power and national identity after World War II*. Cambridge, MA: MIT Press.

Herrera, G. L. (2003). Technology and international systems. *Millennium-Journal of International Studies, 32*(3), 559–93.

Hussain, M. M., & Howard, P. N. (Eds.). (2013). *State power 2.0: Authoritarian entrenchment and political engagement worldwide*. Burlington: Ashgat.

Jackson, P. T. (2011). *The conduct of inquiry in international relations: Philosophy of science and its implications for the study of world politics*. London: Routledge.

Jasanoff, S. (2004). Ordering knowledge, ordering society. In S. Jasanoff (Ed.), *States of knowledge: The co-production of science and social order* (pp. 13–45). London: Routledge.

Jasanoff, S. (2005a). In the democracies of DNA: Ontological uncertainty and political order in three states. *New Genetics and Society, 24*(2), 139–56.

Jasanoff, S. (2005b). *Designs on nature: Science and democracy in Europe and the United States*. Princeton, NJ: Princeton University Press.

Jasanoff, S., & Kim, S.-H. (2009). Containing the atom: Sociotechnical imaginaries and nuclear power in the United States and South Korea. *Minerva, 47*(2), 119–46.

Keohane, R. O., & Nye, J. S. (1998). Power and interdependence in the information age. *Foreign Affairs, 77*(5), 81–95.

Knorr Cetina, K., & Bruegger, U. (2002). Global microstructures: The virtual societies of financial markets. *American Journal of Sociology, 107*(4), 905–50.

Kremer, J.-F., & Müller, B. (2013). *Cyberspace and international relations: Theory, prospects and challenges*. Berlin: Springer.

Latour, B. (2005). *Reassembling the social-an introduction to actor-network-theory*. New York: Oxford University Press.

Law, J. (2004). *After method: Mess in social science research*. London: Routledge.

Linklater, A. (2009). Human interconnectedness. *International Relations, 23*(3), 481–497.

Litfin, K. T. (1994). *Ozon discourses: Science and politics in global environmental cooperation*. New York: Columbia University Press.

Litfin, K. T. (1999). The status of the statistical state: Satellites and the diffusion of epistemic sovereignty 1. *Global Society: Journal of Interdisciplinary International Relations, 13*(1), 95–116.

Litfin, K. T. (2002). Public eyes: Satellite imagery, the globalization of transparency, and new networks of surveillance. In J. N. Rosenau & J. P. Singh (Eds.), *Information technologies and global politics: The changing scope of power and governance* (pp. 65–88). Albany: State University of New York Press.

Luke, T. W. (1994). Placing power/siting space: The politics of global and local in the new world order. *Environment and Planning D, 12*, 613–628.

MacKenzie, D. A. (1993). *Inventing accuracy: An historical sociology of nuclear missile guidance*. Cambridge, MA: MIT Press.

Manjikian, M. M. (2010). From global village to virtual battlespace: The colonizing of the internet and the extension of realpolitik. *International Studies Quarterly, 54*(2), 381–401.
Mann, M. (2008). Infrastructural power revisited. *Studies in Comparative International Development, 43*(3), 355–365.
Mayer, M. (2012). Exploring China's rise as knowledge power. In E. Fels, K. Harmat, & J.-F. Kremer (Eds.), *Power in the 21st century: International security and international political economy in a changing world* (pp. 287–311). Berlin: Springer.
Mayer, M., Carpes, M., & Knoblich, R. (2014). The global politics of science and technology: An introduction. In M. Mayer, M. Carpes, & R. Knoblich (Eds.), *International relations and the global politics of science and technology* (Approaches, concepts, and interdisciplinary conversations, Vol. I). Heidelberg: Springer.
Miller, C. A. (2004). Climate science and the making of a global political order. In S. Jasanoff (Ed.), *States of knowledge: The co-production of science and social order* (pp. 46–66). London: Routledge.
Mitchell, T. (2002). *Rule of experts: Egypt, techno-politics, modernity*. Berkeley: University of California Press.
Mowlana, H. (1997). *Global information and world communication: New frontiers in international relations*. London: Sage.
Mueller, M. L. (2010). *Networks and states: The global politics of Internet governance*. Cambridge, MA: MIT Press.
Mueller, M., & Chango, M. (2008). Disrupting global governance: The Internet whois service, ICANN, and privacy. *Journal of Information Technology and Politics, 5*(3), 303–25.
Mueller, M., Schmidt, A., & Kuerbis, B. (2013). Internet security and networked governance in international relations. *International Studies Review, 15*(1), 86–104.
Nye, J. S. (2004). *Soft power: The means to success in world politics*. New York: Public Affairs.
Nye, J. S. (2004). *Power in the global information age: From realism to globalization*. Abingdon, NY: Routledge.
Oldenziel, R. (2011). Islands: U.S. as networked empire. In G. Hecht (Ed.), *Entangled geographies. Empire and technopolitics in the global Cold War* (pp. 13–42). Cambridge, MA: MIT Press.
Schipper, F., & Schot, J. (2011). Infrastructural Europeanism, or the project of building Europe on infrastructures: An introduction. *History and Technology An International Journal, 27*(3), 245–264.
Simmons, B. A. (2013). Preface: International relationships in the information age. *International Studies Review, 15*(1), 1–4.
Singer, P. W. (2009). *Wired for war: The robotics revolution and conflict in the twenty-first century*. New York: The Penguin Press.
Singh, J. P. (2013). Information technologies, meta-power, and transformations in global politics. *International Studies Review, 15*(1), 5–29.
Skolnikoff, E. B. (1994). *The elusive transformation: Science, technology and the evolution of international politics*. Princeton: Princeton University Press.
Strange, S. (1988). *State and market*. London: Bloomsbury Academic.
Stroeken, K. (Ed.). (2013). *War, technology, anthropology*. New York: Berghahn Books.
Townes, M. (2012). The spread of TCP/IP: How the Internet became the Internet. *Millennium: Journal of International Studies, 41*(1), 43–64.
Urry, J. (2003). *Global complexity*. Cambridge: Polity.
Wall, T. (2011). Surveillance and violence from afar: The politics of drones and liminal securityscapes. *Theoretical Criminology, 15*(3), 239–54.
Warkentin, C., & Mingst, K. (2000). International institutions, the state, and global civil society in the age of the World Wide Web. *Global Governance, 6*, 237.
Whatmore, S. J. (2009). Mapping knowledge controversies: Science, democracy and the redistribution of expertise. *Progress in Human Geography, 33*(5), 587–98.
Wight, C. (2006). *Agents, structures and international relations: Politics as ontology*. Cambridge: Cambridge University Press.
Youngs, G. (1999). *International relations in a global age: A conceptual challenge*. Cambridge: Polity Press.

Part I
Interactional Techno-Politics

A Bullock Cart on Nuclear-Powered Wheels: Nuclear Science, Indigeneity and the National Development Narrative in India

Jayita Sarkar

Abstract The Indian atomic energy program has been historically represented by its elites as a significant step towards a science-led national development project that would emancipate the country from its backwardness. While no national nuclear program is indigenous, the discourse of indigeneity runs deep in India. On one hand, the task of national development in the resource-scarce country necessitated scientific projects that made an immediate contribution to the socio-economic upliftment of the lives of its citizens. On the other hand, the discourse of indigeneity, by underlining the need to establish an indigenous scientific knowledgebase, justified curiosity-driven research or 'science for science's sake.' The trajectory of the Indian atomic energy program manifests a curiosity-indigeneity paradox, which permits the justification of expensive scientific and technological projects without any tangible impact upon national development. This paper analyzes this paradox with respect to (a) the 1951 Franco-Indian nuclear cooperation agreement, and (b) India's thorium research and development enterprise.

Keywords India • France • Beryllium • Thorium R&D • Indigeneity • Curiosity science • Blackett

1 Introduction

Jawaharlal Nehru, India's first prime minister wrote in *The Discovery of India* that 'the scientific approach and temper are, or should be, a way of life... It is the temper of a free man' (Nehru 1948). As independence neared, the Indian nationalist leadership increasingly felt that India could not afford the luxury of pure scientific

J. Sarkar (✉)
Graduate Institute Geneva, Geneva, Switzerland
e-mail: jayita.sarkar@graduateinstitute.ch

M. Mayer et al. (eds.), *The Global Politics of Science and Technology - Vol. 2*,
Global Power Shift, DOI 10.1007/978-3-642-55010-2_2,

research—Indian science had to serve industry (Prakash 1999: 193) and that Indian scientists therefore must strive for the benefit of the community instead of their individual quest for knowledge (Abraham 1998). Science, as the antidote to India's underdevelopment, would be promoting industrialization, and modernizing the economy and the society.

Indigeneity as a discourse was omnipresent in the national development narrative of India, which in the economic sphere took the form of import-substituting industrialization. Independent India would consume only products that it itself could manufacture, instead of importing commodities from foreign countries. Dependence on the outside world for basic commodities was deemed by the Indian political elites, as equivalent to a compromised national sovereignty. Yet, the discourse of indigeneity was paradoxical in the case of atomic energy, as no national nuclear programs till date have been entirely indigenous. The discovery of nuclear fission in 1939 was a continental effort by physicists and chemists in Europe. The Manhattan Project—that produced the world's first atom bombs, was a joint endeavour of the United States, the United Kingdom and Canada, which could not have attained its goals without the efforts of the numerous European physicists who crossed the Atlantic fleeing Nazi persecution. Yet, as the nuclear age progressed, countries strived to engrave their national identities on the nuclear artifacts within their territories, whether reactors or weapons systems. In 1956, when the swimming pool-type research reactor Apsara attained criticality in Trombay, it was the only reactor in Asia outside the Soviet Union. In his speech dedicating it to the nation, Nehru praised the advancement of Indian science and the remarkable feat of Indian scientists, ignoring the fact that it was almost entirely based on British design, and could not have been achieved without the active help of British scientists (Abraham 2009: 110–111).[1]

The discourse of indigeneity with respect to nuclear programs is not unique to India. In France, the French quest to preserve the *francité* or Frenchness of the nuclear program led to a conflict in the 1960s between the French atomic energy commission, *Commissariat à l'énergie atomique*, (CEA) and the national electricity service provider, *Electricité de France* (Hecht 1998). Convinced that French national identity was at stake, the CEA lobbied for retaining the 'French' gas-graphite reactors, while the EDF wanted France to adopt the more economic 'American' light water uranium reactors. Another fitting contemporary example is Iran, which is facing sanctions from the United States and the European Union, on suspicions that it is nurturing a nuclear weapons program. The only operational nuclear reactor in Iran is the aging Bushehr VVER-1000 reactor, for which construction began in the mid-1970s, with West German help. In the 1990s, this reactor received technological assistance from Russia. Yet, Iran calls uranium enrichment

[1] W.W. Harpur of the UK Atomic Energy Research Establishment at Harwell wrote to A.I. Scott of the Collaboration Branch that "it is rather galling that they (the Indians) claim it was done entirely by themselves" and added that "this claim was made before and we appear to have swallowed it and said nothing." Letter from W.W. Harpur to A.I. Scott, 30 August 1956, AB16/1535, Co-operation with India 1954–1957, UK National Archives, Kew.

from this reactor an indigenous feat, and a major sign of success of Iranian science (Vaez and Sadjadpour 2013). Despite the high economic costs of domestic uranium enrichment, especially with stringent international sanctions, Iran takes high pride in it as it stokes its technology-driven nationalism.

Technology-driven nationalism, which is closely related to the discourse of indigeniety, is not specific to contemporary times. According to historian Michael Adas, while in the sixteenth and seventeenth centuries, European travelers viewed their Christian faith as the source of their superiority over non-Western populations, from the early nineteenth century onward, science and technology became the measure of universal modernity with Europe in the lead (Adas 1989: 71). This attitude was reflected in the British colonial enterprise in India, which tended to represent the natives as non-scientific and backward. In response and protest, nationalist Indian elites attempted to portray their country's cultural underpinnings as progressive and modern, and tried to represent Indian traditions as scientific and rational.[2] Nehru's conviction that scientific temper is the legitimate attribute of the free man thus played an important role in his nation-building project of transforming former colonial subjects into citizens of a free country (Prakash 1999:118).

Since power politics is at the core of international relations, defence technology offers higher allure to states than technology that serves only peaceful ends. Owing to its dual-use nature, nuclear technology therefore, occupies an important place in the international power structure of states. Since Indian political elites have always prioritized India's right to keep its option open for the bomb, the indigeneity discourse in India, is closely tied to the question of safeguards, namely, the international measures instituted by the International Atomic Energy Agency (IAEA), to prevent the diversion of nuclear technology and materials from peaceful to military uses.

This chapter studies the discourse of indigeneity in the Indian nuclear program with particular reference to (a) the 1951 Franco-Indian nuclear cooperation agreement, and (b) India's thorium research and development enterprise, with reference to its breeder reactor program. In doing so, this research argues that the indigeneity discourse is used by the political and scientific elites of India to justify expenditures for curiosity-driven science, despite its limited immediate contribution to socio-economic betterment of the country. The result is a curiosity-indigeneity paradox that leads to high resource-allocation for scientific and technological projects that have little tangible impact upon national development.

---

[2] Hindu revivalist groups like the Arya Samaj began to reinterpret ancient Hindu texts in their quest for a 'Hindu science'. Gyan Prakash writes that it was a "historical compulsion that drove the Hindu intelligentsia of British India to negotiate the relationship of classical knowledge with Western science and to represent their traditions as scientific".

2 Franco-Indian Nuclear Cooperation Agreement 1951

In 1951, India became the first country with which France entered into a bilateral cooperation in the field of atomic energy. This nuclear cooperation agreement (NCA) involved the study and construction of a beryllium-moderated low-power reactor.[3] In the first stage, preliminary theoretical studies were to be conducted on the construction of a nuclear reactor of moderate power, which would be beryllium-moderated and use natural uranium as its fuel. This first stage would take place in France. In the second stage, such a reactor would be constructed in India over a period of 5 years. At the end of the 5-year period, it was decided that the CEA would pass on the authority of the reactor to the Atomic Energy Commission of India (AEC). During the same 5-year period, the AEC would supply beryllium to the CEA for the construction of a similar reactor in France. Since the reactor in India would become the property of the AEC, the expenses incurred in its construction must either be directly paid or eventually reimbursed by the AEC. Each party would be responsible for the expenses related to their personnel involved and for the material costs for the studies undertaken in their own countries. The agreement required the CEA and the AEC to cooperate fully in all appropriate scientific and technological domains.

Beryllium-moderated natural uranium reactors had a couple of advantages for India. First, since these reactors used natural uranium, India did not need to master enrichment technology in order to operate them.[4] Second, being beryllium-moderated, these reactors required, at least theoretically, less quantities of natural uranium, therefore allowing India to circumvent the disadvantage posed by its lack of uranium deposits (Mahut 2009: 59).[5] Under the 1951 bilateral agreement, France was responsible for supplying all the uranium required for research and construction of the reactor, while India was the supplier of the beryllium.

However, beryllium-moderated reactors were not a proven technology in the early 1950s, and both the CEA and AEC were aware of it. In fact, after 10 years, the project was jointly abandoned since the technology was found to be infeasible (Goldschmidt 1982: 619). The Indo-French NCA however demonstrated the common difficulties that the AEC and the CEA were facing during the early Cold War, with respect to access to information and resources in the field of atomic energy. In the face of US-led postwar information censorship in atomic energy, and the Anglo-American efforts through the Combined Development Trust, to establish a

[3] Secret envelope from Jules Guéron to Raoul Dautry containing resumé of discussions with Homi J. Bhabha in London, 2 July 1950, 307 AP 225, 1950–1951, Papers of Raoul Dautry, French National Archives, Paris, France.

[4] The author is grateful to M.V. Ramana for this suggestion.

[5] Bertrand Goldschmidt, who headed the CEA mission to India in December 1951, indicated this potential benefit of beryllium-moderated reactors to Indian scientists.

monopoly over strategic minerals, the CEA felt cornered and actively sought international partners.[6] The AEC, which found itself at the receiving end of Anglo-American attempts to control Indian strategic minerals, decided to make an exception for the CEA. By the terms of the NCA, France was granted priority access to Indian beryllium, even though since 1948 New Delhi had imposed an embargo on its exports. Second, despite the ambiguity surrounding beryllium reactors, the Indo-French NCA involved joint research, thus providing the AEC with the possibility of developing a knowledgebase of its own. It also offered the opportunity to cooperate in the field of atomic energy with a developed country as an 'equal,' which was of high significance to a newly independent country like India. As a result, notwithstanding the technological uncertainty and the high economic costs involved in the Indo-French joint cooperation, the AEC embarked on this curiosity-driven science project, in the hope of developing a nuclear R&D base of its own. Needless to mention that the investment made in this endeavor did not yield any production of electricity for public consumption, since this technology was found to be unsuccessful and eventually discarded.

3 Thorium Research and Development

While thorium-related technology was extensively researched by the United Kingdom, the United States, Japan, Russia and France, by the mid-1970s, most of these countries had started abandoning their research since uranium was available in ample abundance in nature, making the search for an alternative nuclear fuel not an immediate concern. India remains, an exception, (together with Russia) which continues a large thorium R&D program till this day. This program is perceived as intrinsically related to India's three-stage program, and therefore to New Delhi's quest for a self-reliance in nuclear energy. Although India lacks uranium reserves, it has large deposits of thorium contained in the monazite sands in Travancore in the south. Self-reliance in atomic energy thus involve developing reactor technology that use thorium as core fuel, instead of uranium, thus curtailing India's dependency on foreign suppliers with respect to nuclear fuel. The AEC therefore underline the importance of thorium-fuelled breeder reactors as vital for the attainment of this goal.

The proponents of the thorium fuel cycle underline several advantages of the cycle, which are not undisputable. First, the thorium fuel cycle is claimed to have

[6] In May 1939, Joliot-Curie tried to conclude an agreement with the Union minière du Haut-Katanga to secure French access to uranium deposits in Belgian Congo. The Nazi invasion of Paris in 1940 stalled all progress. In 1942, the United States signed an agreement with the Belgian firm leading to exclusive US access to Congolese uranium. For a detailed account of these pre-war French attempts see Bertrand Goldschmidt, *Atomic Rivals*, trans. Georges M. Tanner (New Brunswick and London: Rutgers University Press 1990: 45–59). See also Bertrand Goldschmidt, *L'aventure atomique: Ses aspects politiques et techniques* (Paris: Fayard 1962).

low risks of proliferation, since it does not produce plutonium. However, the thorium cycle produces fissile uranium (U-233), which entails a proliferation risk. Second, those in favour of the thorium fuel cycle argue that it produces less waste than the uranium fuel cycle owing to the absence of plutonium-production and the reduced amounts of other radiological waste. This however, does not make a thorium fuel cycle waste-free, because it produces certain long-lived radionuclides that need to be treated in order to prevent a radiological hazard to the environment. Third, the thorium fuel cycle has been the cornerstone of the AEC's proclaimed cost-effective nuclear energy program. Yet, according physicist M.V. Ramana, while 'the hope about thorium was optimistic in the extreme,' no concrete economic ways of using thorium has yet been devised by the Indian atomic energy establishment (Ramana 2013: 160).

Furthermore, the efficacy of breeder reactors has been held suspect for a long time. France, for instance, had relinquished its plans for breeder reactors in the late 1990s as they were found to be neither economical nor reliable (Ramana 2013: 266). This was in spite of the fact that the CEA had constructed three breeder reactors, which included the Superphénix that operated at a commercial scale (Ramana 2013: 255). Ironically, India had obtained the designs for its own fast breeder test reactor from France in the late 1960s (Ramana 2009a) and has continued to cling to its outdated breeder reactor plans.[7]

One of the reasons as to why there has been no public questioning in India of the breeder program and thorium R&D enterprise is because, these have been historically equated with Bhabha's dream projects, which had to be fulfilled at all costs by the Indian Department of Atomic Energy (DAE) (Ramana 2013: 267). Ramana points out that during the negotiations concerning the US-India civil nuclear agreement in the early 2000s, many retired DAE officials had questioned how the agreement might impact Bhabha's three-stage program, especially with respect to India importing light water reactors (LWRs) from foreign countries. The supporters of the US-India nuclear agreement had argued that the import of LWRs is only a temporary measure, to enhance the performance of the nuclear energy program. In other words, while the DAE is almost dogmatic in its defence of the three-stage program, its adherence has not been entirely uniform.

[7] In 1969, the AEC and the CEA signed a bilateral agreement by which, India obtained the design of the French Rapsodie test reactor and the steam generator of the Phénix reactor. Based on the design of the Rapsodie reactor, India's fast breeder test reactor in Kalpakkam, is India's first and only breeder reactor operational at present. M.V. Ramana, "The Indian Nuclear Industry: Status and Prospects," *Nuclear Energy Futures Papers* 9 (Waterloo, ON: CIGI 2009).

4 The Indigeneity-Curiosity Paradox

Despite the efforts for a science-driven national development program and high resource allocation for atomic energy, the performance of the DAE has not lived up to its own promises. At present, nuclear energy meets around 3 % of India's energy needs, which is otherwise largely met by coal. While the technological sanctions that followed New Delhi's first nuclear test in May 1974 hampered the DAE's activities, as the import of nuclear technology and materials became difficult, it must also be accepted that most of DAE's projects were more futuristic than rooted in meeting present pressing needs of the country. For a newly independent country battling poverty and a large population, mobilizing resources for these projects was not easy. Yet, the attractions of developing an indigenous knowledgebase in atomic energy, and the national hubris associated with it, continued to create justifications for these projects.

Several scholars have criticized the DAE for its practice of secrecy, which they argue, has prevented an independent review of its projects and finances (Sharma 1983; Ramana 2009b). The lack of accountability of the DAE helped it to perpetuate its faulty policies of high government investment in basic or pure science research (i.e. science as knowledge) instead of applied research (i.e. science for economic growth). M.V. Ramana noted that the lack of public debate of DAE's plans has permitted the continued importance of breeder reactors. According to Dhirendra Sharma, the problem originated in the organization of atomic energy as directly under the Indian state, which made fundamental research the domain of government-funded research institutes, notably Homi Bhabha's Tata Institute for Fundamental Research (TIFR), instead of public universities (Sharma 1991: 32–36).

In the 1960s, as India spiralled into political chaos and greater economic misery, some wondered if India's woes did not lay in the organization of science in India (Raina and Jain 1997). One such critic was Patrick Maynard Stuart Blackett—a physicist by training and a Cambridge alumnus (like Nehru and Bhabha), who won the Nobel Prize in physics in 1948, and played the role of an informal advisor to Nehru on science and defence policy (Anderson 1999a, b, 2010). In 1961, Blackett was appointed by the Indian minister of scientific research and cultural affairs to conduct a reappraisal of the National Physical Laboratory (NPL) in Delhi. In the report that he submitted within a month, he called for the reorganization of the entire Laboratory, in order to tackle the 'culture of administrative rigidity' at the NPL (Anderson 1999b: 348). He criticized the practice of the Government of India to lure away promising scientists from basic research in universities towards industrial research in government research institutes (Anderson 1999b: 349).[8] He argued that this resulted in basic research or curiosity-based scientific research, which is more suited to universities, being pursued in the government laboratories,

[8] Anderson notes that by this Blackett contributed to the larger debate prevailing in India at the time on the unequal relationship between public universities and government laboratories. Both P.M.S. Blackett and A.V. Hill were avid critics of this practice.

at the cost of large sums of public money. Blackett noted that 'India is finding out that the problem of turning science into wealth. . .is very much more difficult than just doing science.' (Anderson 1999a: 257).

He addressed this issue more fully in his 1971 report to the Leverhulme Trust where he noted that the government laboratories in India have demonstrated a 'tendency toward rather basic research without always having any definite practical goal' of making an immediate contribution to industry and agriculture (Anderson 1999b: 353). According to Blackett, on one hand, many government laboratories were not operating in close collaboration with the industry and therefore were unaware of market requirement of their research, while on the other hand, some others thought that it was important to advance only basic research. This faulty practice, stated Blackett, had prevented India from using science to make substantial contribution to national wealth.

In his last speech in 1966, Bhabha echoed Blackett's criticism but argued that this was only the case with the laboratories of the Council of Scientific and Industrial Research, and not TIFR (Bhabha 1966).[9] In the field of atomic energy, the government strategy was slightly different but the outcome similar. Promising scientists were not taken away from universities to government laboratories, but universities were kept altogether outside the purview of atomic energy research. The TIFR established jointly by the Dorab Tata Trust and the Government of Bombay in 1945, dominated research in this field.[10] Since majority of scientific expertise was concentrated in this single government laboratory, it became the centre for both basic and applied research. Owing to its close relationship with the DAE and the secrecy surrounding atomic energy in India, criticism of its policies remained muted. One of the foremost critics of this setup was physicist Meghnad Saha, who had established the Saha Institute of Nuclear Physics in Calcutta. Until his death in 1956, Saha was a vocal opponent of the organization of the Indian atomic energy apparatus. He argued that it deprived other Indian scientists from appraising the DAE's performance.

5 Conclusion

In 1967, the Indian newspaper *The Statesman* noted in its editorial that 'while economic growth bears no close numerical relationship with R&D spending, it does with investment, and where capital is scarce, as in India, the main hope lies in

[9] Bhabha delivered the speech titled "Science and the Problems of Development" on January 7, 1966 to the International Council of Scientific Unions in Bombay. This was about 2 weeks prior to his untimely death in an air crash over the Swiss Alps on January 24, 1966. The transcript of this speech was later published in the *Science* magazine.

[10] Homi J. Bhabha made a request for a grant from the Dorab Tata trust in March 1944 for the establishment of an institute for fundamental research in Bombay. This led to the establishment of the TIFR. Bhabha himself belonged to the Tata family.

increasing the return on its investment' (The Statesman 1967). The article was reiterating Blackett's views expressed in his Nehru Memorial Lecture delivered in New Delhi that year. Titled, 'Science and Technology in an Unequal World', Blackett argued that while governments tend to support curiosity-based research as a long-term investment and as a source of national pride (all countries want to make their contribution to science), they should not do so, unless they can focus equally on all aspects of the 'innovation chain'.[11] This innovation chain, he said, involved the whole range of activities concerning research, development, design, production, marketing, sales and post-sales service. The first stages 'consume wealth' and the later stages 'produce wealth'. However, the later stages require more able scientists, engineers and managers than the initial ones. As a result, Blackett underlined that unless a country is certain that the final stages of the innovation chain will be looked after, it should not engage in innovation-oriented curiosity-based research. Throughout the 1960s, Blackett criticised government practices in science in India. According to him, India, like Japan, must purchase production licenses from other countries, instead of spending large sums of money in innovating known technology.

The 1960s were a tough period for India with severe drought and famine in Bihar and Uttar Pradesh, the devaluation of the rupee, dispute with the United States over PL-480 food aid, war with Pakistan and border dispute with China over Sikkim, which heightened the calls for self-reliance and import-substituting industrialization. Many laboratory directors in India therefore viewed Blackett's critique of the government's scientific practices and his subsequent recommendations unfavourably (Anderson 1999b: 350–352). Nevertheless, Blackett was able to successfully capture the core predicament of India's science-driven national development program very early on.

Although technologically driven national hubris is not unique to India, a postcolonial country may be more susceptible to it. Andrew Rotter offers the example of Indian monazite to demonstrate this. In the years immediately after independence, India initially sold monazite to U.S. companies that processed it and produced thorium nitrate and sold it back to Indian companies. However, New Delhi embargoed the export of strategic minerals in 1948 (Rotter 2000: 98). According to Rotter, this resulted from India's perception of this relationship with the United States as an unequal and colonial one, where the colonized country exported raw materials to the colonizer country, and bought back finished products from the latter. India eventually set up a monazite processing plant in Alwaye, Kerala with help from France, and thereby acquired the technical knowhow of processing monazite.

Technological capability even without tangible contribution to economic growth can thus be held in high esteem by developing countries, as the Indian case

[11] P.M.S. Blackett, *Science and Technology in an unequal world: Jawaharlal Nehru Memorial Lecture* (New Delhi: Jawaharlal Nehru Memorial Fund 1968). Royal Society Archives, Blackett Papers, PB/7.

demonstrates. The Indian elites thought that the bullock cart of underdeveloped India with its social and economic ills could be pulled out of backwardness through large-scale development of the most advanced form of energy, namely atomic energy. That the DAE has attained few of its goals, points to unaddressed issues—long-term self-reliance goals over immediate socio-economic growth, national technological hubris over tangible contribution to national wealth, and ideology over pragmatism. For a developing country with severe economic constraints, the outcome has been quite damaging.

References

Abraham, I. (1998). *The making of the Indian atomic bomb: Science, secrecy and the postcolonial bomb*. London: Zed Books.

Abraham, I. (2009). Contra-proliferation: Interpreting the meaning of India's nuclear tests. In S. D. Sagan (Ed.), *Inside nuclear south Asia* (pp. 106–135). Stanford, CA: Stanford University Press.

Adas, M. (1989). *Machines as the measure of men: Science, technology and ideologies of western dominance*. Ithaca; London: Cornell University Press.

Anderson, R. S. (1999a). Patrick Blackett in India: Military consultant and scientific intervenor, 1947–72. Part one. *Notes and Records of the Royal Society of London, 53*, 253–273.

Anderson, R. S. (1999b). Patrick Blackett in India: Military consultant and scientific intervenor, 1947–72. Part two. *Notes and Records of the Royal Society of London, 53*, 345–360.

Anderson, R. S. (2010). *Nucleus and nation: Scientists, international networks and power in India*. Chicago, IL: University of Chicago Press.

Bhabha, H. J. (1966). Science and the problems of development. *Science, 15*, 541–548.

Blackett, P. M. S. (1968). *Science and technology in an unequal world: Jawaharlal Nehru Memorial Lecture*. New Delhi: Jawaharlal Nehru Memorial Fund.

Goldschmidt, B. (1982). Les problèmes nucléaires indiens. *Politique étrangère, 47*, 617–632.

Goldschmidt, B. (1990). *Atomic rivals*. Translated by Georges M. Tanner. New Brunswick; London: Rutgers University Press.

Hecht, G. (1998). *The radiance of France: Nuclear power and national identity after world war II*. Cambridge, MA: MIT Press.

Mahut, R. (2009). Entre atomes et diplomatie: Les premiers heures du developpement international du CEA. *Revue d'histoire diplomatique, 123*, 55–74.

Nehru, J. (1948). *The discovery of India*. Delhi: Oxford University Press.

Prakash, G. (1999). *Another reason: Science and the imagination of modern India*. Princeton, NJ: Princeton University Press.

Raina, D., & Jain, A. (1997). Big science and the university in India. In J. Krige & D. Pestre (Eds.), *Science in the twentieth century* (pp. 859–878). London: Harwood.

Ramana, M. V. (2009a). India's nuclear enclave and the practice of secrecy. In I. Abraham (Ed.), *South Asian cultures of the bomb: Atomic publics and the state in India and Pakistan* (pp. 41–67). Bloomington, IN: Indiana University Press.

Ramana, M. V. (2009b). The Indian nuclear industry: Status and prospects. In *Nuclear Energy Futures Papers* 9. Waterloo, ON: CIGI.

Ramana, M. V. (2013). *The power of promise: Examining nuclear energy in India*. New Delhi: Viking.

Rotter, A. J. (2000). *Comrades at odds: United States and India, 1947–64*. Ithaca, NY: Cornell.

Sharma, D. (1983). *India's nuclear estate*. New Delhi: Lancers.

Sharma, D. (1991). India's lop-sided science. *Bulletin of the Atomic Scientist, 32–36*.

Vaez, A., & Sadjadpour, K. (2013). *Iran's nuclear odyssey: Costs and risks*. Washington, D.C.: Carnegie Endowment.

The Politics of European Collaboration in Big Science

Olof Hallonsten

Abstract Intergovernmental collaboration in Big Science has been an important resource for European science since the 1950s, as a means to compete on global level. But interestingly, collaboration in (basic) science has traditionally been left outside of the political integration work of the European Community/Union, which has resulted in a cluttered policy field and a situation where European Big Science collaborations are built on ad hoc solutions rather than a coherent political framework and common regulatory standards. Despite this formal detachment, however, the genesis and development of collaborations, and their political realities once launched, often draw upon and reflect the ordinary (geo)political dynamics of Europe. This chapter reports on four historical and two contemporary cases of European collaboration in Big Science, from CERN in the 1950s to the currently planned European Spallation Source (ESS), all well-documented by previous studies, showing that while scientific and technical preconditions doubtlessly impact the fate of these Big Science installations, the logic and cycles of high-level politics in Europe always plays a role and can, in some cases, be said to have been decisive for the realization of a collaborative effort. Always balancing between national interest and the common good, European collaboration in Big Science is thus no different from the process of EC/EU integration, despite being formally detached therefrom. Using a historical perspective to make justice to the rather small collection of cases to study, the chapter covers a distinct instance of where science and technology is directly affected by international politics.

Keywords Big Science • CERN • Common market • European Atomic Energy Community (EURATOM) • European Organization for Nuclear Research • European Union • Franco-German entente

---

O. Hallonsten (✉)
Department of Education and Social Sciences, Wuppertal University, Gaussstr. 20, D-42119 Wuppertal, Germany
e-mail: hallonsten@uni-wuppertal.de

M. Mayer et al. (eds.), *The Global Politics of Science and Technology - Vol. 2*, Global Power Shift, DOI 10.1007/978-3-642-55010-2_3,

1 Introduction

The end of World War II marked the beginning of a new era in science and science policy, characterized by heavy governmental investment in R&D and institutionalization of the doctrine that science and technology has a major role to play in national defense, development and global competitiveness. A most manifest materialization of this "Marriage between science and the state", as journalist Daniel S. Greenberg (1999/1967) put it, was Big Science—enormously capital intensive complexes at the intersection of fundamental science and military R&D that became the subjects of an Arms Race of its own and signs of strength in the global East–west competition.

In Western Europe, the growth of Big Science was conditional upon the willingness of European governments to cooperate. The resurrection of European science after the devastating war required direct scientific competition with the United States and eventually the Soviet Union, Japan and China, and the road thereto required mobilization of resources beyond that of the European nation states. In the several decades to come, collaborative efforts resulted in the creation of joint organizations, laboratories and centers serving science in a variety of areas, among them science requiring large and costly instrumentation, the topic of this chapter. Interestingly, European collaboration in science did not form part of the mainstream Western European political integration process within the EC/EU project but remained, for the whole twentieth century, an area of formally uncoordinated ad hoc solutions and a myriad of different organizational forms and legal arrangements. In effect, this has meant that each new collaborative initiative have had to rely on the political will of the collaborating countries at the specific time of their genesis and development, which paradoxically enough has made European collaboration in Big Science a mirror of the cycles of European integration in the second half of the twentieth century and beyond.

In this chapter, the history of six collaborative European Big Science projects (two of which are in the making) is reviewed. Analyzing the political processes by which they have come into being, the chapter displays the connection of each of them to particular political/diplomatic conditions at the time of their birth and realization. Big Science is in this context taken to mean large scientific facilities that require a single location and whose realization, for financial/political and/or legitimacy reasons, require the collaboration of at least two European countries agreeing on governmental level. In Table 1, the facilities under study are listed, along with some basic information.

The chapter is structured as follows. In the next section, some fundamentals on the political realm of European intergovernmental collaboration in science are presented. Thereafter, brief historic accounts of the six cases in question are presented along with descriptions of the political embeddedness of the projects and highlights of some specific areas of political sensitivity that have been at the center of the negotiations over projects over the years. A final section summarizes

Table 1 The six cases

Facility	Location	Year of first conceptual idea	Year of start of operation	Type of facility	Number of member countries (at start of operation)
European Organization for Nuclear Research (CERN)	Geneva, Switzerland	ca 1950	1954	Nuclear physics/ high energy physics (accelerators)	12
European Southern Observatory (ESO)	La Silla, Chile/ Garching, Germany	1954	1966	Ground-based astronomy (telescopes)	6
Institut Laue-Langevin (ILL)	Grenoble, France	ca 1965	1972	Neutron scattering (reactor)	2
European Synchrotron Radiation Facility (ESRF)	Grenoble, France	1975	1994	Synchrotron radiation (accelerator)	12
European X-ray Free Electron Laser (XFEL)	Hamburg, Germany	ca 2002	2016[a]	Free electron laser (accelerator)	
European Spallation Source (ESS)	Lund, Sweden[b]	ca 1993	ca 2020[b]	Neutron scattering (accelerator)	

[a]Planned
[b]At the time of writing, the ESS is not formally approved/funded

the findings and the argument, and discusses some specific issues of interest in greater detail.

2 The Politics of European Scientific Collaboration

The 1951 Treaty of Paris, establishing the European Coal and Steel Community (ECSC), marked the beginning of formal postwar political integration in (Western) Europe. The ECSC was the first supranational organization on the continent and the first step towards a common market; its overarching goal was clearly to prevent future wars between foremost France and (Western) Germany but the concrete mission and purpose for the organization was to promote economic prosperity by rationalizing the production and sale of the vital raw products coal and steel. Thus restricted, very practical, and economically oriented, neither the agreements of the Paris Treaty nor the resulting ECSC had any elements of collaboration on the R&D side (Middlemas 1995: 21–22). The 1957 Treaties of Rome, establishing the Common Market as well as the European Atomic Energy Community (EURATOM), instructed member countries to collaborate on very specific areas: coal, steel, agriculture and atomic energy. No collaborative mandate was given in the area of science and technology outside nuclear energy, and EURATOM was furthermore subject to a separate treaty that did de facto not form part of the continuing European political integration process that eventually led to the Single European Act and the treaties of Maastricht and Lisbon.[1] Therefore, it was not until the 1970s that the realm of the European Community was extended to the promotion of science and technology, but then in the shape of industry-oriented programs launched to increase competitiveness in specific sectors, such as ESPRIT (for microelectronics and robotics) and the other various efforts within the so-called Framework Programmes (Grande and Peschke 1999: 45; Papon 2004: 69–70). Active partaking of the European Commission in maintaining and developing a broad research base in Europe began only several decades later, with the most recent Framework Programmes and the creation of a European Research Area (ERA). Nowadays, Framework Programmes funding is available also for initial planning of infrastructure projects, and a strategic body is in place to inform planning and decisions of both collaborative and national research infrastructure projects (Hallonsten 2012: 302–303), but for the whole of the twentieth century, the EC/EU stayed out of (basic) science.

Hence, while the necessity for Europe to collaborate in some sciences to compete on a global scale was acknowledged at an early stage, no political frameworks were put in place to create coherence and establish precedence across disciplines and technologies, and the projects that have been launched have therefore been dependent on ad hoc solutions and the recurring reinvention of legal arrangements and organizational structures. Big Science facilities of this kind are almost always conceived within scientific communities as answers to specific (scientific) needs, and then brought to political level by aggregated scientific lobbying efforts. The ultimate realization of a Big Science project is, naturally, the result of negotiation between these scientific interests and political priority-setting. The cutting edge character of Big Science facilities—mandated by the size of investments—typically makes the scientific side of their realization a complex and challenging process with parallel and interrelated tracks of formulating a credible scientific case, mobilizing a competent user community, and technological design and development at the cutting edge. On the other hand, national political systems typically have well-established procedures and institutions (e.g. systems of national laboratories, and political decision-making processes) for handling initiatives, set priorities between competing projects, and realize them politically (Hallonsten and Heinze 2012). In the case of these European collaborative projects, however, the political side of their realization appears just as messy and complicated as the scientific and technological, involving a vast and complex assortment of political agreements and negotiations on a variety of levels. If successful, this political process leads to the signing of an intergovernmental agreement, sometimes

[1] After some initial alleged troubles, EURATOM evolved into a parent organization for nuclear energy and thermonuclear fusion energy research activities including research but only in these distinct areas (Papon 2004: 64–65; Grande and Peschke 1999: 45).

comparable with an international treaty and sometimes establishing a private company with the member countries as shareholders, whereby the collaboration is founded. The variations with regard to the process and the eventual legal agreement are almost as many as there are collaborations; new shapes and forms have emerged for almost every new project.

Some analysts have named this lack of coherence a factor for the relative success of those collaborations that have succeeded, since it has arguably prevented bureaucracy and institutional inertia to take hold and created dynamisms and efficiency by allowing every specific project to meet the demands of its particular scientific community, at a specific time (e.g. Hoerber 2009: 410; Gaubert and Lebeau 2009: 38; Papon 2004). But the incoherence and lack of framework has also created an unnaturally composite system and an opaque and cluttered policy field where is seems impossible to avoid typical pitfalls and repeated exposure to political strains within and between individual European countries. Countries normally partake in collaborations not as an activity separate from national science policy agendas but rather, from the perspective of an individual country, as "the pursuit of one's interests by other means" (Krige 2003: 900). Most countries realize that collaboration is necessary to achieve goals beyond the reach of any one of them, but strong traditions of sovereignty create constant tension between self-interest and common good, for every partaking country, in every collaboration. In practice, his conflict shows itself in each instance where the relative gain of each (prospective) member country, and the (prospected) ratio between effort and benefit, is on the table. From the perspective of a prospective member country, the decision to participate is therefore perhaps best described as a multi-track cost-benefit analysis, where several possible gains and losses (economic, political, diplomatic, reputational) weigh in but where it would appear that every country in the end will simply seek to maximize its net gain. Whether this is at the expense of others, or coincides with the promotion of the common good and the health of the collaboration as a whole, seems to be a function of several variables and an empirical question, which is a main purpose of this chapter to respond to.

In the planning phase, the issue of site-selection is typically the trickiest, along with the question of the relative financial contributions and the policies for scientific access and procurement of goods and services. Typically, once agreements are made, the signed documents are legally binding, which means that (dis)agreements in the planning and negotiation phase are institutionalized in the facility and let to influence its organizational, technical and scientific performance. These areas of conflict show themselves to varying degrees in the six cases discussed below. The contextualization provided above and the unique scientific, technological and political aspects of a very limited number of cases would probably imply that each collaboration rather constitutes a unique response to a unique historical situation. While this is true, the underlying assumption of the chapter is also that those general patterns that are distinguishable, despite the expectable political and scientific uniqueness of coming collaborations, can and should be analyzed and eventually used as a battery of experiences that might assist policymakers in their quest to avoid pitfalls.

3 The Cases

3.1 The European Organization for Nuclear Research (CERN)

In 1954, the first multilateral European collaboration in Big Science was launched as an international treaty organization. Its overarching political logic was the enormous importance and impact of atomic energy for the ending of World War II, which had made nuclear physics a top priority of most governments, the realization that a joint scientific laboratory would be a favorable foundation for the tedious work to achieve political integration in Europe after the war, and the 'Marshall Plan for Science', i.e. the United States' efforts to assist the rebuilding of Western Europe in accordance with its geopolitical preferences (Krige 2006: 57–67; Krige 2014). CERN was created to complement national nuclear physics programs rather than replace them, and the cost of particle accelerators was still on the level of millions rather than hundreds of millions of dollars, which certainly helped in making CERN a largely uncontroversial feature of European science (and politics) (Pestre 1990: 785). Clearly, its political ramifications in the first decade of existence were straightforwardly simple: a peace project, mobilizing European competitiveness in science, and strengthening the ties to the United States.

This changed dramatically in the 1960s, when the international development in particle physics called for expansions of CERN to keep up in the competition with the United States and the Soviet Union (see, e.g., Greenberg 1999/1967). The proposed upgrade program was large enough to give rise to plans for a new, separate laboratory under the name 'CERN II', possibly located at a new site on the European continent. The very fact that another country could become the host (and thus prime beneficiary) of this large joint investment made member countries openly subjugate the collaboration to their own national interest, and nearly all proposed their own sites. The Federal Republic of Germany and the United Kingdom went as far as issuing ultimatums that they would withdraw completely from the collaboration should the new lab not be located within their borders. An attempt to choose a site on so called "scientific" grounds, i.e. by the work of an independent and "objective" expert committee was buried (Pestre 1996: 73, 77–78). In 1970, the situation was resolved by a decision to build CERN II at the existing site in Geneva. As a compromise solution, also reducing the costs of the project significantly, this apparently convinced member states to go along (Krige 2003: 905).

The reasons for the rupture over CERN II are multiple—significantly higher costs compared to the original CERN laboratory, waning enthusiasm towards big technoscientific projects among the member countries, and fears that CERN would monopolize national science budgets[2]—but overshadowing all was the looming end

[2] A fear that would prove true: CERN II did in fact become the only center for experimental particle physics in Europe, with the exception of the German laboratory DESY in Hamburg (see below).

to the protracted postwar economic boom and the relative political instability on the continent in the late 1960s.

3.2 The European Southern Observatory (ESO)

Before the European economic wonder and technoscientific enthusiasm faded, CERN would be accompanied by yet another big collaborative facility project. Already in 1954, the year of the founding of CERN, leading astronomers from six European countries had issued a statement and recommendation that their home countries embark on a project to establish a joint ground-based observatory. Scientific considerations had the project predestined for the Southern Hemisphere, and originally South Africa (Blaauw 1991: 5), which meant that the European Southern Observatory (ESO), as it would be named, was spared from all potential difficulties associated with choosing a site within Europe.

But ESO was to have its delays anyway. The fact that astronomy "lacked the aura of the nuclear" and was "remote from any practical use" made ESO "bereft of any broader industrial or local, national or global political considerations" which seems to have caused vulnerability for the project, since it made the question of participating a purely financial issue for each member country (Krige 2003: 906). This isolation from a broader policy context may be what caused Britain to hesitate and withdraw from the ESO project in the summer of 1960, in favor of a Commonwealth ground-based astronomy project in Australia (Krige 2003: 906; Blaauw 1991: 8–9), but British general euroskepticism is also a plausible factor. The other member countries (Belgium, Denmark, France, the Federal Republic of Germany, the Netherlands, and Sweden) managed to agree on the financing of the project, but still it could only become reality by a private donation from the Ford Foundation, whose contribution of $1 million eventually had the effect of "pushing the project financially over the threshold" (Blaauw 1991: 11). Between 1963 and 1967, the six member countries ratified the convention one by one, and the signing of France in 1964, which meant that 70 % of the funding was secured, gave effective go-ahead for the project (Blaauw 1991:18). By then, scientific considerations and the diplomatic work of the ESO provisional director Otto Heckmann had caused the change of preferred site from South Africa to Chile, where ESO was subsequently built, opening its first telescope for European astronomers in 1966 and run as an international organization (Krige 2003: 906).

3.3 The Institut Laue-Langevin (ILL)

The Institute Laue-Langevin, an reactor-based neutron scattering facility located in Grenoble, France, was founded in 1967 as a French private company on basis of a bilateral agreement between France and Germany that reflected the strong scientific ambitions of both these countries at the time while not exhibiting the typical quandaries of multilateral collaborations in Europe. Especially Western Germany, doubling its total R&D expenditures between 1962 and 1971, was on a path of investing heavily in both domestic and international large facility projects in the late 1960s, and the ILL was one of them (Trischler and Weinberger 2005: 64). The scientific initiative to the project, and the work to mobilize scientific support for its realization, is attributed to professors Heinz Maier-Leibnitz and Louis Néel, and indeed, it is often claimed that the ILL during its approximately four decades of operation has managed to maintain world leadership in the sciences it serves, hosting hundreds of experiments annually and contributing significantly to the development of neutron scattering techniques for wide utility areas in various parts of foremost materials sciences but also the life sciences (Herman 1986: 141; Tindemans and Clausen 2003).

But the ILL project also had its share of politics. The 1960s was a challenging time in the advances of Britain's relations with mainland Europe's integration process, and the British participation in ILL was consequently postponed until a new attitude towards European collaboration had taken hold in the UK (Judt 2005: 292, 526). That the ILL became reality at all, in 1967, was allegedly helped significantly by the signing of the Élysée Treaty 4 years earlier by German chancellor Konrad Adenauer and French president Charles De Gaulle. The creation of the ILL was not a direct part of the treaty but undoubtedly benefited greatly from the reconciliatory and cooperative sentiment it produced. The location in Grenoble is attributed to cheap electricity and colocation with French atomic energy agencies (Hallonsten 2012: 303).

3.4 The European Synchrotron Radiation Facility (ESRF)

In bright contrast to the strains to European collaboration characterizing the 1960s, the decade thereafter was marked by a general wave of renewed Europeanism centered on the entente between France and Germany that became the historically important "motor of Europe" driving the development towards the Maastricht Treaty and eventually the European Monetary Union (Middlemas 1995). In science, the 1970s and 1980s saw the creation of a number of collaborative organizations,[3] among which perhaps the European Synchrotron Radiation Facility (ESRF) is the most successful example.

Synchrotron radiation is extremely intense electromagnetic radiation produced by circular particle accelerators that, after use in solid state physics since the early 1960s grew rapidly in the 1970s as a technique for a wide variety of studies of

[3] European Science Foundation (ESF) in 1973, the European Molecular Biology Laboratory (EMBL) in 1973, the European Space Agency (ESA) in 1975, and the fusion research center Joint European Torus (JET) in 1977, to name a few (Herman 1986: 150–159; Krige 2003: 899).

materials, including life science applications. The expansion warranted initiative on European collaborative level to secure the supply of synchrotron radiation for European science, and in 1979, a feasibility study was presented by the European Science Foundation (ESF), outlining a collaborative facility that would both complement and supersede national European synchrotron radiation sources in performance (Hallonsten 2012: 305). With no financial or political powers, the ESF had to limit itself to efforts in mobilizing scientific support and mounting a lobby effort to win the support of politicians. Quite naturally, the location of the facility became a problematic issue already at the early stages, and several countries proposed their own sites. A 'scientific' site selection procedure, evaluating proposals on objective criteria, was launched. Nothing decisive happened, however, despite the rather generous pledges of financial support issued by some countries as part of their site proposals.

On 26 October 1984, France and Germany announced their joint decision to build the ESRF in Grenoble and together provide between 50 and 70 % of the construction costs of the facility. Other countries were invited to join. While causing surprise and resentment among the prospective collaborating countries, who felt run over by the two big nations, this decision by the government of France and Germany can in retrospect be identified as the most crucial event for the realization of the ESRF. It is probably no coincidence that it came the same year as the First European Framework Programme for Research and Technological Development (FP1) was launched, and the year before the signing of the Single European Act—the ESRF decision was likely a mere piece in the jigsaw puzzle of Franco-German partnership and renewed Europeanism in this era. The ESRF itself has later been identified as the "second prize" in the package deal between the two countries that also located the European Transonic Windtunnel (ETW) facility to Cologne (Papon 2004: 64; Hallonsten 2012: 314).

The other prospective member countries gradually accepted the Franco-German proposal, and negotiations over budget shares ensued. These mirrored much of the geopolitical situation in Europe in the late 1980s: The United Kingdom, expected to contribute considerably to the ESRF due to its strong scientific communities in fields utilizing synchrotron radiation, got away with a mere 14 % after harsh negotiations. French officials, alluding to past Italian vigor at a time when the Italian economy was in free fall and its government thirsty of (symbolic) restoration, allegedly managed to persuade Italy to pay more than was perhaps motivated (Hallonsten 2009: 220–221).

Construction of the ESRF started in 1989, and in September 1994, the facility opened for users (Hallonsten 2012: 305). The ESRF, run as a French private company, is generally considered a world-leading synchrotron radiation facility, exhibiting particular strength in output in relation to investment as well as high quality in technical and scientific assessments The most often mentioned reason for the strong showing is the comparable generosity of the funding portfolio for ESRF that has enabled the facility to move fast and invest in new areas of utility as they emerge and become popular in scientific communities (Hallonsten 2009: 232).

3.5 The European XFEL

The German research center DESY (Deutsches Elektronen-Synkrotron, German Electron Synchrotron) was among the very few national centers of particle physics left in Europe after the monopolization of particle physics budgets by CERN II in the 1970s. Also undertaking some work in synchrotron radiation, DESY was the main force behind the proposal in the 1990s to build a next-generation linear accelerator for particle physics named TESLA (Terra-electronvolt Energy Superconducting Linear Accelerator), combined with a 'next generation' x-ray source in the shape of a free electron laser. The German Federal Ministry for Education and Research showed greater interest in the free electron laser than in the TESLA machine, arguably not only looking at its smaller price tag but also following global trends of diminishing support for particle physics in favor of more application-oriented big science (e.g. Hallonsten and Heinze 2012). Consequently, in February 2003, the German Federal Ministry for Education and Research announced its plans to go ahead with the XFEL as a European facility located in Hamburg, and to cover for approximately half of the construction costs (Hallonsten 2012: 306). In 2004, France, Greece, Italy, Spain, Sweden, Switzerland and the UK signed a Memorandum of Understanding, and in 2005, the project was joined by China, Denmark, Hungary, Poland and Russia. Meanwhile, the technical design of the facility was given substantial updates but in 2007, funding was still not secured beyond the money already pledged by the German government. In October 2007, Russia announced its level of participation, 23.1 % of the construction costs, which was greeted by XFEL management as a "breakthrough" and "turning point", effectively securing the funding of the project but also empowered Russia with a de facto veto right on issues such as policy decisions regarding access (Hallonsten 2012: 306).

The Russian €250 million contribution to the XFEL is said to have been the result of a high-level agreement between German chancellor Angela Merkel and Russian president Vladimir Putin, made at a summit meeting in October 2007 that was otherwise described as a "cold encounter" ("Kühle Begegnung", Kirschstein 2007), and thus constituting a much-needed symbolic act of unity in the otherwise very tense relations between the countries (Hallonsten 2012: 306), or, as Krige (2003: 904) has put it, "just because it is seen as being a 'non-political' activity, scientific collaboration can be a particularly useful first and tentative step in a politically delicate context of alliance building".

The construction of the XFEL is underway since 2010, under the rule of a German private company, and it is estimated that the first part of the facility can be taken into operation in 2015 (Hallonsten 2012: 306).

3.6 The European Spallation Source (ESS)

The most recent large collaborative facility project in European science is the perhaps most delayed in history, now entering its twentieth year under the name the European Spallation Source (ESS) but yet without binding legal agreements on funding and organization of the project.

Accelerator-based spallation facilities are supposed to deliver enhanced performance to neutron scattering research of the type conducted at e.g. the ILL, and they are nowadays seen as largely complementary to synchrotron radiation and free electron laser facilities. Plans for a European spallation facility were drafted already in the early 1990s, but didn't get any political leverage until almost a decade later, when made part of recommendations for large-scale scientific projects by the Organization for Economic Cooperation and Development (OECD) together with similar projects in Japan and the USA (Kaiserfeld 2013). Work on the Japanese and American facilities promptly began, but even at the time of their completion some 7 years later, Europe still had not reached any decisions. Site contenders had come and gone, Germany and the UK had both declared interest and later withdrawn, and a meeting in Bonn in 2002, supposed to settle the issue and initiate an active phase for the realization of the project, had ended with "a kind of acceptance that it is never going to happen" and effectively buried the project (Berggren and Hallonsten 2012: 24–28; Hallonsten 2012: 307). At the end of 2008, when the cabinet level EU Competitiveness Council took up the issue, three site contenders remained—Lund in Sweden, Bilbao in Spain, and Debrecen in Hungary, all in the midst of an intense political lobbying campaign for their facility proposal (Kaiserfeld 2013). At a meeting in Brussels on 28 May 2009, representatives of countries that had declared interest to participate in the ESS decided that the Lund site would be the preferred choice for the ESS facility. However, the decision came with no financial guarantees, and despite the message in local media and elsewhere that the ESS has been decided upon and will be realized, the outcome of the May 2009 meeting was nothing more than an agreement that *if* the ESS is build, it will be built in Lund (Hallonsten 2012: 307).

As the most recent, and yet not formally approved, European collaborative facility project in science, the ESS is still subject to far-reaching secrecy and the political process of its creation is largely shrouded in mystery. Sweden and Denmark, so far the only shareholders in the ESS company, have pledged to cover approximately 50 % of the construction costs but announced a significantly lower commitment to the eventual financing of the operation of the facility. Thus after 20 years of planning, and almost 4 years after the 'decision' in favor of the Lund, Sweden, site, there are still little signs of an imminent go-ahead decision for the project. Considering the current economic situation in Europe, this might warrant some disillusionment when it comes to the prospects of an ESS facility actually becoming reality.

4 Discussion

While the six cases presented are all scientific user facilities and the products of multilateral European agreements on governmental level, they obviously also differ significantly in character. It is plausible that a common legal framework, developed within the EC/EU, and some precedent in treaties and formal agreements *could have* created greater coherence among the facilities and arguably also reduced uncertainties in preparatory work to the extent that some delays and disagreements could have been avoided. The case of the United States, where Big Science installations are almost exclusively built and operated within the National Laboratories system and where there is greater coherence in the organizations of Big Science labs despite great differences in their science missions and their time of creation (Hallonsten and Heinze 2012), supports such a speculation. Clear is that each and every European facility project described above has had to come into being on basis of ad hoc solutions, which is part of the explanation of their dissimilarity. This is also what provides the foundation for the contribution of this chapter: incoherence and discontinuity in politics of facility projects have made these projects prone to adopting and incorporating other more general trends in European politics at the times of their realization.

The political forces behind the creation of CERN are rather easy to identify. Not only is CERN enormously well documented (e.g. Hermann et al. 1987, 1990) but also a piece of folklore in the modern history of science: The postwar policy of the United States to secure its influence in Western Europe, the first wave of Europeanism after the war, and the general nuclear euphoria at the time seems to have sufficed for guaranteeing the success of almost any project connected to nuclear physics. At least it sufficed for CERN.

The 1960s was a period of strains in the European collaborations, most evidently so between Britain and the continent, showing in ESO, ILL (at first) and in the CERN II controversy. A similar euroskepticism characterizes current British foreign policy, which links well to the decline of UK to join both the ESS and the XFEL. In between, Britain agreed to participate in the ILL and the ESRF, however in the latter case with reluctance to contribute on par with its scientific and economic strength. This stance did not delay the project in itself—by the mid-1980s the shift of the locus of Europeanism to the Franco-German relations appears to have been strong enough to overcome most other political obstacles. The ESRF, on many accounts the world's most successful synchrotron radiation facility (Hallonsten 2013), still rides on the Franco-German entente of the 1970s and 1980s.

There is little hard evidence for the claim that Russian participation in the European XFEL project was determined by a need for a reconciliatory move between Russia and Germany, but in historical light the claim seems plausible. Fortunately for the XFEL project, a convention was signed, and funding pledges made, before the outbreak of the financial crisis and the subsequent Euro crisis. The European Spallation Source (ESS) did not have that fortune, and the deepening crisis in the European economy is one possible explanation for the fact that 4 years after the 'decision' in favor of Lund, Sweden, no legally binding agreement and/or funding plan has been presented.[4]

A few other topics on the level of details of the politics of the collaborations deserve mentioning. Troubles in connection with site selection have been mentioned a couple of times, and these troubles have generally been due to ample expectations of socio-economic benefits brought to the host country and region by an international research facility, in addition to potential benefits for the local scientific community and the risk of disadvantages of not hosting. This is, in other words, an incarnation of the basic tension between national interest and common good that always plague European collaboration (cf. Krige 2003). In the cases discussed, the issue of site selection has been resolved either by a kind of default solution or by politics. CERN II ended up in Geneva, at the original CERN site, as a compromise. The ESO was predestined for outside Europe and was spared from conflict over location. ILL had only two partner countries at the time of its launch and economic factors were allowed to rule. ESRF was a piece in a larger game between France and Germany, which also settled the matter of location. In contrast to ESO, the XFEL's predefined site in Hamburg seems rather to have been a liability to the host country, because other countries appear to have less interest in participation if the site is already agreed upon. And for the ESS, it is not a stretch too far to claim that Sweden could emerge as winner only after Germany and the UK had withdrawn their site bids. Generally, it seems site-selection is a key piece in the negotiations and lobbying that produce European collaborative Big Science projects. A clear lesson from history, for policymakers to build future efforts on, is that it is only after a viable site has been agreed upon that the real conditions for the process of realization of the project in question are enough known and countries can start evaluating what their level of commitment should be.

But also after a site has been decided upon and the finances secured, struggles over investments and benefits continue. A mechanism put in place to counter the imbalance effects of investment and return that may make hosting of a facility a major economic boost for the local region is so called *Fair Return* (or *Juste Retour*) on procurement, applied in all collaborations under study here, and essentially securing that the collected value of contracts awarded to firms in a member country on long term reflects that country's relative contribution to the budget. An interesting political reality facing coming collaborations such as XFEL and ESS is the apparent illegality of Fair Return under the rules of the common market of EU (Leonard 2010), and new policies have hence been invented to secure return for investment for collaborating countries. Most extensive is the use of *in-kind contributions* by member states, which is the opportunity for member countries to substitute direct financial investment in a facility for the delivery of goods and

[4] Another contributing factor for this might be that the 'decision' was made without a complete technical design and scientific case for the facility. These have reportedly been under development since 2009, and only after their finalization will it be possible to set a price tag for the future facility, for the prospective member countries to negotiate over.

technology and thus spend their money domestically. The policy has some drawbacks; restricting the call for tenders to the participating countries might exclude competitive alternatives, and there is also a risk that at the time of delivery, which might be several years after the in-kind agreement was made, the best qualified company may no longer be in the country providing the particular in-kind contribution. Both the ESS and the XFEL projects will rely heavily on in-kind contributions—approximately half of the total investment, according to estimations (Hallonsten 2012: 309). The challenges associated with this seem new in historical perspective; perhaps lessons learned from the experiences of Fair Return can provide policymakers and lab administrators with some guidance.

The scientific use of a facility, so far unnaturally uncovered in this chapter, has also been the subject of Fair Return-like arrangements, most famously in the case of ILL and ESRF where the allocation of experimental time at the facilities is corrected after the ordinary peer review process to reflect nations' relative financial contributions (Hallonsten 2009: 244–246). In the case of XFEL, it seems the legal documents invite a similar policy to be implemented once the facility is in operation. There are signs that Russia, having obtained strong influence over the governance of XFEL by their large financial contribution (see above), intend to secure access to the facility for its domestic scientific community through the implementation of a strict Scientific Fair Return policy: The November 27, 2009 press release announcing Russia's signing of the XFEL convention stated that "beam usage time *will be* shared proportionally to each country's contribution to the project" (Russian Corporation of Nanotechnologies 2009, emphasis added). Clearly, Russia sees their participation in the XFEL project merely as "the pursuit of [their] interests by other means".

To Russia's defense, it could of course be reiterated that this is the normal procedure for European collaboration in Big Science; this is, furthermore, also a key lesson of this chapter that perhaps could instill some caution among those directly involved in setting up the scientific organization for the XFEL and other projects. One might indeed speculate, as Papon (2004: 70) and Hallonsten (2012: 311) does, that there is a re-nationalization trend in European scientific collaboration, showing itself in individual countries' increased guarding of national interest at the expense of the common good. Hallonsten (2012: 311) mentions a few recent signs of such a development, first Britain's and Italy's 2010 lowering of their contributions to ESRF, and then the apparent strategy of prospective member countries in the XFEL to enter at the lowest level possible, realizing that their influence in the collaboration will be marginal anyway, given the size of the shares of Germany and Russia. Arguing, on scientific grounds, for the abolition of Scientific Fair Return, these countries have even less of a reason to increase their contribution—their domestic scientific communities will have to compete scientifically for access anyway.

Combining the speculation of a re-nationalization trend with the other overall conclusion of this chapter—that European collaborative projects in Big Science tend to mirror the cycles of European integration and geopolitics—amounts to a rather gloomy realization. It should be emphasized that the argument here is not that high-level politics is the only deciding factor for the achievement of a Big Science collaboration in Europe, only that it seems politics trumps most other motives in the *final decisions*. In this perspective, and considering generally the history of Europe in the last 70 years, it is somewhat ironic if this re-nationalization is the overruling trend, given the success of European integration on other arenas. One would then perhaps be prone to refurbish one of the most famous quotes of the former President of the United States Bill Clinton: "It's the economy, stupid!"

References

Berggren, K.-F., & Hallonsten, O. (2012). Timeline of major events. In O. Hallonsten (Ed.), *In pursuit of a promise: Perspectives on the political process to establish the European Spallation Source (ESS) in Lund, Sweden* (pp. 21–30). Lund: Arkiv Förlag.

Blaauw, A. (1991). *ESO's early history: The European southern observatory from concept to reality*. Garching: European Southern Observatory.

Gaubert, A., & Lebeau, A. (2009). Reforming European space governance. *Space Policy, 25*, 37–44.

Grande, E., & Peschke, A. (1999). Transnational cooperation and policy networks in European science policy-making. *Research Policy, 28*, 43–61.

Greenberg, D. S. (1999/1967). *The politics of pure science* (2nd ed.). Chicago: The University of Chicago Press.

Hallonsten, O. (2009). *Small science on big machines: Politics and practices of synchrotron radiation laboratories*. Ph.D. dissertation, Lund University

Hallonsten, O. (2012). Continuity and change in the politics of European scientific collaboration. *Journal of Contemporary European Research, 8*(3), 300–318.

Hallonsten, O. (2013). Introducing 'facilitymetrics': A first review and analysis of commonly used measures of scientific leadership among synchrotron radiation facilities worldwide. *Scientometrics* 96(2): 497–513.

Hallonsten, O., & Heinze, T. (2012). Institutional persistence through gradual adaptation: analysis of national laboratories in the USA and Germany. *Science and Public Policy, 39*(4), 450–463.

Herman, R. (1986). *The European scientific community*. Essex: Longman.

Hermann, A., Krige, J., Mersits, U., & Pestre, D. (1987). *History of CERN. Volume I: Launching the European organization for nuclear research*. Amsterdam: North-Holland.

Hermann, A., Krige, J., Mersits, U., & Pestre, D. (1990). *History of CERN. Volume II: Building and running the laboratory, 1954–1965*. Amsterdam: North-Holland.

Hoerber, T. C. (2009). The European space agency and the European Union: The next step on the road to the stars. *Journal of Contemporary European Research, 5*(3), 405–414.

Judt, T. (2005). *Postwar: A history of Europe since 1945*. London: Pimlico.

Kaiserfeld, T. (2013). ESS from neutron gap to global strategy: Plans for an international research facility after the cold war. In T. Kaiserfeld & T. O'Dell (Eds.), *Legitimizing ESS: Big science as collaboration across boundaries*. Lund: Nordic Academic Press.

Kirschstein, G. (2007). Kühle Begegnung zwischen Merkel und Putin. *Die Welt,* 16 October.

Krige, J. (2003). The politics of European scientific collaboration. In J. Krige & D. Pestre (Eds.), *Companion to science in the twentieth century* (pp. 897–918). London: Routledge.

Krige, J. (2006). *American hegemony and the postwar reconstruction of science in Europe*. Cambridge, MA: MIT Press.

Krige, J. (2014). Technological collaboration and nuclear proliferation: A transnational approach. In: Mayer M, Carpes M, Knoblich R (Eds.), *The global politics of science and technology* (Vol. 1, Concepts from international relations and other disciplines). Heidelberg: Springer.

Leonard, D. (2010). *Guide to the European union: The definitive guide to all aspects of the EU*. London: Profile Books.

Middlemas, K. (1995). *Orchestrating Europe: The informal politics of the European Union 1973–95*. London: Fontana Press.

Papon, P. (2004). European scientific cooperation and research infrastructures: Past tendencies and future prospects. *Minerva, 42*, 61–76.

Pestre, D. (1990). Some characteristic features of CERN in the 1950s and 1960s. In A. Hermann et al. (Eds.), *History of CERN. Volume II: Building and running the laboratory, 1954–1965*. Amsterdam: North-Holland.

Pestre, D. (1996). The difficult decision, taken in the 1960s, to construct a 3–400 GeV proton synchrotron in Europe. In J. Krige (Ed.), *History of CERN* (Vol. III, pp. 65–96). Amsterdam: North-Holland.

Russian Corporation of Nanotechnologies. (2009). *European XFEL project is set to start*. Press release. Accessed February 14, 2013, from http://en.rusnano.com/press-centre/news/88183

Tindemans, P., & Clausen, K. (2003). ESS on hold: Europe urgently needs a strategy'. *Neutron News, 14*, 2.

Trischler, H., & Weinberger, H. (2005). Engineering Europe: Big technologies and military systems in the making of 20th century Europe. *History and Technology, 21*, 49–83.

Of Red Threads and Green Dragons: Austrian Sociotechnical Imaginaries About STI Cooperation with China

Ruth Müller and Nina Witjes

Abstract In this chapter, we address intersections of Science and Technology Studies (STS) and International Relations Theories (IR) by looking at how Austrian science, technology & innovation (STI) policy makers and related stakeholders envision and enact a close relationship between China and Austria in the field of green technologies. Analytically, we draw on the concept of 'sociotechnical imaginaries' as proposed by Jasanoff and Kim, which attempts to grasp how visions of social order and technological development become entangled in the processes of science and technology policy and politics. China is currently arising as one of the key global players in STI. It is intensely courted by numerous countries seeking collaboration and market access. China's most recent Five-Year-Plan has introduced a focus on environmental sustainability to complement economic growth. Based on interviews, participant observation and document analysis, we show how Austrian STI actors connect their national positioning activities to this recent focus on sustainability. We trace how they attempt to find common ground for relating to the rising and ever so dynamic People's Republic by drawing on a nationally accepted narrative about Austrian technopolitical history and culture that casts the country as pioneer of environmental awareness and green technologies. We understand and analyze this process as part of a broader Austrian sociotechnical imaginary in the making that constitutes a situated response to an increasingly globalized STI system, while building on and reaffirming central elements of Austria's local technopolitical culture and history.

Keywords Socio-technical imaginaries • Techno-political culture • Austria • China • Green technologies • Discourse Analysis

R. Müller (✉)
Research Policy Group, Lund University, Lund, Sweden
e-mail: ruth.mueller@fpi.lu.se

N. Witjes
Science and Technology Policy Research Group, Austrian Institute for International Affairs, Vienna, Austria
e-mail: Nina.Witjes@oiip.ac.at

M. Mayer et al. (eds.), *The Global Politics of Science and Technology - Vol. 2*, Global Power Shift, DOI 10.1007/978-3-642-55010-2_4,

1 Introduction

To commemorate the 40th anniversary of diplomatic relations with China in 2011, the Austrian Economic Chambers commissioned a short film entitled "The red thread".[1] Drawing on a Chinese mythological saying that explains that everything and everyone in life that belong together are bound together by an invisible but powerful and unwithering red thread, the film attempts to invoke a longstanding and somehow 'destined' connection between Austria and the People's Republic of China. The film centers on the unfolding love story of two young middle-class Chinese people, who are in their lives not only connected by the magical red thread but also surrounded by Austrian products, culture and opportunities. Austria, it seems, is not a small, far away country, but already and almost indispensably part of growing Chinese middle class wealth and China's future economic and social development.

As we watch Xue and Jianguo slowly fall in love with each other, the suggested omnipresence of Austrian goods in Chinese daily life is pointed out through pop-ups of Chinese characters, giving information about the products that are visible in the scene or the Austrian invention that gave rise to them—sun-roofs, fire trucks or the fact that the AB0 blood group system was developed by Austrian scientist Karl Landsteiner. Just as Xue warms up to Jinguo, the young lovebirds are separated for a year, when Jinguo is awarded a fellowship to study at a technical University in Austria. Luckily, they meet again, when both receive awards at the Austrian Embassy in Beijing, Xue for her performance of a Mozart sonata and Jinguo for his achievement as an exchange student in Austria. When receiving his award, Jinguo tells us that being back in China now, his mission is "*to help my country with clean, renewable and efficient solar energy*", which he aims '*to bring [. . .] to China*" by working for an Austrian company. The film ends, when Jinguo proposes to Xue on a ski slope, after he has tested that she is adventurous enough to try this new sport and hence, in his eyes, adventurous enough to take on this changing world and its global connections, challenges and transformations with him.

Why are we telling this story? We are leading our paper about the ways in which science, technology & innovation (STI) collaboration with China is currently imagined among key Austrian STI stakeholders with this brief recount of 'The Red Thread', because this promotional film already includes many of the elements that are significant for the ongoing efforts to expand into and work with China in this sector. It entails the invocation of a long-standing history and exchange; it shifts the attention from a focus primarily on Austria's culture and history towards its technoscientific present and future; and within this area, it particularly emphasizes Austria's knowledge and innovation capacities in the field of 'green' technologies; technologies that enable sustainable societal development—something that, as the

[1] Muh, Victor (dir.) (2011). The Red Thread. 30 min. short film, commissioned by the Austrian government to commemorate the 40th anniversary of diplomatic relations with China.

movie suggests, China needs and wants. What we are interested in this paper is how this narrative is constructed. What are the strings that Austrian STI stakeholders use to spin this 'red thread'? Which images of China and of Austria are created, contested and engaged within this process? Which aspects of Austria's techno-political history and culture are mobilized to craft this connecting yarn? The paper takes the example of 'green technologies/sustainable development' (a key aspect of China's current Five-Year-Plan) as a case for how Austrian policy makers envision and enact alignment between the two otherwise very different nation states of China and Austria. It analyses this process as part of a specific Austrian sociotechnical imaginary (Jasanoff and Kim 2009) in the making. An imaginary that constitutes a situated response to an increasingly globalized STI system as it builds on, expands and reaffirms central elements of Austria's techno-political culture and history (Felt et al. 2010; Felt and Müller 2011). This culture is characterized by attempts to "*keep technologies out*" (Felt 2013) that are perceived to be 'unnatural', such as nuclear power or agricultural biotechnologies. It is grounded on past and present discursive practices that have, over nearly four decades, since the initial rejection of nuclear power plants on Austrian territory, added to creating an understanding of Austria as a nation committed to preserving 'nature' as well as acting as a pioneer in biocompatible technologies, e.g. in the area of 'green energy'. It is based on this assumed role as a 'bio-pioneer' that Austrian STI stakeholders find common ground for 'approaching China together'.

2 Theoretical Perspectives

2.1 *Bringing IR and STS Together*

This paper builds on theoretical perspectives from both International Relations Theories (IR), and Science & Technology Studies (STS). In IR it is particularly constructivist approaches (Wendt 1992; Debrix 2003; Ulbert 2005) that allow for integration with key assumptions of most contemporary STS scholarship, which we subsume on the heading of material-discursive (Barad 2007) or, more simply, 'co-productionist' (Jasanoff 2004) approaches.

Social constructivist approaches in IR can be understood as a "critical reflection on the limits of neo-utilitarism" (Ruggie 1998: 856), which dominates the mainstream of IR in the form of neo-realism (e.g. Waltz 1979; Keohane 1986). As Herrera (2003) points out, the relations between neo-realism and constructivism in IR are somewhat similar to what the relations between technological determinism and social constructivism have been in the shaping of the field of STS. While neo-realism understands states as pre-existing entities, whose identities and interests are exogenous and given or can at least be clearly derived from the assumed conditions of anarchy between them, technological determinism (e.g. Heilbroner 1967) assumed technology as 'outside' the social, but acting upon it in certain given ways. In both fields the 1980s brought a critique of these approaches, arguing that both states (Onuf 1989) and technology (Pinch and Bijker 1984) are constructed in social processes and are constantly subject to social interpretation. In both fields, constructivism proposed a shift towards the social becoming of structures, entities and artifacts formerly assumed as pre-existent. In Foucauldian terms, they proposed a shift towards the study of the multiple discursive practices through which states and technologies come into being and remain in existence. In STS this proposition was soon met with a call for symmetry: yes, technology is constructed and related to in social processes, but its own agency cannot be neglected either. The material world, and with it technology, is not just constructed and acted upon, but it is itself an important actor co-constituting the social. Today, a number of articulations of this basic stance exist in STS: Actor-network-theory (Latour 1987, 2005; Law 1992) the material-semiotic/material-discursive approaches of feminist science studies (Haraway 1988; Barad 2007) and the co-productionist framework (Jasanoff 2004). Similarly, in the field of IR, today many "constructivists hold the view that the building blocks of international reality are ideational [i.e. discursive] *and* material"[2] (Ruggie 1998: 879). When it comes to understanding the role of science and technology in international relations, there is hence potential for fruitful synergies between those two approaches in IR and STS.

Today, technology is at the same time a condition, a means and a topic of international relations (Herrera 2003). For the analysis of these complex relationships, we believe one more perspective shared between constructivism in IR and STS to be of great importance. That is a commitment to specificity and context. Analysis in these framework understands that no argument can be made about the role of technology *in general*, but only about a *specific* technology in a *specific* social and historical context (Hecht and Allan 2001). Today, these contexts are increasingly shaped by both local and global elements, even if the context studied is referred to as 'national' (Martello and Jasanoff 2004). With regard to the national, constructivist approaches in IR shift the attention away from a focus primarily on states and their governments towards analyzing the agencies of multiple government and non-government actors that are contributing to the formation of what is then called "the state" and its "international relations". These actors might or might not be located within the limits of the particular nation state under analysis. An analysis of international relations then hence always becomes an analysis of processes on the multiple local and global levels (Rosenau and Czempiel 1992; Messner and Nuscheler 2003; Dingwert and Pattberg 2006). Thus, although we draw on the specific socio-technical histories of *Austria* as well as the perception of China by *Austrian* actors in the field of science, technology and innovation policy, our analysis necessarily oscillates between the national (local) and the transnational (global) level—or between what Martello and Jasanoff term simultaneous processes of *localization* and *globalization* (2004). In our cases, this means that while processes of Austrian STI strategy coordination are governed by national

[2] Emphasis by the authors.

institutions and echo the interests of specific national government and non-government actors, they are at the same time responding to and integrating a variety of transnational initiatives and discourses.

2.2 *Socio-Technical Imaginaries*

Given this multitude of actors, how then do seemingly 'national' positions in STI politics and policy emerge? Positions, which might be contested, but are nevertheless shared and supported by enough heterogeneous actors to allow for action? To address these questions we turn to a conceptional approach outlined by Jasanoff and Kim (2009). Drawing on their comparative work on the roles of nuclear technologies in different national contexts, they propose the framework of 'socio-technical imaginaries' as a tentative answer to the above posed questions. Jasanoff and Kim define

> national sociotechnical imaginaries as 'collectively imagined forms of social life and social order reflected in the design and fulfillment of nation-specific scientific and/or technological projects.' Imaginaries, in this sense, at once describe attainable futures and prescribe futures that states believe ought to be attained (2009: 120).

Their notion of state is—as in constructivist IR—one that assumes that it is being co-constituted by a multitude of actors. To understand the relationships between a specific kind of technology and a specific national context, they propose to attend to the multiple negotiation processes between these actors, to distinct national socio-technical histories and discourses as well as to global dependences and interdependencies. Importantly, the concept of socio-technical imaginaries attempts to grasp how visions of *social* and *technological* progress and order become entangled within the processes of science and technology politics and policy. In their empirical work, Jasanoff and Kim trace how specific imaginaries about the relationship between the social and the technological and their futures become dominant in a specific national context. They show how narratives about the types of knowledge and technologies that should be embraced or rejected, and about how they should be employed, promoted and regulated, are always also narratives about specific social forms and not others, about specific versions of societal organization, governance and well-being. Sociotechnical imaginaries are value-laden constructs that envision possible societal and technological futures, attempt to order their contested presents and in the process both draw on and recreate a nation's socio-technical past.

As STI systems become increasingly global, national sociotechnical imaginaries include increasingly complex narratives about other countries as well, about their techno-political pasts, presents and futures; narratives, which can be contested and ambivalent but nevertheless meaningful within national attempts to achieve shared positioning towards other actors in order to be able to move forward. In this sense, national sociotechnical imaginaries have to be understood as participating in processes of both globalization and localization, as they are forms of discursively relating the local and the global. In the case of China, Pan (2004) argues that the idea of China as a threat to Western countries' political and economic dominance and welfare is a particularly powerful framework shaping national positioning work towards the People's Republic. While we find that the narrative of the 'China-threat' is clearly also the background against which our story about China, Austria and green technologies develops, in this context is appears to somewhat break down and translate into a narrative about China as an 'unavoidable opportunity' (Bruijn et al. 2012: 16). While the notion of China being 'unavoidable' still contains elements of threat and suggests that not engaging with China would come at (potentially high) costs for the reluctant nation state, at the same time the focus shifts to questions of opportunity and hence of how to relate beneficially. It is in this context and as a response to it that Austrian stakeholders develop the narrative of the metaphorical 'red thread' instead of 'threat' to invoke a grown, benign, progress-oriented, mutually beneficial and almost self-evident connection between Austria and China.

2.3 Technopolitical Histories and 'Culturally Appropriate' Technologies

To understand the knitting of this thread in the specific Austrian context, we turn to the few available studies of Austrian technopolitical culture. In conversation with Jasanoff's and Kim's work Felt and co-authors draw our attention to the importance of detailed analysis of the distinct *technopolitical history and culture* of a country for understanding contemporary positioning work towards new technologies and within global STI systems (Felt 2013; Felt and Müller 2010; Felt et al. 2010). Their work offers a historically grounded analysis of how 'Austrian' positions towards novel technologies, such as nanotechnology, build on prior negotiation processes regarding nuclear power and agricultural biotechnologies. In her most recent work on this topic, Felt (2013) particularly shows how discussions about new technologies in Austria, e.g. in focus groups with lay people, draw strongly on a version of Austrian technopolitical history to which attempting and achieving to '*keep [certain] technologies out*' is elementary:

> In all [focus] groups, [...] the strong opposition to nuclear power plants on Austrian territory since the 1970ies (e.g. Nowotny 1979; Torgersen 2002) as well as the rejection of genetically modified food/crops about two decades later (e.g. Seifert 2003) served as a shared frame of reference. A tacit agreement seemed to exist about their importance for Austrian (technopolitical) culture. [...] Keeping these specific technologies out created the imagination of a well delimited Austria, in its sociotechnical practices different from 'the others.' [...] with it emerged a particular kind of "imagined community": of Austrians as being "naturally" opposed to nuclear energy, as refusing certain interventions into nature and food, short as being close to specific values (Felt 2013: 4, 16).

Connecting this specific Austrian "*repertoire of sociotechnical resistance*" (2013: 6) to Benedict Anderson's (1983) work about *imagined communities* as being foundational to the formation and cohesion of nation states, Felt shows how the rejection of certain technologies gets "*woven into partly mythical narratives that participate in the formation of a specific kind of national technopolitical identity*" (2013: 3). This identity however is only partly defined by the absence of certain technologies; rather the emergent narrative space becomes filled with counter-images, which, as Felt indicates, constitute an "*alternative innovation space with at the core a clean energy production through hydroelectric and solar power as well as an organic food culture caring for its environment.*" (2013: 17) These 'green' technologies can be understood as "*culturally appropriate technologies*" (Martello and Jasanoff 2004: 8) in a country that likes to think of itself as a pioneer in and ambassador of environment-friendly technologies and lifestyles; an "*organic food nation*" (Felt 2013: 4), in which the only nuclear power plant that was ever built never went online due to a public vote, its ruins serving today as a symbol of citizens' power to reject.

In Austria, 'green technologies' hence offer the opportunity to promote mid- and high-tech Research and Development (R&D), while at the same time keeping the likelihood of public dissent and controversy very low. In their study of 'sustainable energy regions' in Austria, Späth and Rohrracher (2010) show how initiatives to create such regions met with surprisingly little resistance both within local communities (including local entrepreneurs) and on different levels of government.[3] Combined with the simultaneous basic and applied character of this research field, 'green technologies' thus present an attractive case for policy makers and others, who are interested in getting many actors aboard their enterprise of promoting Austria as a country of science, technology and innovation. For this endeavor it is not necessary that Austria 'is' a pioneer in green technologies or a particularly environment-friendly country; indeed Austria has been repeatedly criticized by environmental activist groups, such as Greenpeace, for upholding the "*myth*" of being environment friendly, but at the same time not meeting the Kyoto Protocol climate protection goals (Graber 2013). While such criticism can gradually crumble a sociotechnical imaginary,[4] in the Austrian case, to date, the myth manages to outshine parts of the reality. It is based on this well rooted narrative that Austrian STI stakeholders are constructing an imaginary on how to relate to China at this moment of change and uncertainty.

---

[3] However, they also show how this striking level of consent is based on avoiding too controversial topics in the first place, e.g. the sensitive topic of switching from cars to public transport in rural regions. Felt et al. (2010) also indicate that Austria's technopolitical culture shows strong conflict-adverse elements.

[4] See e.g. changes in the US imaginary regarding nuclear technologies (Jasanoff and Kim 2009).

3 Material and Methods

3.1 Project Background

While Felt and co-authors built their work on Austrian sociotechnical imaginaries mainly on focus group and interview work with lay citizens, this paper offers insights into ongoing discussion processes from a different end of the spectrum of involved actors. The analysis in this paper draws on interview work, participant observation and document analysis in two projects at the Austrian Institute for International Affairs (OIIP), funded by and performed in cooperation with the Federal Ministries for Transport, Innovation and Technology (BMVIT) and for Science and Research (BMWF).[5] In both projects, the project team accompanied and analyzed multi-stakeholder-processes of discussion and coordination pertaining to Austrian STI-internationalization beyond the EU and its associated states. While one of the projects was focused on STI-internalization more generally, the second project was specifically focused on the possibilities and challenges of STI-cooperation with China.

3.2 Material

As part of the analytical input for stakeholders within the Federal Ministry for Transport, Innovation and Technology concerned with strategy development regarding China, the project team conducted a series of interviews with a broad range of individuals with different types of expertise pertaining to China. The team conducted 13 semi-structured interviews with experts from the fields of basic research, applied research, economy/business and diplomacy. All participants either had experiences of working with/in China in the field of STI and/or were in charge of the broader internationalization activities of their institution, for which China is a reoccurring topic. Further, the project team was present at, contributed to and documented a number of stakeholder and working group meetings, some of which engaged with STI-internationalization activities in general, while others focused on STI-cooperation with China in particular. In addition, the project team conducted a document analysis of a range of written texts. These included (1) media articles concerning the relationship between Austria and China in the fields of green

[5] Project (1) *Wissenschafts- und Technologiekooperationen mit China: Hoffnungen, Möglichkeiten und Herausforderungen*. Project team: Heidenwolf, C./Müller, R./Rampke, B./Sigl, L. Project duration: 01.2012-03.2013. Project (2) *Analytische Unterstützung der FTI-Arbeitsgruppe 7a Internationalisierung*. Project team: Müller, R./Heidenwolf. C./Sigl, L. Project duration: 03.2012-06.2013.

technologies and sustainability in three different Austrian newspapers[6] from January 2006 to March 2013: (2) policy documents and public accessible official statements by Austrian ministers and other relevant actors on the topic and (3) announcements, promotional material, and reports from Austrian export-oriented business actors.

3.3 Method of Analysis

While Jasanoff and Kim propose an intriguing conceptional framework, their work offers less insights into how to engage with sociotechnical imaginaries methodologically. Here, we turn to Hajer's approach of *argumentative discourse analysis* and particularly to his work on discourse coalitions and the institutionalization of discourse. In line with constructivist approaches he proposes to analyze discourses at a concrete-political level (Ulbert and Weller 2005) and aim at understanding how and under which conditions a specific discourse becomes dominant within a national setting. In a Foucauldian tradition, Hajer understands discourses as "ensemble[s] of ideas, concepts and categories produced, re-produced and transformed within a specific set of social practices and thereby giving meaning to the physical and social reality." (Hajer 2003: 279). Discourse coalitions are groups of actors that participate in the formation and propagation of a specific discourse. As they might be quite diverse, 'storylines', simplified and condensed articulations of the key elements of the discourse, serve as media to organize the actors themselves and to further 'impose their view of reality on others' (Hajer 1995: 47).

Hajer's methodological approach is helpful for analyzing complex processes of knowledge negotiation between multiple actors, in our case in order to understand the formation of national sociotechnical imaginaries. Multiple stakeholders within ministries, academia, business and diplomacy take part in the processes of negotiating a common position towards China, contributing different sets of knowledge and at times diverging agendas from their specific areas of expertise and activity. In order to create mutual understanding and political relevance, knowledge from these specialized discourses has to be transferred to the level of broader social understanding (Hajer 1995: 61f.). As in other cases (Hecht and Allan 2001; Jasanoff and Kim 2009), this is achieved by drawing on familiar elements of Austria's national identity and socio-technical history.

[6] Analyzed newspapers (online versions) were "Der Standard" (Vienna), "Die Presse" (Vienna) and "Profil"(Vienna).

3.4 Confidentiality and Consequent Limitations

Austria is a small country. The number of individuals with STI expertise regarding China is hence obviously limited. Thus, while this article builds equally on the analysis of expert interviews, field notes from participant observation, policy documents and media articles, we decided to only quote directly from the latter two kinds of official documents to ensure the anonymity of our interview partners. These quotes however are indicative of broader discursive storylines pervading the entirety of the collected material.

4 The Discursive Knitting of the Red Thread

4.1 The Red Dragon Rises

In this next chapter we will trace how Austrian STI actors discursively enact green technologies as an obvious and timely connection between Austria and China. While a number of countries, such as India and Brazil are currently emerging as increasingly influential players in the global STI system, no other country has undergone remotely as radical a transformation in the field of Science, Technology and Innovation as China. Having been considered mainly a center for the industrial production of low-level technological goods for a long time, China is swiftly and constantly developing its STI sector. Being known for copying innovation in this process, it increasingly shifts to original homegrown research and creation. China is now claiming a place amongst the leading nations in a growing number of scientific and technological fields. As a consequence, the interest in cooperating with China has increased significantly throughout the world, and with it did attempts to establish new relationships and intensify existing ones.

This is the case in Austria, too, where the interest in economic and technological cooperation with China has grown rapidly over the last years. As in most countries, this rising attention is motivated partially by the objective of establishing stronger ties to the growing Chinese academic excellence, but more significantly by China's potential as a rapidly growing market for export and local adaptation of mid- to high-tech products. With its steadily expanding infrastructure projects and the rising consumption of its burgeoning middle class, China currently depends significantly on imports from other countries to keep up with the pace of its own development and its people's desires. Yet, Chinese policy makers emphasize that while they seek out collaboration and are interested in importing technologies, China is also striving for advancing its economic and technological independence (Schwaag Serger and Breidne 2007).

In this short intro, we already see a number of temporalities at play that are pervading and structuring the narratives about and images of China that emerge in the discussion about STI cooperation with China among Austrian stakeholders.

China is sketched as a formerly backwards country, a country that has damaged its scientific institutions and STI labor force severely during the Cultural Revolution (see also Benner et al. 2012: 3ff) and has for the last decades at best been the 'work bench' of the world, but clearly not a center of technoscientific development and excellence. It further discursively emerges as a country that at some point has realized its 'mistakes' and set foot on a path of emulating the STI development of today's post-industrial countries.[7] Yet, once on this path China started to travel it without necessarily attending to all the rules its predecessors had set out and allegedly obliged to (e.g. regarding intellectual property rights). Devoting significant resources to this endeavor, China is now perceived to be a country on the fast track to innovation leadership that threatens to outpace its prior role models. Its centralistic, non-democratic structure and non-compliance to international rules is perceived to enable both the speed and the determination of this development. Where this development will exactly lead in the longer run, appears as unclear to most Austrian STI actors. With its growing STI sector, its huge domestic market and strategically ensured access to the natural resources of many developing countries of the world, it becomes less and less clear to which degree and for how long China will remain dependent on what is referred to as the Western world.[8]

Austrian stakeholders' imaginaries about China reflect and remake the speedy temporalities of China's recent past, bustling present and probably effusive future. Take this quote by an Austrian Professor of Agricultural Sciences, who is remembering his earlier and present academic collaboration with China:

> There have been times when people were working in laboratories and offices without windows, wearing seven underpants and scarfs in the wintertime. Nowadays they have the most modern PCs and better equipment than we have at [our] institute [...].

This quote is illustrative of images of China's past—China as technological backwards country—and of its current and future direction—China as an ever more dominant technological (and political) power—that are shared in virtually all stakeholder statements.[9] They are almost always connected not only to a narrative of China catching up, but pushing ahead of what is perceived as its former role models. However, while in this narrative the past and the future appear rather settled, the question becomes *how to relate to China in the present*; in a present that seems utterly dynamic and multiple; in which the 'old' China still exists in so many parts and aspects of the country, while the 'new' China is arriving at what seems to be warp speed; and in which technological progress does not necessarily entail assimilation to 'Western' culture[10] and hence reduction of difference, complexity and necessity for more than lingual translation.

---

[7] See Grunwald (2009) for a discussion of the history of the Chinese STI system.

[8] See Woods (2008) for a discussion about China as an emerging donor in development assistance.

[9] In interviews as well as in media articles.

[10] We are highly aware of the complexity and difficulty of the term "Western culture" (as of "West" and "culture" in general). We use it here, provocatively, to indicate the implicit assumption of many stakeholders that there is but one trajectory of (technological and social) progress, which

In the discussion among STI stakeholders, a key question thus becomes whether or not a country like Austria *can* profit from China's rise or will inevitably be part of the countries that are being ousted by its growing power. This question is the background, the foundational uncertainty against which the discursive positioning work of Austrian STI actors needs to be understood. The current situation is imagined as one in which China is highly courted and basically picks whom they want to collaborate with or give access to their markets. Why then should China pick Austria?

4.2 The Dragon Goes Green

The discursive response to this question hinges on interpretations of political, social and economic developments in China that culminated in the release of China's 12th Five-Year-Plan (2011–2015). Since 1953, the Five-Year-Plan (FYP) has been China's most important instrument for domestic planning, as it is defining and explicating the country's main aims for social and economic development within the next 5-year period. The 12th FYP defines social security as well as the ecological and sustainable development of the national economy as its key goals. In contrast to former plans it shows moderate growth targets, which are 4.5% below the average growth of the last years. The plan emphasizes the importance of reducing both China's excessive export dependence by strengthening the national potential for genuine innovation and its dependence on external resources (e.g. fossil fuels) by developing a „low-carbon economy". The acquisition, creation and use of 'green' and 'low-carbon emission' technologies are core aims of the plan. The 12th FYP thus indicates a significant shift in China's official narrative of what constitutes 'development', moving towards a definition that does not only include the accumulation of economic wealth, but also environmental sustainability. Clean water, fresh air, non-toxic soil and seemingly undisturbed landscapes become symbols of progress rather than relicts of a rural past and so do the technologies that promise to enable creating and sustaining such kinds of environments.

Austrian actors relate to these developments by indicating that China is now moving onto a path Austria has already been travelling for a long time—longer than most other countries. Its nature—air, water, landscapes—becomes proof and

will lead to China becoming more like 'the West' in its values, economic and political model, forms of communication and social etiquette as well as cultural and lifestyle practices. We understand 'the West' not as geographical location, but as a symbolically and politically meaningful construct in the historically grounded and ongoing negotiations of power and dominance – more simply put negotiations about "who lives, who dies, and at what price", political questions that are always also "inside of technology" (Quotes from the Essay "Your are cyborg" by Hari Kunzru (1996), citing Donna Haraway. Available online: http://www.wired.com/wired/archive/5.02/ffharaway.html. Accessed February 2014).

symbol of this history and ongoing practice as well as the resulting expertise. Austria's nature is, so they argue, *enabled by a certain relationship with and to technology*. Sophisticated waste disposal systems, energy from waterpower, well-developed systems of public transport, energy-efficient construction of buildings, all are cast as contributing to the creation and maintenance of this kind of nature. The quote below by the Austrian Federal Minister of Economy, Family and Youth about the Austrian pavilion at the World Expo in Shanghai in 2010 illustrates this argument.

> We do not only present Austrian history and culture at the Expo, but also our clean environment and the high quality of life. Together with the numerous personal contacts we make, we are creating an important platform for closer cooperation in the field of eco-innovations (BMWFJ 2010).

In this discursive element, nature and technology are not offered as opposites: rather, the narrative suggests that in the 3rd millennium, nature increasingly depends on the *right* technologies, sustainable and green. Technologies, which as the narrative further elaborates have a long tradition in Austria, as they have a 'natural' fit with Austrian mentality and its people's relation to nature. The quote below by the Austrian Minister of Science and Research is indicative for this unfolding storyline,

> The ecological consciousness is, compared to other countries, particularly pronounced in Austria. To take full advantage of this is not only a touristic, but an ecological and scientific chance (BMWF 2012).

Thus, though there has been cooperation in the field of STI between Austria and China since 1984,[11] the FYP 2011–2015 offers particularly suitable attachment sites for Austrian narratives about scientific and technological pasts and futures and for many of its key STI actors. In a situation of rapid change and multiple ambivalences, in which it is hard to know how to relate in the present, drawing on a culturally established narrative of Austria as a 'green' country allows imagining a vantage point from which to relate to China. Summing up a broadly shared consensus and impetus among Austrian STI actors concerned with China, the Austrian Federal Minister of Agriculture, Forestry, Environment and Water Management announced already in 2010: "We want to be there right from the beginning, when China's dragon goes green!" (Lebensministerium 2012) That it makes sense to be there, or that China could be particularly interested in Austria's presence above other nations, is argued through the narrative about Austria's deeply rooted expertise in this field, but also through a narrative of longstanding, rather uncontroversial connection between Austria and China.

---

[11] The *Science and Technology Agreement* (April 1984) builds the foundation for bilateral STI-cooperation between Austria and China. Both partner agreed on the funding of joint research projects.

4.3 The Red Thread

China has entered the green path Austria is already travelling. This narrative creates similarity between two otherwise very different countries. It mobilizes specific pasts, presents and futures of both countries and rearranges them in ways invoking connection. To bolster this emerging narrative of relatedness, Austrian STI actors additional emphasize on the one hand a presence of a (not necessarily STI related) *positive* history between the countries and on the other hand an *absence* of negative history. Particularly political representatives emphasize the "*longstanding diplomatic relations*"[12] between Austria and China, whose 40th anniversary has recently been celebrated; their shared status as 'nations of culture'; and the 'positive image' Austria enjoys in China (Xu and Yafang 2006). While these platitudes are of course part of the standard repertoire of any diplomatic verbiage geared at improving relations, it is interesting that in the Austrian case these accounts of positive presences are complemented by what Felt calls "*imaginaries of the absent*" (Felt 2013: 3). For example, a former Austrian Minister of Economy and Work[13] stated that Austria was benefiting from a "*lack of negative historical experiences*" of China with Austria. He further argues, rather mysteriously, that "*due to our history we Austrians are well able to empathize with and understand others*",[14] which he considers a key skill for succeeding on the Chinese market. In a global network of colonial and post-colonial connections, Austria is sketched as a country devoid of such histories, a harmless country, politically neutral; its people cooperative, accommodating and adaptive; an ambassador of green technologies, who is again characterized by being *free of* something rather than by strong presences (Felt 2013: 9ff.). While Felt shows us how the narrative of 'freedom' in Austrian technopolitcal culture entangles "*the material absence of a given technology [with] the imagination of political freedom*" (Felt 2013: 17)—emphasizing the people's ability to *choose* not to let certain technologies in—here we see how in a global context this nexus becomes further linked to the absence of certain 'negative' histories of global relations. As Wodak et al. (2009) point out, this implies fading out certain histories, e.g. the existence and agencies of Austro-Fascism, a trope that becomes particularly visible when Austria's international image is compared to what is invoked as the much more charged history and image of Germany. Similarly, while focusing on sustainable development and problematizing the current state of nature in China and its global consequences, questions of human rights violations in contemporary

---

[12] Homepage Austrian Ministry of Foreign Affairs: Bilaterale Beziehungen - Österreich und China http://www.bmeia.gv.at/botschaft/peking/bilaterale-beziehungen/bilaterale-beziehungen-oesterreich-und-china.html (Accessed February 2014).

[13] Austrian Ministries tend to change names quite often; areas of competence are renamed, fused or split. The former Ministry of Economy and Work was roughly what now is the Ministry of Economy, Family and Youth.

[14] Austrian Minister of Economics, cited in *Wirtschaftsblatt* (2007) (Link is not available anymore).

China are hardly a topic of discussion in the STI collaboration discourse.[15] One could argue that one of the many 'benefits' of the discursive focus on green technologies is that it allows postponing (or avoiding) the human rights discourse altogether, as one is attending to other urgent matters.

This is but another aspect that makes green technologies such an attractive medium for relating to China discursively, strengthening the emergent discourse coalition and their storylines, and drawing new members to their cause. What are the current impacts and achievements of this coalition? First, it has to be noted that one primary achievement is the establishment of the coalition itself. It is remarkably heterogeneous, including members of virtually all relevant ministries, industrial leaders and members of their representative organizations, scientists and R&D developers. While the technology and export oriented actors are clearly pursuing relations with China more adamantly than actors related to academia, who are still hesitant regarding their benefit for strengthening Austria's academic excellence, none of the actors interviewed objected to the basic storyline that China should be engaged and that this should happen with a focus on green technologies. The storyline of green technologies as a connective thread with China was part of each interview with virtually any of the above mentioned stakeholder groups. The Austrian Ministry of Transport, Innovation and Technology is currently pushing ahead by drafting a 'China Strategy' for the resort, a move that has received attention by other stakeholders. The green tech focus of the Austrian pavilion at the 2010 Shanghai World Expo is another example for the emergent physical manifestations of the ongoing discourse. As are a growing number of memoranda, agreements and to some extent contracts between Austrian and Chinese stakeholders in the area of low carbon construction and infrastructure. Whether these efforts will lead to more stable manifestations of the socio-technical imaginary currently constructed by Austrian stakeholders remains to be seen. This will depend on a number of factors, prominently among them China's ambitions to increasingly achieve technological self-sufficiency. As the Science Counselor at the Embassy of the People's Republic of China in Vienna states "*We are facing a strategic turn: Whereas foreign technologies were required for the previous economic recovery, Research and Development will be self-sufficient in the near future*".[16]

If and when this emergent socio-technical imaginary of STI independency, which currently still co-exists with China's "*extended hand*" (Bruijn et al. 2012) seeking cooperation and transfer, gains momentum, it will demand new positioning work on the side of Austrian (and other) STI stakeholders, requiring the adaptation of current imaginaries or the creation of new ones.

---

[15] We would like to emphasize that it is not only in STI negotiations with China that human rights questions are not being raised; yet there is a stronger expectation that China needs to be confronted. Hardly anyone however would expect Austria to address e.g. the existence of Guantanamo in STI negotiations with the US.

[16] http://www.profil.at/articles/0543/560/124491/china-drachensteigen (Accessed February 2014).

5 Conclusions: Waltzing with the Dragon: The Technopolitics of the 'Natural'

In this paper, we have tried to sketch the discursive positioning work of Austrian STI stakeholders towards cooperation with China; in other words we have attended to how they are crafting a specific socio-technical imaginary as a shared basis and strategic focus for ongoing and future STI engagements with China. We have shown how in the discursive negotiations we analyzed Austria emerges as bio-pioneer, a green country, in which 'nature' is enabled rather than disturbed by (the right) technologies. Building on the historical fact and fiction of successfully keeping certain technologies out, Austria has—at least discursively—become an "*alternative innovation space*" (Felt 2013: 17) devoted to green technologies and sustainable energy. In this narrative, its 'intact' nature becomes a symbol of its technological capacities. Critical voices explicitly refer to this narrative as myth, and point to Austria's failure to meet the Kyoto protocol goals. Austria further emerges as a harmless country, *free of* negative (imperial, colonial) history. Austrian STI actors engaged with internationalization draw these aspects together to sketch Austria as an experienced, neutral and engaged ambassador of green technologies, willing to travel east to promote its nature-enabling technologies and adapt them to the local contexts. While China and Austria are narrated as always already connected countries (like Xue and Jianguo, whom we met in the introduction to this paper), it is the current moment in time, the moment of 'green revolution' that brings them, almost naturally, closer together. Through this imaginary, a moment of uncertainty in the face of China's rapid development is recreated a moment of obvious opportunity.

We understand this very specific imaginary as part of a broader Austrian technopolitical discursive landscape and as related to other, topic-specific imaginaries. Together they constitute a specific, historically grounded national technopolitical culture that needs to be better understood, both *for and through* the analysis of international relations in the area of science, technology and innovation. While analysis that primarily focuses on the national is worthwhile, analysis that focuses on the imagined and enacted relations of more than one nation allows for understanding how national sociotechnical imaginaries and their processes of constitution are at once situated responses to increasingly globalized STI systems as well as building on, contesting and reaffirming very local elements of a country's technopolitical culture and history. They are hence simultaneously participating in processes of both *localization* and *globalization* (Martello and Jasanoff 2004). These entanglements between the local and the global, and particularly their historicity and genealogy, need to be taken into account for understanding the becomings of international STI relations, e.g. in the area of science diplomacy or international cooperation. Understanding national socio-technical imaginaries, the ongoing discursive positioning work of the complex networks called nation states, requires attention to global connections and local differences and promises to be a dynamic field of contemporary research. We understand this type of analysis as contributing to both knowledge production in the constructivist tradition in IR and to research in STS that is interested in mapping transnational phenomena or doing comparative work, as well as politically important for understanding the ever more prominent and complex role of science and technology in international relations.

References

Anderson, B. R. O'G. (1983). *Imagined communities: Reflections on the origin and spread of nationalism*. London: Verso.

Barad, K. (2007). *Meeting the universe halfway: Quantum physics and the entanglement of matter and meaning*. Durham: Duke University Press.

Benner, M., Liu, L., & Serger, S. S. (2012). Head in the clouds and feet on the ground: Research priority setting in China. *Science and Public Policy, 39*(29), 1–13.

BMWF (Austrian Ministry of Science and Research). (2012). *Alpbach - Karlheinz Töchterle am Tiroltag: Ökologisches Bewusstsein ist Chance für die Europaregion Tirol*. Press release. August 19 2012. Accessed February 12, 2014, from http://www.ots.at/presseaussendung/OTS_20120819_OTS0026/alpbach-karlheinz-toechterle-am-tiroltag-oekologisches-bewusstsein-ist-chance-fuer-die-europaregion-tirol

BMWFJ (Austrian Ministry of Economy, Family and Youth). (2010). *Mitterlehner: Österreich punktet in China mit ÖkoInnovationen*. Press release. July 5 2010. n.p. Accessed February 2014, from http://m.bmwfj.gv.at/Presse/Archiv/Archiv%202010/Seiten/Mitterlehner ÖsterreichpunktetinChinamitÖko-Innovationen.aspx

Bruijn, J. A., Adriaans, P.W., Hooymans, C. M., Klasen, E. C., Morley, P. et al. (2012). *China's extended hand. How Chinese and Dutch knowledge can strengthen each other*. Dutch Advisory Council for Science and Technology Policy (AWT). Accessed February 2014, from http://www.awt.nl/upload/documents/publicaties/engels/a78uk.pdf

Debrix, F. (2003). *Language, agency, and politics in a constructed world*. New York: Armonk.

Dingwert, K., & Pattberg, P. (2006). Global governance as a perspective on world politics. *Global Governance, 12*(2), 185–203.

Felt, U. (2013). *Keeping technologies out: Sociotechnical imaginaries and the formation of a national technopolitical identity*. Pre-Print; Published by the Department of Social Studies of Science, University of Vienna, February 2013. Accessed February 2014, from http://sciencestudies.univie.ac.at/publications

Felt, U., & Müller, R. (2011). Tentative (Id)entities. On technopolitical cultures and the experiencing of genetic testing. *BioSocieties, 6*(3), 342–363.

Felt, U., Fochler, M., & Winkler, P. (2010). Coming to terms with biomedical technologies in different technopolitical cultures. A comparative analysis of focus groups on organ transplantation and genetic testing in Austria, France, and the Netherlands. *Science, Technology, and Human Values, 35*(4), 525–553.

Graber, F. (2013). *Die Kyoto-Ziele waren überehrgeizig*. In: Wirtschaftsblatt: 26.04.2013. Vienna: Online edition. Accessed February 2014, from http://wirtschaftsblatt.at/home/nachrichten/oesterreich/wien/1394369/Die-KyotoZiele-waren-uberehrgeizig

Grunwald, R. (2009). China. In D. Simon, A. Knoe, & S. Hornbostel (Eds.), *Handbuch Wissenschaftspolitik* (pp. 518–533). Wiesbaden: VS Verlag für Sozialwissenschaften.

Hajer, M. A. (1995). Discourse coalitions and the institutionalization of practice: The case of acid rain in Great Britain. In F. Fischer & J. Forester (Eds.), *The argumentative turn in policy analysis and planning* (pp. 43–67). Durham/London: Duke University Press.

Hajer, M. A. (2003). Argumentative Diskursanalyse. Auf der Suche nach Koalitionen, Praktiken und Bedeutung. In R. Keller, A. Hirseland, W. Schneider, & W. Viehöver (Eds.), *Handbuch Sozialwissenschaftliche Diskursanalyse* (pp. 271–297). Opladen: Leske und Budrich.

Haraway, D. (1988). Situated knowledges: The science question in feminism and the privilege of partial perspectives. *Feminist Studies, 14*(3), 575–599.

Hecht, G., & Allan, M. (Eds.). (2001). *Technologies of power*. Cambridge: MIT Press.

Heilbroner, R. (1967). Do machines make history? *Technology and Culture, 8*(3), 335–345.

Herrera, G. L. (2003). Technology and international systems. *Millenium: Journal of International Studies, 32*(3), 559–593.

Jasanoff, S. (Ed.). (2004). *States of knowledge: The co-production of science and social order*. London: Routledge.

Jasanoff, S., & Kim, S.-H. (2009). Containing the atom: Sociotechnical imaginaries and nuclear power in the United States and South Korea. *Minerva, 47*(2), 119–146.

Keohane, R. O. (Ed.). (1986). *Neorealism and its critics*. New York: Columbia University Press.

Kunzru, H. (1996). *You are cyborg*. Accessed February 2014, from http://www.wired.com/wired/archive//5.02/ffharaway.html?person=donna_haraway&topic_set=wiredpeople

Latour, B. (1987). *Science in action: How to follow scientists and engineers through society*. Cambridge, MA: Harvard University Press.

Latour, B. (2005). *Reassembling the social: An introduction to actor-network theory*. Oxford: University Press.

Law, J. (1992). Notes on the theory of the actor-network: Ordering, strategy, and heterogeneity. *Systems Practice, 5*(4), 379–393.

Lebensministerium (Federal Ministry of Agriculture, Forestry, Environment and Water Management). (2012). *Ausgewählte highlights der exportinitiative umwelttechnologien*, Wien 21. Feb. 2012. n.p. Accessed April 2013, from http://www.lebensministerium.at/umwelt/green-jobs/exportinitiative/exportinitiative-umwelttechnologien/Highlights.html

Martello, M. L., & Jasanoff, S. (2004). Introduction: Globalization and environmental governance. In S. Jasanoff & M. L. Martello (Eds.), *Earthly politics: Local and global in environmental governance* (pp. 1–29). Cambridge: MIT Press.

Messner, D., & Nuscheler, F. (2003). Das Konzept global governance: Stand und perspektiven. *INEF Report, 67*, 1–56.

Muh, Victor (dir.) (2011). *The Red Thread*. 30 min. short film, commissioned by the Austrian government to commemorate the 40th anniversary of diplomatic relations with China.

Nowotny, H. (1979). *Kernenergie. Gefahr oder Notwendigkeit*. Frankfurt a. Main: Suhrkamp.

Onuf, N. (2012). *World of Our Making: Rules and Rule in Social Theory and International Relations*. Reissue edition, London: Routledge.

Pan, C. (2004). The "China Threat" in American self-imagination: The discursive construction of other as power politics. *Alternatives: Global, Local, Political, 29*(3), 305–331.

Pinch, T. J., & Bijker, W. E. (1984). The social construction of facts and artefacts: Or how the sociology of science and the sociology of technology might benefit each other. *Social Studies of Science, 14*(3), 399–441.

Rosenau, J. N., & Czempiel, E. O. (1992). *Governance without government: Order and change in world politics*. Cambridge: Cambridge University Press.

Ruggie, J. G. (1998). What makes the world hang together? Neo-utilitarianism and the social construction challenge. *International Organization, 52*(4), 855–885.

Schwaag Serger, S., & Breidne, M. (2007). China's fifteen-year plan for science and technology: An assessment. *Asia Policy, 4*, 135–164.

Seifert, F. (2003). *Gentechnik-Öffentlichkeit-Demokratie. Der österreichische Gentechnik-Konflikt im internationalen Kontext*. Wien: Profil Verlag.

Späth, P., & Rohrracher, H. (2010). 'Energy regions': The transformative power of regional discourses on socio-technical futures. *Research Policy, 39*(4), 449–458.

Torgersen, H. (2002). Austria and the transatlantic agricultural biotechnology divide. *Science Communications, 24*(2), 173–183.

Ulbert, C. (2005). Sozialkonstruktivismus. In S. Schieder & M. Spindler (Eds.), *Theorien der internationalen beziehungen* (pp. 409–440). Opladen & Farmington Hills: Verlag Barbara Budrich.

Ulbert, C., & Weller, C. (2005). *Konstruktivistische analysen der internationalen politik*. Wiesbaden: VS Verlag für Sozialwissenschaften.

Waltz, K. N. (1979). *Theory of international politics*. New York: Waveland.

Wendt, A. (1992). Anarchy is what states make of it: The social construction of power politics. *International Organization, 46*(2), 391–425.

Wodak, R., de Cillia, R., Reisig, M. & Liebhart, K. (2009). *The discursive construction of national identity*. Edinburgh: Edinburgh University Press.

Woods, N. (2008). Whose aid? Whose influence? China, emerging donors and the silent revolution in development assistance. *International Affairs, 84*(6), 1205–1221.

Xu, L., & Yafang, L. (2006). *Österreichs Botschafter in China über die bilateralen Beziehungen*. Interview with the Austrian Ambassador about bilateral relations. Accessed February 2014, from http://german.china.org.cn/german/235726.htm

Liberal Preferences as an Explanation for Technology Choices. The Case of Military Robots as a Solution to the West's Casualty Aversion

Niklas Schörnig

Abstract When it comes to choices regarding military technology, many scholars of international relations favor a realist perspective, arguing with "technological imperatives", "adaptation of technology" and the "security dilemma". When it comes to military robotics in general and drones in particular, the "technological imperative" seems to be a valid and powerful explanation for the worldwide proliferation of drone technology. What is missing from a realist explanation, however, is why drones become so popular in the first place and why especially Western democracies have been implementing and fostering the development from the very beginning. The text argues that drones and other unmanned military systems offer Western democracies a unique opportunity to implement what has been termed the "New Western Way of War", i.e. warfare with minimized casualties on their side and reduced civilian casualties on the side of their opponents. So rather than understanding "technology push" and "mission pull" as two contradicting explanations of technology development, the text argues that rather a combination of both offers a comprehensive picture.

Keywords Liberal theory • Drones • Military robots • Weapon technologies • War

1 Introduction[1]

In July 2013, military history was written. For the first time ever an unmanned aerial vehicle (UAV), Northrop Grumman's X-47B demonstrator, successfully landed on the flight deck of the US aircraft carrier George H.W. Bush without any human

[1] Some of the arguments presented in this text resume thoughts already presented in earlier works, especially ideas presented in Müller and Schörnig (2010), Sauer and Schörnig (2012) and Schörnig (2013).

N. Schörnig (✉)
PRIF, Frankfurt am Main, Germany
e-mail: schoernig@hsfk.de

M. Mayer et al. (eds.), *The Global Politics of Science and Technology - Vol. 2*, Global Power Shift, DOI 10.1007/978-3-642-55010-2_5,

interference. Amongst jetfighter pilots, arrested landing on a moving carrier is considered to be one of the most daring maneuvers. After the first landing the UAV started and landed again but aborted a third approach due to unspecified internal problems and was diverted to its inland base where it returned safely. The performance of this UAV demonstrates the enormous development unmanned military systems have seen over little more than a decade, becoming what can be considered the single most important asset in Western warfare. Ten to fifteen years ago, UAVs were remote controlled and only capable of timely reconnaissance. Today, many current drones are not only capable of carrying missiles and a variety of sensors which communicate intelligence, surveillance and reconnaissance (ISR) signals in real time but are on the verge of making their own decisions based on ever increasing computing power and artificial intelligence, rendering them semi-autonomous—with the operator controlling rather than actually "flying" the UAV. This development is reflected in the tremendous increase in their sheer numbers in military arsenals. In 2002, for example, the US-military had roughly 50 drones in its arsenals in contrast to approximately 8,000 today (Singer 2013: 9). In a 2012 report the US Government Accountability Office stated that 76 states worldwide were using military drones (US Government Accountability Office 2012: 10) but only one year later expert Peter Singer came up with the number of 87 (Singer 2013: 9).

While drones have been in the spotlight of public awareness for some time now, due to their controversial use by the United States in Pakistan, Yemen or Somalia, it is, however, safe to say that they are only the frontrunner of the probably most fundamental change in military affairs since the introduction of nuclear weapons: the robotization of the military as a whole. If current trends continue, it will take only years rather than decades until military robots will dominate the air, the ground and the seas. But where does this tremendous dynamic steam from, what are its driving forces? Is it the inevitable result of current technological developments, as proponents of the technological imperative would suggest (e.g. Buzan 1987)—that is, whatever is technologically possible, will be build? A cursorily look might support that impression: Robots have not only become important for the military in recent years but have become relevant as a civilian consumer product as well. Robots vacuum private homes; they clean windows or mow the lawn. Cars have already demonstrated the ability to navigate in human traffic while the ability to self-park has become almost a standard at least in high-class limousines. From this perspective, the triumph of robots seems rather obvious: the technological development of more powerful microchips, better algorithms and enhanced sensors is the independent variable pushing both civilian as well as the military developments in the robotic realm and changes the physical conditions and the social environment, i.e. military postures, strategies and tactics, alike. This deterministic notion is not totally false. As MacKenzie and Wajcman put it, "[t]echnological determinism contains a partial truth. Technology matters." (MacKenzie and Wajcman 1999: 3). However, in the same text they warn us that "[a]s a simple cause-and-effect theory of historical change, technological determinism is at best an oversimplification" (MacKenzie and Wajcman 1999: 4). Many authors would agree that there is an interchange between technology and society in general and the military in particular rather than a one-way connection. While the availability of a certain technology might have an impact on military, there is a political and military shaping of certain technologies as well. Usually, military technologies do not spring up out of thin air but are the result of deliberate investment choices in the light of broader strategic decision: states that will focus on maritime defense will invest in other technology than a classical land power. In addition, the natural limitation of state resources forces states to make choices between competing solutions to military ends. In 2004, for example, the US Army cancelled its advanced helicopter project, the fast and stealthy RAH-66 Comanche, and opted for slow and non-stealthy drones instead (Fulghum and Wall 2004). On the other hand, considerations about domestic jobs or even pork barrel politics might actually prevent the military from procuring the most advanced technology and settle for second best solutions, while bureaucratic resistance or "incompatibility" to existing doctrine might prevent the military to procure certain weapon types at all (see Rosen 1991). In sum, when it comes to military technology there is a lot more "wiggle room" for "agency" than the technological imperative suggests.

This leads me to my research question. While experts claim that at the end of 2012 more than 80 countries had some kind of drone technology in their military arsenals (Singer 2013), the 2011 edition of the Military Balance yearly published by the International Institute for Strategic Studies, listed 34 countries holding either medium or heavy sized UAVs out of which roughly two thirds were democratic states (IISS 2011: 24–26).[2] In addition, it is safe to say that Western states, with the United States and Israel being the most active, have actively been pushing the development of military robots and drones in recent years. So, the question is, whether military robots and drones do have a specific appeal to Western democracies and where this appeal might come from vis-à-vis alternative weapon systems.

To answer this question I will, first, take into account classical theories of international relations (neorealism, constructivism, liberalism/democratic peace theory) to gain a broader perspective on the interplay between technology and the military. Second, I will argue that a liberal perspective, informed by recent works on the so-called democratic peace is especially useful to understand the enormous pace with which the robotization of especially Western militaries is happening. Not being threatened in their very existence anymore, military engagements for almost all Western states, including Israel, have become "wars of choice" (Freedman 2006/2007) rather than of necessity.[3] In consequence, minimizing casualties amongst their own troops has become one major objective to Western states, almost on par with mission fulfillment. Investing heavily in military robots can thus be understood as the most viable strategy amongst several to minimize losses on one's own side. In other words: to fully understand the current dynamics in Western armament, one

[2] Depending on which democracy index one uses, either 22 (Economist Intelligence Unit Democracy Index; =65 %) or 24 (Polity IV; 71 %) out of 34 countries can be seen as democratic.

[3] For the case of Israel fighting wars of choice rather than of necessity see Levy (2010a, b).

has to paraphrase former US-President Bill Clinton—"It's the casualty, stupid!". But while, as Frank Sauer and I have argued elsewhere (Sauer and Schörnig 2012), drones might look as a silver bullet of Western warfare in the short run, the heavy investment into unmanned systems is not without danger in the long run. These potential dangers might include the hacking and spoofing of unmanned systems, a lowered threshold when it comes to military engagements due to the near-perfect protection of Western soldiers or the development of fully autonomous systems with the capability to search and kill without human interference. While many of these potential long term problems are just warranted speculations, the political impact of some might be tremendous. Research should therefore focus on the evaluation of the likelihood of certain repercussions and what could (and should) be done to prevent those worst-case scenarios.

2 Military Research, Development and Procurement: Power and Prestige

What are the driving forces behind the current drone hype? Different theories of international relations point to different answers, all claiming (more or less) prima facie plausibility. From the perspective of theories of international relations, the idea of technological determinism is closely related to Kenneth Waltz' neorealism (1979). According to (neo)realists, the anarchy of the international system forces states to compare their own military capabilities vis-à-vis potential challengers as power is, by definition, a relational concept. While there is still debate amongst realists about how much power is enough [which is the basic issue of the well-known debate between defensive and offensive realists; see, amongst others, Taliaferro (2000)], all realists agree that falling behind one's opponent is a recipe for disaster and has to be avoided at all costs to ensure national survival. From this perspective implementation of new technology into the military becomes a dominant strategy no matter how the opponent acts. If the opponent has access to the new technology it seems unlikely that he will hesitate to implement it to his or her advantage—one's own implementation only levels the field then. And if he does not implement it, one's own implementation offers the chance to gain an edge over him. In other words: the implementation of new technology into the military happens instantly no matter what the other side does—it is the "technological imperative" driving development and procurement (for a concise summary of the concept see, for example, Reppy 1990: 102–103). Once a military technology owned by one state has proven its worth on the battlefield, other states are compelled to develop and/or procure this particular capability for their own military. As Kenneth Waltz argues: "Contending states imitate the military innovations contrived by the country of greatest capability and ingenuity. And so the weapons of major contenders, and even their strategies, begin to look much the same all over the world" (Waltz 1979: 127). João Resende-Santos supports this notion of rapid emulation: "States, like firms, emulate successful innovations of others out of fear of the disadvantages that arise from being less competitively organized and equipped" (Resende-Santos 1996: 196). From this perspective, military robots should tempt all militaries worldwide and the spread of this technology, in general, is well explained by neorealist assumptions. Especially when it comes to the so-called D^3-tasks—military assignments that are dull, dirty or dangerous—robots are seen as the better alternative compared to humans.

The second perspective, which does not offer a different explanation for the current trend but rather an additional element, is informed by the "softer" constructivist notion of norms, status and prestige. According to social constructivists, the purpose of a specific weapon system is not ontologically given but depending on the specific meaning states attribute to the system. The question becomes not one of strategic advantage but of the symbolic nature of specific weapon systems. This connection is most obvious in the nuclear realm where "membership" to the club of the "haves" has tremendous implications beyond the sheer strategic implications (van Wyk et al. 2007: 26) but is not limited to nuclear weapons. In consequence, the proliferation of specific weapon systems might not be influenced by military needs but rather its immersion in specific normative systems. Dana Eyre and Mark Suchman argue that certain conventional weapons do have a similar, although smaller effect. They explain: "It is quite common for developing nations to maintain only a single 'squadron' of four or five advanced aircraft—too few to offer any substantial strategic or tactical benefits ... but enough to constitute a reasonable air show" (Eyre and Suchman 1996: 93), stressing the "symbolic nature of weaponry" (ibid.). Given that unmanned aerial vehicles and military robots are at the technological forefront of weapon development at the moment, it is no surprise that, for example, many states have started either procuring drones or developing and manufacturing their own. Iran, for example, has not only run its very own drone program for some years now but presented an allegedly fieldable armed drone in 2010, called the "ambassador of death", exactly when a broader public started to debate the American drone strikes in Pakistan (Shachtman 2010). While experts doubt that the Iranian drone is much more than a bomb with wings, the signal Tehran wants to send out is that they can compete with the US-superpower on a technological level, thereby strengthening their claim for regional supremacy. In other word: while having some form of drone is evidence for having a state-of-the-art military, claiming to operate an armed drone is the signal to play in the upper league, with both creating a proliferation pull for drones independent from sheer capability-based arguments.

In sum, both a realist and a constructivist perspective on military robotization help to explain at least parts of the accelerating dynamic of the overall phenomenon. While neorealism assumes uniform and rapid emulation as the "primary behavior expected of units" (Goldman and Andres 1999: 83), i.e. all states, constructivism would describe the general trend but refrain from general predictions which states would have a more intense desire for prestige than others.

What is left unexplained, however, is why democracies in particular—especially the United States and Israel—have been in the driving seat of the drone-boom so far and have accepted quite some trouble and investment to keep their edge in the military robot business. This question is even more puzzling as Western states have not been facing an enemy similar to the Soviet Union since 1989, rendering explanations based on a security dilemma whatsoever obsolete. In addition, Western states have—in contrast to autocracies—always been relatively reluctant when it comes to military swaggering, leaving the pursuit of prestige a potential but not a likely candidate for an explanation.

3 The Liberal Perspective: Democracies and Military Technology

In contrast to neorealists, liberals focus on the preferences and specific interests of states and their individual political system. In this context, Democratic Peace Theory has come up with the notion that democracies are distinct as a group from all other possible groups of states. The most important finding in this respect has been what has become to be known as the "democratic peace" or "separate peace", i.e. the fact that mature democracies have not fought a war against each other since at least 1815 (e.g. Doyle 1983) despite being involved in military conflict as often (or only slightly less often) as comparable autocracies are. This empirical finding was deemed so robust in regard to different definitions of war and democracy as well as various statistical methods applied, that Jack Levy (1988: 662) called it "as close as anything we have to an empirical law in international relations". While the inner-democratic peace is still the most important feature of democracies, researchers have found that their behavior differs significantly from that of other states in other realms as well.[4] So the question to be asked here is whether there is something particular in (Western) democracies which make them especially interested in unmanned military systems.

3.1 Is There a Puzzle? Democracies, Technological Innovation, and the Military

When one looks at the connection between democracies and military robotization, the answer seems rather obvious. Western states possess a vivid high tech information technology industry and have gained experience in the realm of civilian robotization (e.g. in the field of industrial robots, but also in the field of civilian applications as presented in the introduction). From the perspective of the technological imperative described above, Western democracies can rely on competitive

[4] For a short overview of specific democratic behavior see, for example, Sauer and Schörnig (2012: 365–369).

civilian markets to bring up innovative solutions, generating a broader range of technological choice for consumers as well as the military (Müller and Becker 2008). Based on Cold War experiences, Matthew Evangelista has argued that democracies seem to be more innovative while it seems easier for authoritarian states to channel massive resources into refining specific systems once the basic designs have been reengineered (Evangelista 1988). This observation might help explain the advantage of democracies in the realm of robotics and military robotics, as well as the delayed but fast catching up of potential contenders like China or Russia. On the other hand, allocation of defence related resources is a contested field, especially in democratic states where competing interests between "butter and guns" tend to constrain excessive defence budgets (see e.g. Fordham and Walker 2005). In consequence, choices have to be made and not every available technology finds its way into the military arsenals.

In addition, the military is—generally speaking—a conservative institution, and usually quite averse to rapid technological change (Rosen 1991), at least when this change threatens to lead to significant organizational changes. When the army, for example, abandoned its advanced helicopter project RAH-66 Comanche in 2004, thereby swapping a fast, stealthy and *manned* high-tech helicopter for technologically inferior, slow, non-stealthy and hardly battle-proven drones, this was a rather unusual move (Fulghum and Wall 2004). This general reluctance to change should hold even more in the case of military robotics. Despite the abstract advantages described above by the three D's, military robots raise suspicion within the military ranks. First, as Bradley Strawser who teaches philosophy at the US Naval Postgraduate School in Monterey puts it, "flying a UAV from a desk . . . is not nearly as exciting or sexy as being, say, a traditional fighter pilot" (Strawser 2013: xviii). Related to that, there is, second, not much glory and heroism in remotely controlling a drone or ground robot while being in a safe container thousands of miles away. Efforts by the Pentagon to propose a medal for drone- and cyber-operations has stirred criticism by veterans and Congress alike (Shinkman 2013). Third, if there is something military commanders do not like, it is not remaining at the helm. When it comes to military robots and the increasing amount of automated (or even "autonomous") behavior, however, they literally have to hand over control to computer algorithms they most likely do not even understand—at least in detail. The computerization of the military has already increased operational speed significantly and many—if not most—decision at higher command levels are aided by computer programs anyway. With military robots, there is the danger that the warfighter might be thrown out of the decision making-loop altogether.

All this taken together makes one doubt that the militaries of Western democracies have either been the most eager advocates of the robotization of their services. The major driving forces for the particular interest of Western states in military robots have to be located somewhere else.

3.2 The Importance of Casualty-Aversion After the Cold War

Many scholars of strategic studies as well as military practitioners argue that since the end of the Cold War Western states have entered into what by now has been termed a "post-heroic-phase which can be described by an increasing aversion against losses on their own side (Luttwak 1995). Not having to fight "wars of necessity" (Freedman 2006/2007), i.e. wars of self-defence or for a generally accepted national interest anymore, military engagements became a matter of choice, usually humanitarian or state-building missions or missions to enforce international law with a low tolerance regarding casualties amongst their own ranks. The underlying supposition is fed by considerations that can be traced back to Kant's essay on the Perpetual Peace. The starting point is Kant's idea that in a republic (or today, a democracy), there is congruence between the people who decide in favor of or against a certain act of foreign policy and those who will enjoy the benefits of that decision or bear its costs. That is, people make decisions about themselves. This property is particularly crucial in matters of war and peace. Kant came to this conclusion: "If . . . the consent of the subjects is required whether there shall be war or not, nothing is more natural than that they weight matters well, before undertaking such a bad business. For in decreeing war, they would of necessity be resolving to bring down the miseries of war upon their country" (Kant 1984 [1795]: 12). In today's parlance: Citizens answer for their decision with property, life and limb. By contrast, an undemocratic ruler can shift the costs of foreign policy decisions to his subjects (in the language of modern microeconomic theory, externalize them), while he alone enjoys their benefits.

In addition to this rather utilitarian explanation there are other normative factors at play as well. Almost all modern democracies are based on particular norms dating back to the enlightenment, especially that every individual human being is the bearer of innate, indefeasible rights, e.g. equality before the law, physical integrity, personal freedom and so on (e.g. Owen 1994). This offers an additional rationale for Western casualty aversion: democracies value the individual over the collective; they grant their soldiers particular human rights (at least more than most armies of authoritarian states) and they externalize these norms even when dealing with the population of the opponent's state as the intensive debates about how to reduce what has—euphemistically—been called "collateral damage" show. Finally, as democracies are based on the rule of law, they tend to abide international law to a higher degree than non-democracies—which is not to say that they *always* abide international law (e.g. Morrow 2007).

While most of the norms described would let a casual observer expect a more peaceful behavior (respect for human life; rule of law etc.) at least at first glance, recent research has shown that this conclusion might be premature as they at least have the potential to allow both peaceful as well as bellicose behavior (Geis et al. 2013). While, for example, Western democracies might shy away from military action because they do not want to expose their military personal to physical danger, the same emotion of compassion might be triggered by humanitarian crises, especially when the so-called CCN-effect, i.e. intense media coverage, kicks in, resulting in the demand that "something has to be done". This leaves the political decision makers between a rock and a hard place as he or she has to square the circle of conflicting preferences, i.e. engage militarily but ensure minimal losses on the Western states side and reduced civilian casualties when possible (Levy 2010a).[5] This tension was clearly visible, for example, during NATO's Kosovo campaign, where the aim to stop ethnic cleansing was hampered by the effort to avoid NATO casualties, e.g. by relying on planes attacking from extreme heights to avoid enemy anti-aircraft fire.

The pressure on the political leadership gets even worse when unanticipated losses occur while there is no visible progress regarding mission accomplishment. As Gelpi, Feaver and Reifler have shown, Western publics react strongly when they think that their soldiers die in vain (Gelpi et al. 2006). That means that in addition to a more general "casualty aversion" in wars of choice, political decision maker have to factor in sudden changes of the public mood due to "loss aversion" in particular military engagements as well. This might ultimately lead to what I have termed "casualty trap" elsewhere (Schörnig 2009): In order to avoid additional losses (and the potential tipping of public opinion from favor to opposition against a specific engagement), troops are called back into the relative safety of their camps thereby compromising the mission goals, leading to a stalemate position where the mission is not aborted but cannot be fulfilled anymore as well.

In a similar but more general fashion, the British sociologist Martin Shaw came up with the concept of the "New Western Ways of War" (Shaw 2005). According to Shaw, Western states have to transfer many of the risks associated with warfare away from their soldiers, i.e. they have to wage what he calls "risk-transfer war", or "the new Western way of war". On the other hand, modern wars are fought under "global media surveillance", limiting military options. Shaw describes 15 "rules" of risk-transfer warfare, many of which are similar to the conditions derived from liberal democratic peace theory. Wars must, for example, anticipate the problems of global surveillance (rule 4), they have to "rely on precision weaponry to sustain their legitimacy" (rule 11), minimize the "risks of 'accidental' civilian casualties" (while "small massacres must be regarded as inevitable"; rule 10) and wars "must, above all, minimize casualties to Western troops" (rule 7). Only and only if these (and the other) rules are followed, Western publics will accept the use of deadly force in the international realm.

[5] James Igoe Walsh suggests that civilian casualties are even more important to the public than those amongst American servicemen or –women (Walsh and James 2012).

3.3 Waging the "New Western Way of War"

One option to solve the dilemmas described above would be to avoid any military confrontation at all. This, however, would significantly limit a state options for action in the international realm (e.g. to participate in alliance-warfare or an UN-mission) and might even contradict demands by one's own public as described above. In consequence, since the end of the cold War Western states have tried several ways to avoid the casualty trap while keeping military options open—with varying degrees of success (see also Schörnig 2013). There is more than one way to wage a "risk-transfer war".

The first alternative, which might be called "division of labor", relies on providing air power by the Western states while the actual fighting on the ground is left to local forces, supporting the fraction which seems most in line with the Western interest. At best, Special Forces or intelligence agents in theater support the fraction of choice with tactical advice, technology or intelligence. This approach was used during the first phase of the Afghanistan War in 2002/3 as well as the Libya War in 2011. The problem with the "division of labor" approach is that there is no guarantee that—at least in the long run—the ally on the ground will do as he is told as local fighters usually have an agenda of their own which might not correspond with the interest of the supporting power. Without having own ground forces in theater, control is virtually impossible to establish. In addition, local troops might interpret international humanitarian law in rather "unorthodox" ways (or simply ignoring it), opening up gateways for criticism on the international arena.

The second possibility, which might be called "outsourcing" uses so-called "private military companies" (PMCs) rather than one's own soldiers for fighting. The basic rationale is that the public does not react strongly (if at all) to losses amongst "mercenaries", reducing pressure on the political elite in case of high casualty rates (Schooner 2008). However, when it comes to actual fighting rather than support (i.e. bodyguards, guarding facilities etc.) using PMCs has shown similar problems as relying on local groups on the ground (Singer 2008: 234–242): neither is it clear whether they follow orders as regular troops nor do they always follow the laws of war or the rules of engagement. Finally, being civilians rather than soldiers, their legal status in war is far from clear. While reliance on PMCs as a fighting force rather than a viable support in logistics and guard duty actually seems a thing from the past as no Western states puts PMCs rather than soldiers to the frontline, the third approach to avoid the casualty trap can be seen by many Western democracies at the moment: the use of Special Forces like the American Delta Force, the British SAS or German KSK. This approach, which might be labeled "professionalization", rests on specially trained elite soldiers, capable to execute intensive missions with a minimum of soldiers in a very precise way. Due to their effectiveness—as well as the clandestine nature of their work, guaranteeing a minimum of public scrutiny—most Western countries have heavily invested in their special forces. But not every type of mission is of course suited for Special

Forces. The political fallout of failure might be dramatic either because infiltration is still too dangerous, or because Special Forces alone do not command enough firepower. Finally, even with Special Forces being the most discreet military units, once casualties occur amongst them the fact cannot be hidden from the public for long.

3.4 *Robotization of the Military as the Silver Bullet?*

It is obvious that a fourth option to deal with the necessity to conduct military operations with minimal casualties is to rely heavily on military robots. They combine many of the advantages of the other approaches without the respective drawbacks and seem to be a "silver bullet" (Sauer and Schörnig 2012). First, it is obvious that no soldier is in danger when a robot is doing his or her job. When there is no pilot in the plane, no pilot can be harmed. In addition, military robots like drones help to protect human soldiers in the field where their ability to loiter for an extended time is crucial. While human airborne assets can only stay for a short period of time and are usually called in a case of emergency, a U(C)AV can stay aloft for 24 h, escorting a patrol or guarding a mission from the very first minute, thereby providing real time intelligence (when unarmed) and enhanced situational awareness or by offering *close air support* (when armed) in the case on an ambush (Harrington 2010).

This ability to loiter and wait for the perfect time to engage a target with high precision is brought forward by proponents of drone procurement in the context of civilian casualties as well (Partlow 2010). In contrast to manned systems which only have a small window of engagement before they have to return to base, they argue that a drone which is about to attack a military target in a basically civilian environment (a village or a city) can wait until the expected harm to civilians is minimized. It is interesting to see that the increasing relevance given to civilian casualties of military engagements which has been noticeable since the end of the Cold War has led to specific technological innovations as well. Drones usually use what has been termed "smart munitions", i.e. laser-guided missiles which are said to allow very precise engagements. When drones started to be armed in 2001/2002, the American military usually relied on Hellfire air-to-surface-missiles (ASM), originally designed as an anti-tank weapon but one of the smallest ASM in the military's arsenals. As it turned out, the Hellfire was still too powerful when used against individual humans in urban scenarios. In order to reduce civilian casualties and appease international public opinion, current research aims at minimized warheads to reduce the lethal radius of the blast (Harrington 2010). According to proponents all this helps to stick to fundamental principles of international humanitarian law like the discrimination between combatants and civilians as well as the principle of proportionality.

Finally, with robots the question of reliability is of course solved, at least as long as fully autonomous systems are not introduced but rather systems where a human stays “within” or at least “on the loop” and where the room for “decision making” is well defined. But one thing is sure: with robots on the front line no military commander has to worry that his troops will withdraw once a certain amount of casualties has been suffered. Not knowing the concept of self-preservation, robots can be ordered to engage to the last without any question or hesitation. This programmed obedience, however, might be one of the major political reasons for authoritarian rulers to develop autonomous military robots in the future as robots would not hesitate to take violent measures against “their” own (civilian) population while regular troops might refuse and revolt. However, the political nexus between authoritarian regimes and the robotization of the military has not been in the focus of systematic research yet.

It is obvious that in comparison to the other three alternatives described above, robotization seems to have the greatest appeal for Western states as it seems to offer most advantages but—at least at first glance—almost no disadvantages. It is no wonder then that Western politicians (especially American and Israeli ones) rather than the military have pushed for a more extensive use of military robots—as it, for example, can be seen in the extensive use of armed drones both in Afghanistan as well as in Pakistan, Yemen or Somalia by President Obama[6] when the rising numbers of US casualties in Afghanistan and Iraq restricted the Presidents room of maneuver. And it is no wonder that many Western states which have been involved in actual combat missions and have experienced the volatility of their respective public opinion once body bags returned home are now pushing for armed drones in particular and military robots in general.

4 The Dark Side: Potential Long Run Effects

This paper is not the right place to dwell into the potential long-run problems of Western robot armament as I have done this extensively in other publications (see Müller and Schörnig 2010; Sauer and Schörnig 2012; Schörnig 2013). However, the enthusiasm with which Western states push in the direction of modern high-tech armament and military robotization has led to blindness towards shortcomings which might lead to problems and dangers in the medium and long run. Just to name a few: If risk-free warfare is a real possibility for Western states, many scholars fear that the threshold to actually engage militarily will be lowered and that the use of military force will lose its status as a last resort. Second, as described above, non-democracies will be tempted follow suit but for more realist reasons rather than fear of casualties. This, however, might lead to increased arms races in already tense regions. As the recent history of “targeted killings” has shown, drones are particularly suited for stealthy surprise attacks out of the literally blue sky,

[6] For the drone strikes in Pakistan, Yemen and Somalia see the numbers presented by the independent Bureau of Investigative Journalism (Bureau of Investigative Journalism 2013).

thereby fueling first-strike fantasies of successful decapitation-strikes. Third, the more dual-use high-tech goes into the development and production of ever more sophisticated military robots, the more complex the necessary software code gets. This, however, increases the vulnerability of the robotic systems against cyber-attacks, spoofing and hacking. In his drone-thriller "Kill Decision" Daniel Suarez describes, for instance, an attack by what seems to be an American MQ-9 Reaper drone against a praying crowd in Karbala, Iraq (Suarez 2012). In the book the drones has been reengineered and is part of a plot to blame the United States for the attack. In reality, creating a similar scenario might be even easier when a real drone is manipulated by a virus or hacking, decreasing accuracy and leading to unintentional but high civilian casualties and a fatal political fallout. Finally, up to now Western states have been using drones only in very asymmetric situations where the airspace was hardly contested as drones like the MQ-1 Predator or MQ-9 Reaper are propeller-driven, relatively slow and not too stealthy. They are easy prey for modern anti-aircraft systems. If Western states want to extend the use of drones and military robots to more contested scenarios in order to avoid casualties not only in some of them but to wage a "bloodless war" (Mandel 2004) on their side at least, they cannot rely on remote control via satellite anymore. Especially when it comes to dogfights or the attack of antiaircraft defense, decisions have to be made in fractions of a second. A signal propagation delay of one or two seconds between the drones, a satellite and the drone's operator might already be too long. Therefore, more and more functions have to be "decided" by the drone, including the perfect moment to release fire. This trend towards more and more autonomy will most likely lead to computers and algorithms making the ultimate decision about life and death—a tremendous challenge, to say the least, for international law as well as morality and ethics (see, amongst others, Sparrow 2007; Human Rights Watch 2012).

Proponents of military robots argue that these are solvable technological problems. Roboticist Ron Arkin, for example, believes that it will be possible to program an "ethical governor" which might ultimately lead to robots which are "more humane in the battlefield than humans" (Arkin 2009: 30). Other observers are more pessimistic: Wendell Wallach and Colin Allen, for example, come to the conclusion that "... the prospect of reducing ethics to a logically consistent principle or set of laws is suspect, given the complex intuitions people have about right and wrong" (Wallach and Allen 2009: 215). So, while current research by Arkin and other proponents of autonomous military robots is driven by the motive to avoid casualties and reduce human suffering, it is far from clear whether these "ethical barriers" can be implemented. In the wake of this research, however, the arms race in the realm of autonomous robots is fueled tremendously, as states like—for example—China and Russia are very likely to react to the Western approach as predicted by the security dilemma, reducing the chances for regulation and arms control (e.g. Altmann 2013).

5 Conclusion

Military robots are booming and they will continue to do so. As it has been described above, to fully understand the current dynamics especially on the side of Western democracies, several explanations have to be combined. It is true that the ongoing technological advancement in the realm of information technology and the related proliferation of drone technology can be understood as a classic case of the realist "technological imperative". However, technological progress and fear of insecurity are only part of the story as other approaches are better suited to help us understand the specific interest of certain states in the development of specific technologies in the first place. In consequence I have argued that Western political decision makers have been very willing to support the robotization of their respective armies as they understood it as a solution to the short term problem of potential backlash at the ballot box due to too high casualty rates in military engagements. The preference to minimize casualties amongst one's own can therefore be understood to be one very important factor driving the development of military robotics in Western democracies. As early as 2000, the National Research Council of the United States argued that the interest in UAVs was fueled by a "combination of 'mission pull' (e.g. risk avoidance and cost avoidance), which require that systems be developed for certain missions, and 'technology push' which is fueled by advances in particular technology" (National Research Council 2000: 22), supporting the argument by MacKenzie and Wajcman presented in the introduction. This is not to say that casualty aversion is the only "pull"-factor in the current process of the robotization of the military. As I have argued, other factors—e.g. prestige—might come into play in other groups of countries.

Finally, technology as such needs to have reached a certain level before one can actively accelerate certain trends and developments. In the case of military robots the processing power of microchips and a certain minimum amount of wireless bandwidth to allow for a certain degree of automation and real-time surveillance was needed in order to kick-start the current development. However, taking the sub-systemic liberal perspective of International Relations which focuses on which individual preferences key actors have has proven to be a valuable source of additional understanding when it comes to certain technology choices of democracies and their status as "innovators" in the drone-hype.

What has to be focused on now is whether the influence of democratic preferences on military technology development is a one way street or whether there are backlashes of the technology on democratic norms and institutions as well. While some authors, like Peter Singer, fear that the casualty reducing effects of military robots might undermine democracy (Singer 2012), others see the threshold to wage war lowered (Sauer and Schörnig 2012). But more research is needed to proof that these theoretical deductions are not only valid form a logical point of view but empirically true and relevant as well.

References

Altmann, J. (2013). Arms control for armed uninhabited vehicles: An ethical issue. *Ethics and Information Technology, 15*(2), 137–152.

Arkin, R. C. (2009). Ethical robots in warfare. *IEEE Technology and Society Magazine, 28*(1), 30–33.

Bureau of Investigative Journalism. (2013). *Drone strikes.* Accessed December 18, 2013, from http://www.thebureauinvestigates.com/category/projects/drones/

Buzan, B. (1987). *An introduction to strategic studies: Military technology and international relations.* London: Macmillan.

Doyle, M. W. (1983). Kant, liberal legacies, and foreign affairs (I). *Philosophy & Public Affairs, 12*(3), 205–235.

Evangelista, M. (1988). *Innovation and the arms race. How the United States and the Soviet Union develop new military technologies.* Ithaca, NY: Cornell University Press.

Eyre, D. P., & Suchman, M. C. (1996). Status, norms, and the proliferation of conventional weapons: An institutional theory approach. In P. Katzenstein (Ed.), *The culture of national security. Norms and identity in world politics* (pp. 79–113). New York: Columbia University Press.

Fordham, B. O., & Walker, T. C. (2005). Kantian Liberalism, regime type, and military resource allocation: Do democracies spend less? *International Studies Quarterly, 45*(1), 141–157.

Freedman, L. (2006/2007). Iraq, liberal wars and illiberal containment. *Survival, 48*(4), 51–65.

Fulghum, D. A., & Wall, R. (2004, February 29). Comanche Helicopter program killed. *Aviation Week.*

Geis, A., Müller, H., & Schörnig, N. (Eds.). (2013). *The militant face of liberal democracies: Liberal forces for good.* Cambridge: Cambridge University Press.

Gelpi, C. F., Feaver, P. D., & Reifler, J. (2006). Success matters. Casualty sensitivity and the war in Iraq. *International Security, 30*(3), 7–46.

Goldman, E. O., & Andres, R. B. (1999). Systemic effects of military innovation and diffusion. *Security Studies, 8*(4), 79–125.

Harrington, C. (2010, August 18). Unmanned unbound. *Jane's Defence Weekly*, 22–26.

Human Rights Watch. (2012). Losing humanity. The case against killer robots. Washington, DC: Human Rights Watch.

IISS. (2011). Unmanned aerial vehicles: Emerging lessons and technologies. In IISS (Ed.), *The military balance* (pp. 20–26). London: Routledge.

Kant, I. (1984) [1795]. *Zum ewigen Frieden.* Stuttgart: Reclam (Translation taken from *Perpetual peace: A philosophical essay*, translated with Introduction and Notes by M. Campbell Smith, London: George Allen and Unwin, 1917).

Levy, J. (1988). Domestic politics and war. *Journal of Interdisciplinary History, 18*(4), 653–673.

Levy, Y. (2010a). The hierarchy of military death. *Citizenship Studies, 14*(4), 345–361.

Levy, Y. (2010b). The gap of legitimacies syndrome: A conceptual framework. *International Political Science Review, 31*(1), 77–94.

Luttwak, E. N. (1995). Toward post-heroic warfare. *Foreign Affairs, 74*(3), 109–122.

MacKenzie, D., & Wajcman, J. (1999). Introductory essay and general issues. In D. MacKenzie & J. Wajcman (Eds.), *The social shaping of technology* (2nd ed., pp. 3–27). Trowbridge: Redwood Books.

Mandel, R. (2004). *Security, strategy, and the quest for bloodless war.* Boulder: Lynne Riener.

Morrow, J. D. (2007). When do states follow the laws of war? *American Political Science Review, 101*(3), 559–572.

Müller, H., & Becker, U. (2008). Technology, nuclear arms control, and democracy: Reflections in the light of democratic peace theory. In M. Evangelista, H. Müller, & N. Schörnig (Eds.), *Democracy and security: Preferences, norms and policy-making* (pp. 102–119). London: Routledge.

Müller, H., & Schörnig, N. (2010). Drohnenkrieg: Die konsequente Fortsetzung der westlichen Revolution in Military Affairs. *Aus Politik und Zeitgeschichte, 50*(2010), 16–23.

National Research Council. (2000). *Uninhabited air vehicles. Enabling science for military systems* (NMAB-495). Washington, DC: National Academy Press.
Owen, J. M. (1994). How liberalism produces democratic peace. *International Security, 19*(2), 87–125.
Partlow, J. (2010). 13 die as U.S. uses drone to it suspected Afghan insurgents. Accessed December 18, 2013, from http://articles.washingtonpost.com/2010-01-13/world/36822928_1_taliban-safe-house-helmand-insurgents.
Reppy, J. (1990). The technological imperative in strategic thought. *Journal of Peace Research, 27* (1), 101–106.
Resende-Santos, J. (1996). Anarchy and the emulation in military systems. Military organization and technology in South America, 1870–1914. *Security Studies, 5*(3), 193–260.
Rosen, S. P. (1991). *Winning the next war. Innovation and the modern military*. Ithaca, NY: Cornell University Press.
Sauer, F., & Schörnig, N. (2012). Killer drones – The silver bullet of democratic warfare? *Security Dialogue, 43*(4), 363–380.
Schooner, S. L. (2008). Why contractor fatalities matter. *Parameters, 38*(3), 78–91.
Schörnig, N. (2009). In der Opferfalle. Die Bundeswehr und die zunehmenden Gefallenen der Bundeswehr in Afghanistan. *Standpunkte* 2/2009. Frankfurt: HSFK.
Schörnig, N. (2013). Unmanned warfare: Towards a neo-interventionist era? In G. Kümmel & B. Giegerich (Eds.), *The armed forces: Towards a post-interventionist era* (pp. 221–235). Wiesbaden: Springer.
Shachtman, N. (2010). Iran's Rebotic 'Ambassador of Death' is more envoy of annoyance (Updated). Accessed December 18, 2013, from http://www.wired.com/dangerroom/2010/08/irans-ambassador-of-death-drone-is-more-envoy-of-annoyance/
Shaw, M. (2005). *The new Western way of war*. Malden: Polity.
Shinkman, P. D. (2013). Pentagon strikes 'drone medal'. Accessed October 10, 2013, from http://www.usnews.com/news/articles/2013/04/15/pentagon-strikes-drone-medal
Singer, P. W. (2008). *Corporate warriors: the rise of the privatized military industry* (Updated edition). Ithaca, NY: Cornell University Press.
Singer, P. W. (2012). Do drones undermine democracy? Accessed December 18, 2013, from http://www.nytimes.com/2012/01/22/opinion/sunday/do-drones-undermine-democracy.html?pagewanted=all&_r=0.
Singer, P. W. (2013). Die Zukunft ist schon da. Die Debatte über Drohnen muss von der Realität ausgehen. *Internationale Politik, Mai/Juni*, 8–14.
Sparrow, R. (2007). Killer robots. *Journal of Applied Philosophy, 24*(1), 62–77.
Strawser, B. J. (2013). Introduction: The moral landscape of unmanned weapons. In B. J. Strawser (Ed.), *Killing by remote control. The ethics of an unmanned military* (pp. 3–24). Oxford: Oxford University Press.
Suarez, D. (2012). *Kill decision*. New York: Dutton.
Taliaferro, J. W. (2000). Security seeking under anarchy. Defensive realism revisited. *International Security, 25*(3), 128–161.
US Government Accountability Office. (2012). *Nonproliferation, agencies could improve information sharing and end-use monitoring on unmanned aerial vehicle exports* (US Government Accountability Office Report 12-536). Accessed December 18, 2013, from http://dronewarsuk.files.wordpress.com/2012/09/us-gao-_-noproliferation-of-uavs.pdf
van Wyk, J.-A., Kinghorn, L., Hepburn, H., Payne, C., & Sham, C. (2007). The international politics of nuclear weapons: A constructivist analysis. *South African Journal of Military Studies, 35*(1), 23–45.
Wallach, W., & Allen, C. (2009). *Moral machines: Teaching robots right from wrong*. Oxford: Oxford University Press.
Walsh, J. I. (2012). Do drones change Americans' views on the use of force? Accessed December 18, 2013, from http://themonkeycage.org/blog/2012/08/28/do-drones-change-americans-views-on-the-use-of-force/.
Waltz, K. N. (1979). *Theory of international politics*. Reading: Addison Wesley.

The "Tight Oil Revolution" and the Misinterpretation of the Power of Technology

Susanne Peters and Werner Zittel

Abstract The "tight oil revolution" is a prominent example of the misleading belief in the power of technology. We identify the "domination of nature paradigm", which originated in the sixteenth century and developed into an influential ideology of Western society, as a root cause for the optimism that technology will always push the geological limits and provide us with an abundance of fossil fuels. As a case study, our discussion of the "tight oil revolution" will demonstrate (1) that natural factors like steep depletion rates are alarming symptoms of its unsustainability; (2) that the emergence of this revolution is only partially linked to the "power of technology", but also to deregulation and Wall Street economics. The article concludes that we need to accept that there are insurmountable limits to technology's "domination of nature". Theoretically, this recognition should be reflected in International Relations, allowing for the incorporation of natural factors such as "peak oil" in order to provide us with sustainable models for a post-fossil era.

Keywords Tight oil • Domination of nature • Laws of physics • Peak oil • Resource depletion • Fracking

The belief in the limitless power of technology has been a dominant feature of the Western world's thinking since the sixteenth century and has further intensified with the age of globalization. "Technology" is supposed to provide the ultimate solution to one of the most pressing issues of today's world: the problem of how to

S. Peters (✉)
Kent State and Webster University Geneva, Geneva, Switzerland
e-mail: Speters9@kent.edu

W. Zittel
Ludwig-Bölkow-Systemtechnik GmbH (LBST), Ottobrunn, Germany
e-mail: Werner.Zittel@lbst.de

M. Mayer et al. (eds.), *The Global Politics of Science and Technology - Vol. 2*,
Global Power Shift, DOI 10.1007/978-3-642-55010-2_6,

supply enough energy to maintain our level of civilization for an ever-increasing number of people on the globe. The focus of the problem is oil, because oil so far is irreplaceable for transportation and petrochemical industries, two key areas that are indispensable for the sustainability of economic activity. As oil prices have skyrocketed over the last decade, long thought forgotten terms from the 1970s such as "energy security" and "resource scarcity" have returned to the public debate with an ever-pressing urgency. In particular, one group of scientists and oil experts had warned that oil production would peak around 2005, and that with the anticipated decline of global oil production after the peak, we would need to begin transforming our fossil fuel dependent societies into "post-fossil" ones. Into these debates on "peak oil" broke the news of the "tight oil revolution", which—according to the IEA (2012) WEO—will turn the Unites States into a "Saudi America" and a potential OPEC member, thus contributing significantly to the increase of the global reserve base. Technological progress is supposed to be the driving factor behind this revolution—a development that the deniers of "peak oil" have always counted on.

In contrast to this new euphoria, this chapter argues that the "tight oil revolution" is a preeminent example of Western society's misleading belief in the power of technology. We identify the "domination of nature" paradigm as a root cause for this core belief. The belief stipulates that with the help of the power of technology mankind can dominate nature in such a way as to exploit it for man's own needs. The basis for this paradigm is the perception of an unbridgeable dichotomy between "nature" and "society"—a view that has been accepted and sustained by most social sciences, while rejecting the incorporation of natural and material components in their ontologies and epistemologies.[1] A second important obstacle to acknowledging the limits of nature is the lack of the recognition of the "Laws of Entropy" which have challenged the mechanistic view of the sixteenth century scientific revolution. Still today, we are led by the paradigm that we have to "dominate" nature—an ideology that has led us to an unsustainable estrangement from nature. Since it is important to understand the implications of a lacking recognition of the limits of nature for peace and cooperation among states and nonviolence within societies, a second theoretical part will discuss to what extent political science, and in particular its subfield of "international relations" has started engaging these questions by incorporating the environment and ecology as independent variables in its studies. This analysis will conclude that there are some new interesting and promising approaches in international relations, but that their analytical power will be limited so long as the field has not begun developing the theoretical tools needed to engage with the material world.

In a third empirical part, we will use the case study of the "tight oil revolution" to show how the prevailing "domination of nature" thinking is still, once again,

[1] Deudney (1999a: 42) reminds us that important exceptions to this rule are being found in the disciplines of archeology, geography, economics, and history, all of which have contributed to the "study of the effects of the natural environment on politics".

misleading us into believing that technology could offer a sustainable solution to our energy problems. While acknowledging that the "tight oil revolution" will provide us with a deferral of a few years, the following will employ this much celebrated "game changer" in global oil supply as a case study to show that it was not only technology, but rather "Wall Street-economics", politics and deregulation which resulted in this "revolution". Within a few years time we will most probably observe that it will indeed not be sustainable. The chapter concludes that natural scientists and social scientists alike should eventually overcome the thinking of the "mechanistic age" and acknowledge that the laws of nature bestride and set the frame in which politics, technology and economics can evolve, and not vice versa. We would also like to encourage social scientists—in particular those working in the field of international relations—to integrate the analysis of both the material and social worlds in order to not only leave critical questions like the future of our energy to economists and engineers. We need to better understand how our societies can learn to adopt energy systems that rely on efficiency, savings, and renewable energy.

1 Modern Ideology and Physical Laws

1.1 Domination of Nature

Western society's unbounded belief in the power of technology can be tracked down to the "domination of nature" paradigm that originated in the sixteenth century, which over time became an influential ideology of modern Western society. This new ethic stipulated that nature had to be "conquered" and exploited to satisfy our "insatiable" human needs (Leiss 1974: xvi). It is connected with the "scientific revolution" of the sixteenth and seventeenth centuries, which intended "to mechanize and to rationalize the world view" and to bring order into a lawless nature by subjugating it to the technological innovations brought forward by scientists (Merchant 1989: 2). While modern scientific thinking is associated with the work of Galileo and Newton,[2] the "mastery of nature" paradigm is mostly associated with Newton and has been developed by Bacon and Descartes. Jeremy Rifkin, in his 1981 analysis "Entropy—A New World View", by referring to these protagonists of the mechanical paradigm concluded that "(a)fter three hundred years we are still living off their ideas" (Rifkin 1981: 19). Currently, the "mastery of nature" paradigm "is shared by almost every society on earth" (Pattberg 2007: 1).

[2] Galileo's credo that theory must be confirmed by experimental observations and Newton with his mathematical formulation of the laws of nature have revolutionized the classical understanding of the world establishing the fundamentals of natural sciences.

The scientific revolution changed the Renaissance image of the earth as a "nurturing mother" and "living organism" (Merchant 1989: 3)[3] into one of a machine that is "composed of interchangeable atomized parts that can be repaired or replaced from outside." (Merchant 1989: 193) A second important image was that of nature as being in "disorder", to be fixed in the mechanical world by redefining order as "to mean the predictable behavior of each part within a rationally determined system of laws" (Merchant 1989: 193).

This view of nature as dead, passive and disorderly matter opened up and sanctioned the new ideology which required that nature needed to be submitted to new technologies, in particular for mining and metallurgy (Merchant 1989: 164). What the protagonists of the "mastery of nature" ideology actually did was to construct a "detached natural scene" to be observed from the outside with the conviction that this scene belonged to a different world (Bramwell 1989: 30 citing Keith Thomas). Thus the idea of the "mastery of nature" was based on the "methodical separation of 'nature' and 'society'" (Leiss 1974: xi), leading to the notion of an unbridgeable dualism between the two. This notion ultimately was responsible for a development that led to men's estrangement from nature.

A specific example of that dualism is expressed by the belief that technological progress and economic incentives help to exceed the limits set by geology. This belief finds special application in the debate about fossil fuels. Though everybody agrees that physical resources—fossil fuels in particular—are finite, most people believe that these limits have no influence on today's behavior and might be altered by technological progress in the unspecified future.

1.2 The Laws of Physics

The ultimate consequence of the mechanistic and deterministic view culminated in La Place's "demon",[4] first published in 1814, with the belief of being able to make precise predictions into the future. Only in the early twentieth century, Bohr and Heisenberg could argue with the "Copenhagen Interpretation"[5] that exact predictions

[3] For a critique of Merchant's radical gender orientation in her interpretation see Bramwell (1989: 24–26).

[4] LaPlace stated that once the initial and boundary conditions of all atoms in the world are known, their movements could be predicted into any future. Since such knowledge can only be theoretical and virtual, this statement became famous as "La Place's demon".

[5] The uncertainty principle states that it is impossible to know a priori all relevant physical data of any physical system exactly. This result of mathematical calculations led Bohr and Heisenberg to the "Copenhagen interpretation", which identifies any physical prediction at an atomic level as a probabilistic statement (Born 1957). This interpretation was fundamental in relating modern physics to macroscopic observations, while also allowing the integration of neuronal aspects of thinking and even esoteric views in the broader physical frame without logical contradictions (Dürr and Österreicher 2001).

of future developments are impossible due to fundamental laws of physics and should be replaced by probability statements. Though almost never realized outside of philosophical circles, this theoretical discussion was essential in its consequences as it severely reduced the "mastery of nature" paradigm in its claims.

The Entropy concept became another threat to the mechanistic view of the world after it was proposed by the mathematician Lazare Carnot in 1803 and finally formulated by Rudolf Clausius about 150 years ago within the frame of thermodynamics. The laws of thermodynamics have general consequences reaching far beyond classical physics. This amounts to an alternative worldview that—as Rifkin hoped some three decades ago—"will eventually replace the Newtonian world machine as the organizing frame of history" (Rifkin 1981: 6).

The first law of thermodynamics stipulates that the total energy of an isolated system is constant, and therefore only its form can be transformed. Indeed, any physical movement is only possible by the conversion of energy. The second law of thermodynamics prescribes that energy can only be converted in one direction, that is from "useable to unusable, or from available to unavailable, or from ordered to disordered." (Rifkin 1981: 6) Each conversion process increases the state of disorder. "Entropy" is the measure of the state of disorder of a given physical system; and there is no escaping from the physical reality that the "entropy" of a system always rises, unless external energy—previously not included in the system under consideration—is added. Thus, the entropy concept also destroys one of the key legends of our modern times, which is that science and technology bring more order and through more order, more progress to the world.

The key here is that these processes are irreversible. Nicholas Georgescu-Roegen concludes from this observation that "(i)f the entropic process were not irrevocable, i.e. if the energy of a piece of coal or of uranium could be used over and over again ad infinitum, scarcity would hardly exist in man's life" (Georgescu-Roegen 1971: 6).[6]

Increasing order of a system is only possible by adding higher ordered energy from external sources. In that perspective, the evolution of the industrial society was only possible by adding chemical or nuclear energy (coal, oil, gas, uranium) from resources, which previously were external to the considered system. Indeed, most efficiency gains in the economy can be traced back to the substitution of human labor by technical work from fossil fuels. Kümmel et al. (2009) challenge the common assumption that growth is driven by innovations and efficiency improvements, arguing that the dominant driver of economic growth in the past was a substitution process to fuels with higher energy density, e.g. from coal to oil.

In the last decade of the twentieth century it seemed that the second law of thermodynamics had finally found some recognition and had begun taking over the mechanistic interpretation of nature as advanced by the "scientific revolution". In combination with a rise of oil prices in the mid-2000s, an increasing number of analyses on energy and energy security started from the assumption that fossil fuels

---

[6] To illustrate the consequences of the processes of entropy Homer-Dixon (2007) uses the example of the Roman Empire's collapse.

are finite and that we are close to a situation of "peak oil", urging us to take a sharp turn in the way we organize our oil dependent societies. These studies were mainly provided by natural scientists (and very few social scientists), who connected each under the umbrella of the "Association of the Study of Peak Oil" (ASPO). But in general—apart from some new approaches that will be discussed below—it is still safe to assume that most modern society's natural and social scientists (including economists) have difficulties acknowledging this problem as they still feel challenged by recognizing our limitations in relation to space, time, matter and energy. They try to ignore the laws of nature by assuming that money and technology drive the world instead of acknowledging that it is being subjected to the laws of physics with the conversion and deterioration of energy reservoirs from higher to lower order. However, in the last 20 years some notable and interesting new approaches have emerged in the field of international relations, which intend to link the environment and the ecology to human outcomes in our society. We will discuss below to which extent these approaches have developed into feasible and viable new alternatives in overcoming the traditional dichotomy of nature and society in the social sciences.

2 New Approaches in International Relations

In the field of international relations theory it was first and foremost Daniel Deudney who as early as 1999 demanded in a seminal article to "bring nature back in". His goal was a revitalization of the much discredited tradition of "geopolitics" and to place an emphasis on the critical role that environmental factors such as, for example, "climate" and "arable land" might play for certain political outcomes (Deudney 1999a: 27). As a remedy for this lacuna, he advertised a return to a "functional-materialist theory" (Deudney 1999a: 50). We should also mention Karen Litfin's "evolutionary idealism", which intends to integrate constructivism and materialism (2003). But both, Litfin and Deudney's theoretical reflections need to be further developed in order to make them operational for empirical study in this area. However, it might also be possible to ignore this theoretical deficit in international relations altogether. Timothy Mitchell's (2011) strictly empirical study on "Carbon Democracy", which links the availability of oil with the existence of democratic structures, justifiably received widespread attention because it enlightens the understanding of the power of the material world in determining political outcomes.[7] Although, in order to make Mitchell's results applicable to further case studies, they would need to be embedded in a theoretical conceptualization. In response to new awareness of the environmental degradation of a globalized world which "coincided with a turbulent period in the field of conflict

[7] Though Mitchell (2011: 238) does not want to be misunderstood as belonging to the camp of determinists and Malthusians.

studies" (Klem 2003: 10), two new subfields within "international relations" have emerged: "environmental security" and "political ecology". In the following section, we will explain why these two subfields fall short of our request to incorporate nature and its limits in studies on peace, conflict and the sustainability of our societies.

The subfield "environmental security and conflict" emerged at the beginning of the 1990s with the intention to explain the connection between violent conflicts and environmental degradation (Klem 2003). Thomas Homer-Dixon, the most prominent representative of this school, made the concept of "environmental scarcity" the focal point of his analysis by claiming that it contributes to ethnic violence, civil strife, inter-state war, as well as to the North-South conflict in general (Homer-Dixon 1999: 5). His study exclusively covered renewable resources[8] such as "cropland, forests, river water, and fish stocks" (Homer-Dixon 1999: 8). Environmental degradation is expected to affect the developing more than the developed world and therefore will subject the former to much more violent conflicts. Homer-Dixon discerns three causes for the scarcity of renewable sources: environmental degradation, population growth and an unequal distribution of resources. As a further problem Homer-Dixon diagnoses an "ingenuity gap", with the "groups most affected by ecological scarcity" being the "least capable of innovation" (Klem 2003: 13). His much discredited neo-Malthusian approach and his conjuring up of environmental disaster in the third world has been criticized widely in the field of political science for legitimizing the North's intention to control the South (Dalby 1999).[9] For the purposes of our discussion it is sufficient to note that "environmental security" studies are generally reduced to the "environment" in their identification of the causes of the outbreak of violence without adequately taking into account political and societal factors of the conflict. While the method promotes a disciplinary approach, it still needs to develop a comprehensive theory to capture sufficiently the dynamics between the environment and societies under examination.

Parallel to the evolution of "environmental security", the subfield of "political ecology" developed in sharp distinction to the assumptions of the Toronto school of "environmental security". In their 2001 publication "Violent Environments", several of the authors start out their chapters with an intensive and comprehensive criticism of Homer-Dixon's research project.[10] By integrating "Marxian political economy with cultural studies and the new ecology" while claiming not to neglect the "causal powers inherent in Nature itself" (Peluso and Watts 2001: 24–25), political ecologists intend to provide "tools for thinking about the conflicts and

---

[8] Non-renewable resources were excluded from any analysis because they were considered to be abundant in the 1990s. This changed in the mid 2000s when questions of the availability of fossil fuels became an integral part of Homer-Dixon's analysis. See Homer-Dixon (2007) and Homer-Dixon and Garrison (2010).

[9] See also Deudney (1999b) and Peluso and Watts (2001: 15).

[10] Peluso and Watts are extensively criticizing Homer Dixon in their introductory chapter as well as Hartman, Fairhead, Richards, and Watts in their contributions to the volume (Peluso and Watts 2001).

struggles engendered by the forms of access to and control over resources" (Peluso and Watts 2001: 25). But political ecologists' explanations for violence and civil wars usually suffer from a serious shortcoming: they focus explicitly on the politico-economic factors which might have caused this conflict, while material or environmental factors are usually ignored as a possible trigger for this outbreak of violence. In response to their Marxist heritage, as far as natural resources are concerned, they are rather treated like dependent variables that gain relevance exclusively by a transformation process through labour. Accordingly "political-ecological" studies focus on questions of the link between violent conflict and the special nature of the labour process involving natural resources. This approach has, for example, strong analytical power when applied to "climate change", which seems to be a very conducive field for demonstrating the incompatibility between capitalism and sustainable development (Peet et al. 2011).[11]

But if applied to non-renewable resources and, for example, to the phenomenon of "peak oil", the limits of this approach become obvious. To begin with, peak oil stands for the point of the global maximum oil production rate and expresses the phenomenon that the use of crude oil—as of any finite natural resource—is limited due to the finiteness of its reservoirs. This finiteness implies that in-between the start of production and its endpoint a maximum of extraction is followed by a declining supply rate long before the resource is exhausted. Based on the detailed analysis of many empirical observations, the geologists Colin Campbell and Jean Laherrere stated in the 1990s that the world's crude oil production will peak in-between 2000 and 2010 and that this will have serious consequences on world markets and society (Campbell and Laherrere 1995). However, in his article on peak oil, Gavin Bridge critiques "peak oil's core claim that geological limits are now the primary driver of oil's availability" (Bridge 2011: 308). As a political economist, he focuses on "above ground" conditions on oil's availability", such as "cartelization, the investment strategies of international oil companies, and resource nationalism", as a cause for the "oil crunch" (Bridge 2011: 314 ff.). He concludes that the "creation of scarcity" is at the "heart of capitalist society" (Bridge 2011: 316). Thus according to his analysis there are no limits to nature which have to be factored in. As a consequence, he is in line with the "domination of nature" paradigm and when emphasizing the "(s)ocial mastery of the earth's energies" as "one of the most potent of modernity's ideological tropes" (Bridge 2011: 308). Thus, political ecologists seek to explain any limitations of the availability of fossil fuels with the political economy of the process of resource extraction. They would consider it as a misleading and detrimental to social science analysis to strive to integrate "nature" as an independent variable in social science analysis.

The polarization of political views between the proponents of peak oil and political economists is also reflected in the debate between economists on the "fixed stock paradigm" and the "opportunity cost paradigm". The "fixed stock

[11] For further theoretical developments to be utilized for "energy affairs" within "International/Global Political Economy" see Stoddard (2013) and Coole (2013).

paradigm" relies on physical measurements of availability, suggesting that mining in the long run is inherently unsustainable due to its finiteness (Prior et al. 2010; Gordon et al. 2007; Gagnon et al. 2009; Hall and Cleveland 1981). The "opportunity cost paradigm" assesses "resource availability by what society has to give up producing for another unit of a mineral commodity" (Tilton and Lagos 2007). It claims that while over time depletion tends to drive the opportunity cost of mineral production up, new technology and other forces can offset this upward pressure. At the heart of this discussion stands the question of whether or not depleting resource use becomes a problem for societies: If so, scarcity drives up prices with unprecedented consequences to technology, the economy and society. If not, new technologies will be phased in, and in time continuously balancing this dependence without severe consequences for the economy and society. Due to these fundamentally contradicting consequences, these discussions are polarized and mixed with "stake-holder" argumentations: the "opportunity cost paradigm" fits well into any growth oriented economic thinking while the "fixed stock paradigm" requires adequate economic and political measures, which would unavoidably shake present life style patterns.

3 Exploring the "Tight Oil Revolution"[12]

Until 2012, the "peak oil" debate had made it into some of the mainstream publications on energy security. In its 2008 WEO Report, even the IEA issued a strong warning that fossil fuel dependent economies are not sustainable for much longer, while others pointed to possible conflicts and violence. But all of a sudden a new phenomenon—the "tight oil revolution"—seemed to vindicate those analysts—most of all the managers of the oil industry—who had always argued that, once again, a new technological step would result into mastering nature's finiteness of fossil. For observers, this new development was not surprising. In its latest report, the IEA weakens hopes on its one year earlier announced energy revolution as "light tight oil shakes the next ten years, but leaves the longer term unstirred...but this does not mean that the world is on the cusp of a new era of oil abundance." (IEA 2013: 25)

The US shale revolution seems to have entered the world of energy like a "phoenix from the ashes". The big bang of public awareness on the oil revolution came with the publication of the International Energy Agency's "World Energy

[12] To avoid confusion, the oil which is subject to our analysis is usually called "light tight oil" in contrast to shale oil, which is synthetic oil produced with high energy input from oil shales (so called kerogene is transformed into liquid oil by being heated up to 500 °C). Both are extracted from almost impermeable shales, the latter still chemically bound in organic matter, while the first (light tight oil) is already conventional oil but trapped in fine rock pores or cracks where it cannot escape without stimulation.

Outlook" in November 2012, which stated that the United States was about to turn into the largest global oil producing nation, and subsequently overtaking Saudi Arabia by the mid-2020s.[13] By around 2030, the IEA expected the US to become a net oil exporter, and energy self-sufficient, though only in expectation of a reduction of consumption by 1.4 % annually.[14] The U.S. Energy Information Administration estimates "technically recoverable resources of 345 billion barrels" of world shale/tight oil resources and some 58 billion alone in the US (EIA 2013: 2–3). Apart from some noticeable exceptions like Daniel Yergin, nobody seemed to be aware of the advent of this "revolution", though the deposits have been known for decades. But after the 2012 IEA report, huge media resonance transported the message of an energy revolution without almost any critical assessment, echoing the main message of IEA's press release. Daniel Yergin assures us that "(t)his is like adding another Venezuela or Kuwait by 2020, except these tight oil fields are in the United States" (Yergin, quoted in Ghouri and Ghouri 2012: 3).

3.1 Reasons Driving Depletion

The development of light tight oil deposits in North Dakota (Bakken Formation) and South Texas (Eagle Ford Shale Formation) is treated like a real "game changer" (Bremmer and Hersh 2013) and is expected to have dramatic geopolitical consequences by changing the energy market and the need for a US presence in the most oil sensitive area: the Gulf region. But given these geopolitical stakes, what are the reasons for this new hope of an US energy revolution? Is it really based on access to new resources by means of new technology shifting the peak oil problem far into the future or is it based on hopes and wishful thinking of the US elites during times when new countries such as Russia and China continue to gain power at world level, threatening the pretense of the US as being the only superpower worldwide? Understanding this heated debate of a new "revolution" first requires a short explanation of some of the fundamental trends driving resource depletion.

The general pattern of resource extraction can be presented irrespective of the nature of the resource. First, each promising region is screened and mapped for possible reservoirs, which later on are tested by exploration and appraisal wells. With increased geological knowledge, engineers make estimates about the number and size of the reservoirs and together with economists, rank them according to their economic and technical properties. The subsequent phase of development and production mirrors this internal ranking, as the most promising assets - those

[13] The big bang introduction of the "tight oil revolution" was only preceded by very few analyses referring to the Shale oil revolution (e.g. Mills 2012).

[14] Certainly, energy self-sufficiency is not only a result of the tight oil revolution, but also because of the expected impact of new fuel-efficiency measures in transport (International Energy Agency 2012: 1 and table 3).

providing the highest earnings - being developed first. However, as soon as these are depleted, the next valuable assets will substitute them and so forth. The transition from initial production to depletion can take decades or in some cases even centuries. But hydrocarbon reservoirs are unique due to the high mobility of oil and gas they accumulated in traps under impermeable rock structures. As production is spurred by high reservoir pressure, the production profile has a typical shape: In the beginning, production rises rapidly when further wells are added. However, the extraction process reduces the pressure and increases the water level that limits most reservoirs at the bottom side. A once pure oil well will steadily convert into a well that produces water. At the end of their lifetime, some wells produce up to 95 or more percent water mixed with only a small amount of oil. This increases cost and extraction energy demand. In addition the production rate declines when the reduced oil flow can no longer be compensated by the addition of new wells, irrespective of rising cost.

This transition from "easy to extract at high flow rates" to "technically challenging to extract at low flow rates" takes place in each oil or gas field and in each region, irrespective of geological and geographical details. It is accompanied by rising energy input. The energy return on energy investment (EROEI) accelerates the rate of depletion during later stages when energy becomes scarce, and in regards to a net balance, even less oil is extracted (Hall and Cleveland 1981; Hall and Klitgaard 2012). For instance, for the process of conventional oil extraction less than 1 % of the obtained energy is needed to achieve the extraction. Tar sand extraction and upgrading require about 20–50 % of the gained energy for the extraction and preparation processes. Further, 10–15 % are consumed by refining and transport. Declining crude oil quality requires increased refining effort. A worldwide historical analysis and trend extrapolation indicates that around 2030 the energy return on world oil production might become smaller than the energy needed to extract, refine and transport that oil (Gagnon et al. 2009). As mentioned above, nearly 20 years ago, geologists—based on the findings of detailed analyses—recognized with high probability, that in the first decade of the twenty-first century the world aggregate of oil production would reach its maximum, followed by an unavoidable period of declining supply rates (Campbell and Laherrere 1995; Banks 1999; Bowlin 1999). This analysis confirms much older work from the late 1970s (Speth 1980: 413–419), which predicted a peak in global oil production around the turn of the century. The recent extrapolation included longer time records from production profiles. Increasing prices followed by economic and financial problems were some of the expected consequences.

In short, the depletion process counteracts the above-discussed belief on technological progress and everlasting economic growth, while bearing severe economic and political risks. In the following section, the International Energy Agency's understanding of these processes as the most powerful authority of the energy establishment is shortly reviewed. This will offer illuminating insights on the agency's ambivalent messages, which often are in need of interpretation.

3.2 The IEA Forecasts and Oil Companies Under Pressure

In its 1998 World Energy Outlook, the International Energy Agency (IEA) reported the problem of peak oil for the first time, by addressing "the pessimists' view" (IEA 1998: 95ff). In 2000, world oil prices started to rise at an unprecedented speed never before observed, with oil analysts commenting on the ascent as unexpected, but not based on fundamentals. From US$12/bbl in 1998, the Nymex crude oil price rose to US$40 during the beginning of the financial downturn, soon rising further to US $140 in mid-2008, and eventually leveling off around US$80–100 following the economic recession, a level not seen during any previous period of economic growth before. The IEA never repeated or extended the peak oil analysis of 1998, but instead exchanged the responsible staff (Bourdaire 2011) and each year focused on different regions which might increase production.

In 2011, the IEA assessed the light tight oil potential for the first time. Just one year later, in 2012, this assessment concluded that "the global energy map is changing", as already pointed out above (IEA 2012: 23). However, in the report these statements are considerably weakened, with the IEA reporting that "none of these scenarios [are] a forecast. Each is intended to demonstrate how markets could evolve under certain conditions." (IEA 2012: 38). Furthermore, at the release of the 2013 report, the IEA executive director, Maria van der Hoeven, dampened the high expectations by predicting a "peak around 2020" which will then plateau (Wald 2013).

The trend of worsening supply conditions is reflected in the difficulties of big oil companies to maintain a high level of reserves, which is an important aspect for shareholders. In the late 1990s, many powerful Western oil companies merged: Exxon and Mobil, Chevron and Texaco, Conoco and Phillips, BP, Amoco and Arco. The companies acquired reserves and production capacities by mergers instead of their own exploration initiatives. Only one "big sister" was missing at that time: Shell. Consequently, despite its exaggeration of reserves in 2003—the CEO had to leave when this became public (Donovan 2012)—Shell's oil production started to decline. Between the end of 2002 and 2013, Shell's oil production decreased by almost 40 % according to company reports. Simultaneously, exploration and production expenses increased from just below US$5 billion to US$30 billion in 2012. The combined production of large companies listed by US and European stock markets declined from its all-time high in 2004 by 2.6 % annually until 2013, while the expenses of exploration and production increased in a similar proportion to Shell's example (EWG 2013).

New discoveries were disappointing. Deepwater areas failed, which 10–15 years earlier were seen as increasingly important supply regions: Gulf of Mexico production started to decline around 2010, production in Mexico has been in decline since 2005, Angola has passed peak production, and the enthusiastically celebrated discovery of the Kashagan oil field in the Caspian Sea in 2000 is still at least 8 years behind schedule, even after spending around US$60 billion. Just a few years earlier, all of these regions were presented as essential parts for the future of world oil supply.

3.3 *The "U.S. Energy Revolution": Producing Hot Air*

Today, attention has been refocused on the "energy revolution" in the USA, where oil and gas production has increased: Since 2005, gas production has risen due to shale gas developments. Oil production rose, mainly due to tight oil developments. Almost 6,000 new oil wells were drilled in North Dakota alone, increasing production from 4 kb/day in 2005 to 800 kb/day by July 2013. The two largest companies involved in the tight oil and shale gas business with 25 % of the share—XTO and Chesapeake—grew tremendously. In 1993, both were relatively small sized companies with several ten million US$ in turnover. By 2010, both had increased their gas sales to several billion US$ each.

In the face of depleting options, the shale developments can be seen as the next step in the life cycle of oil and gas production. But the basic formula of "easy to difficult" also applies to shales: The success of shale gas developments combined with the original high gas price resulted in a price collapse, which then became too low to generate profit. Better returns could only be gained by transactions of claims and drilling rights instead of investing into drilling activities themselves.

After the collapse of U.S. gas prices in 2008, the economic situation for shale gas and tight oil producers became anything but lucrative. Many asset sales were addressed to improve the financial situation of the sellers and the reserve replacement situation of the buyers. In 2010, more than US$50 billion worth of assets changed owners. These purely financial transactions established close ties between the oil and gas business with the investment banking institutions (Rogers 2013). Many positive production reports—further echoed by the media—were exaggerated, as shown by an NYT investigation which over years collected and finally published e-mails and reports received via whistle blowers within companies (New York Times 2011, 2011a). However, investments declined steadily to less than US$4 billion in 2013 (IHS Herold 2014). These developments fit well into the concept of a life cycle of resource extraction when the focus has to shift from the vast crude oil and gas deposits to the less favorable unconventional reservoirs. All these unconventional oil and gas sources are only viable at a higher technical and economic effort. Among them are the tar sands and extra heavy oil deposits in Canada and Venezuela, as well as oil shales and light oil in tight hydrocarbon reservoirs.

3.4 *The Development of Gas and Oil Fields*

In order to extract gas (or oil) from tight reservoirs, a well is drilled and stimulated by creating fractures that give access to the oil or gas in the pores and cracks of the rock formation. The stimulation process usually requires the exertion of high pressure (up to 1,000 bar), and the injection of water (several million liter per well), propping agents (e.g. sand), and chemical additives such as gelling agents, friction reducers, surfactants, breakers, corrosion inhibitors and biocides. Due to its nature, the stimulation process is usually known as hydraulic fracturing or in short "fracking". Most of the wells show steep depletion rates, which in the first year frequently reach up to 30 % for oil wells and up to 80 % for gas wells. A continuously rising number of new wells are required to compensate the depletion of the older wells.

The growth of oil and gas production in 2005 was due to several factors which set incentives: Rising prices attracted investments, improved technology allowed for horizontal drilling, and the lifting of environmental laws constrains reduced cost and development time. The most promising gas shales, such as Barnett, Fayetteville, Haynesville, Eagle Ford and Marcellus Shale were developed over these years. As of now production in most of these fields is about to peak or has already peaked. As of the end of 2013, total US shale gas production has plateaued since 2 years. In all likelihood, total U.S. gas production will start to decline within the next few years—despite the hopes set by exaggerated reports.

Fracking technology gained strong political support during the 2000s– otherwise it could not have happened at that scale. However, time will limit its effects. First, political incentives for well stimulation were set in force in the 1970s when politicians and experts defined a tight gas reservoir as the expected value of permeability to gas flow as being less than 0.1 Millidarcy. This political definition was being used to determine which wells might receive tax credits for producing gas from tight reservoirs (Holditch 2006). Strong support for the promotion of this technology came from the 2001 National Energy Policy developed by the Energy Task Force under the direction of Dick Cheney. For this report, Cheney rebuffed requests from the Environmental Protection Agency (EPA) to include some mentioning of fracking's environmental concerns. Cheney's office was involved in the report's portrayal of fracking as being safe for the environment although at the time he was Vice-President of Halliburton, a company heavily engaged in this extractive industry.

On December 12, 2003 the US EPA, BJ Services, Halliburton and Schlumberger signed a Memorandum of agreement to eliminate the diesel fuel injection used with hydraulic fracturing activities. This agreement was a prerequisite for EPA to include in its environmental impact study in 2004, "that the process of hydraulic fracturing did not pose a threat to drinking water." (Lustgarten 2010) Again, the conclusion of the EPA study was the prerequisite for the exemption of gas and oil drilling from the Safe Drinking Water Act (SDWA) in the Energy Policy Act in 2005. (EPA 2005) The gas industry was no longer obliged to disclose the chemicals used for fracking activities, as the exemption from the SDWA removed the legal basis for the EPA to monitor corresponding activities unless violations were proven. Since then, Vice-President Dick Cheney became instrumental in achieving this success because of his connection to Halliburton, with the passage having been labeled the "Halliburton loophole" (New York Times 2009). Various attempts to remove the exemption paragraph from the Energy Policy Act were started, though without success, as it could be proven that the industry frequently violated its 2003 "diesel agreement" (e.g. H.R. 7231 in the 110th Congress and H.R. 2766 in the 111th Congress, Waxman and Markey 2010).

4 Conclusion

The theoretical discussion on the evolution of the laws of physics showed that mankind's thinking has not yet fully overcome the "domination of nature" paradigm as developed by the scientific revolution in the sixteenth and seventeenth centuries. While the laws of thermodynamics, and in particular the second "Law of Entropy", have slowly overtaken the supremacy of the "mechanistic age", they still have not become the status quo of social scientists who need to realize that only by the addition of external energy, degradation of energy quality and scarcity can be beaten. While there is no question about the benefits of technology to our societies, we need to acknowledge that there are insurmountable limits to technology's ability to dominate nature. In order to prepare for the post-fossil fuel era, we need to come to grips with this aspect of physical reality.

The depletion of geological energy reservoirs starts with the first well drilled. There is no doubt that technology improvement can postpone the declining share of easy to develop resources. These resources are substituted during the extraction period by the development of more complex and expensive geological reservoirs. This race between depletion and technology gradually shifts the situation to the worse. There is general agreement that conventional easy to develop crude oil extraction has already peaked and is more and more substituted by the use of unconventional resources. Following the period of polar oil and deep water offshore production, the presently called "Energy revolution" in the US, focuses on light tight oil and gas extraction from tight reservoirs. These stocks were known for decades but only exploited when rising prices, improved technology, favorable political framework (which steeply reduced environmental policy hurdles) and economic incentives such as changing SEC[15] rules, and huge profits for the financial sector (credits) paved the way.

From that point of view, the presently stated "US energy revolution" should be further interpreted as a step towards the depletion of fossil fuels. The engagement of companies in that field is an admission that conventional assets are getting exhausted—otherwise these would be utilized. It is an attempt to carry on "business as usual" a bit longer during a time when long-term thinking would be challenged in the face of an obvious structural change in the fundamentals of energy supply, affecting all structures of public life, including economics and politics.

[15] In recent years, the Security and Exchange Commission (SEC) modified the reporting rules for gas and oil companies by allowing the inclusion of unconventional reserves. Without this change, most companies would no longer be able to report reserve additions larger than the decline due to the produced volumes during the year.

In order to engage in this debate with relevant and significant prognosis and guidance, social scientists and scholars of International Relations in particular, need to overcome the discipline's traditional inherent dichotomy between nature and society, and develop theories and concepts explaining the dynamics between energy and political outcomes. For a peaceful transition from the fossil to the post-fossil world, we urgently need to develop models, new ideas and new comprehensive theoretical approaches integrating nature and society in order to foster the feasibility and acceptance of a post-carbon life style.

References

Banks, H. (1999). ENI's Franco Bernabe says cheap oil won't last long. *Forbes, 161*(12), 84–86.

Born, M. (1957). *Physik im Wandel meiner Zeit*. Braunschweig: Vieweg Verlag.

Bourdaire, J.-M. (2011). Biographical notes. In C. Campbell (Ed.), *Peak oil personalities* (pp. 53–69). Skibbereen: Inspire Books.

Bowlin, M. (1999). ARCO chairman says last days of oil age have begun: Calls on U.S. energy industry to meet clean fuel challenge. Accessed January 30, 2014, from http://www.prnewswire.com/news-releases/arco-chairman-says-last-days-of-oil-age-have-begun-calls-on-us-energy-industry-to-meet-clean-fuel-challenge-74862562.html

Bramwell, A. (1989). *Ecology in the 20th century. A history*. New Haven, CT: Yale University Press.

Bremmer, I., & Hersh, K. A. (2013, May 23). When America stops importing energy. *International Herald Tribune*.

Bridge, G. (2011). Past peak oil: Political economy of energy crisis. In R. Peet, P. Robbins, & M. Watts (Eds.), *Global political ecology* (pp. 155–186). London: Routledge.

Campbell, C., & Laherrere, J. (1995). *The world's oil supply 1930–2050*. Geneva: Petroconsultants.

Coole, D. (2013). Agentic capacities and capacious historical materialism: Thinking with new materialisms in the political sciences. *Millennium: Journal of International Studies, 41*(3), 451–469.

Dalby, S. (1999). Threats from the South? Geopolitics, equity and environmental security. In D. H. Deudney & R. Matthew (Eds.), *Contested grounds. Security and conflict in the new environmental politics* (pp. 155–185). Albany, NY: State University of New York Press.

Deudney, H. D. (1999a). Bringing nature back in Geopolitical theory from the Greeks to the global era. In D. H. Deudney & R. Matthew (Eds.), *Contested grounds. Security and conflict in the new environmental politics* (pp. 25–60). Albany, NY: State University of New York Press.

Deudney, H. D. (1999b). Environmental security: A critique. In D. H. Deudney & R. Matthew (Eds.), *Contested grounds. Security and conflict in the new environmental politics* (pp. 187–219). Albany, NY: State University of New York Press.

Donovan, J. (2012, December 18). Chris Finlayson and the Royal Dutch Shell reserves scandal. Accessed January 30, 2014, from http://royaldutchshellplc.com/

Dürr, H.-P., & Österreicher, M. (2001). *Wir erleben mehr als wir begreifen – Quantenphysik und Lebensfragen*. Freiburg: Herder Spektrum.

Energy Information Administration. (2013). Technically recoverable oil and shale gas resources: An assessment of 137 shale formations in 41 countries outside the United States. Accessed January 30, 2014, from http://www.eia.gov/analysis/studies/worldshalegas/

EPA. (2005, August 8). *Energy Policy Act 2005*, Sec 322 (Hydraulic Fracturing).

EWG. (2013). *Fossil and nuclear fuels – the supply outlook*, Energy Watch Group. Accessed January 30, 2014, from http://www.energywatchgroup.org

Gagnon, N., Hall, C. A. S., & Brinker, L. (2009). A preliminary investigation of energy return on energy investment for global oil and gas production. *Energies, 2*, 490–503.
Georgescu-Roegen, N. (1971). *The entropy law and the economic process.* Cambridge: Harvard University Press.
Ghouri, S., & Ghouri, A. (2012, June 18). The US unconventional oil revolution: Are we at the beginning of a new era for US oil? *European Energy Review.*
Gordon, R. B., Bertram, M., & Graedel, T. E. (2007). On the sustainability of metal supplies: A response to Tilton and Lagos. *Resources Policy, 32*, 24–28.
Hall, C. A. S., & Cleveland, C. J. (1981). Petroleum drilling and production in the United States: Yield per effort and net energy analysis. *Science, 211*, 576–579.
Hall, C. A. S., & Klitgaard, K. (2012). *Energy and the wealth of nations: Understanding the biophysical economy.* Heidelberg: Springer.
Holditch, S. A. (2006). Tight gas sands. *Journal of Petroleum Technology, 58*(6), 86–93.
Homer-Dixon, T. (1999). *Environment, scarcity, and violence.* Princeton, NJ: Princeton University Press.
Homer-Dixon, T. (2007). *The upside of down. Catastrophe, creativity, and the renewal of civilization.* London: Souvenir Press.
Homer-Dixon, T., & Garrison, N. (Eds.). (2010). *Carbon shift. How peak oil and the climate crisis will change Canada.* Toronto: Vintage Canada.
IHS Herold. (2014). *Global upstream performance review 2013.* Accessed January 30, 2014, from http://www.herold.com/research/ihs_herold.page?p=42
International Energy Agency. (1998). *World energy outlook 1998.* Paris: OECD/IEA.
International Energy Agency. (2012). *World energy outlook 2012.* Paris: OECD/IEA.
International Energy Agency. (2013). *World energy outlook 2013.* Paris: OECD/IEA.
Klem, B. (2003). Dealing with scarcity and violent conflict (Clingendael Conflict Research Unit Working Paper 24). Accessed January 30, 2014, from http://www.clingendael.nl/sites/default/files/20031000_cru_working_paper_24.pdf
Kümmel, R., Ayres, R., & Lindenberger, D. (2009). The laws of physics and the methods of economics. Accessed January 30, 2014, from http://ufn.ru/en/pacs/all/ and http://www.ewi.uni-koeln.de/fileadmin/user_upload/Publikationen/Working_Paper/EWI_WP_08-02_Cost-Shares-Output-Elasticities.pdf
Leiss, W. (1974). *The domination of nature.* Boston, MA: Beacon Press.
Litfin, K. (2003). Towards an integral perspective on world politics: Secularism, sovereignty and the challenge of global ecology. *Millennium: Journal of International Studies, 32*(1), 29–56.
Lustgarten, A. (2010, March 18). EPA launches national study of hydraulic fracturing. *Pro Publica.* Accessed January 30, 2014, from http://www.propublica.org/article/epa-launches-national-study-of-hydraulic-fracturing
Merchant, C. (1989). *The death of nature. Women, ecology and the scientific revolution.* New York: HarperOne.
Mills, R. (2012, April 17). Cheer up: The world has plenty of oil. *European Energy Review.* Accessed January 30, 2014, from http://www.europeanenergyreview.eu/site/pagina.php?id=3641
Mitchell, T. (2011). *Carbon democracy. Political power in the age of oil.* London: Verso.
New York Times. (2009, November 3). *The Halliburton Loophole* (Editorial). Accessed January 30, 2014, from http://www.nytimes.com/2009/11/03/opinion/03tue3.html
New York Times. (2011). *Leaked industry documents.* Accessed January 30, 2014, from http://www.nytimes.com/interactive/us/natural-gas-drilling-down-documents-4-intro.html?_r=0
New York Times. (2011a). Drilling down series. Accessed January 30, 2014, from http://www.nytimes.com/interactive/us/DRILLING_DOWN_SERIES.html?_r=0
Pattberg, P. (2007). Conquest, domination and control: Europe's mastery of nature in historic perspective. *Journal of Political Ecology, 14*, 1–9.
Peet, R., Robbins, P., & Watts, M. (Eds.). (2011). *Global political ecology.* London: Routledge.

Peluso, N. L., & Watts, M. (Eds.). (2001). *Violent environments*. Ithaca, NY: Cornell University Press.

Prior, T., Giurco, D., Mudd, G., Mason, L., & Berisch, J. (2010, 22–25 August). Resource depletion, peak minerals and the implications for sustainable resource management. *Paper presented at the International Society for Ecological Economics* (ISEE) 11th Biannual Conference, Oldenburg/Bremen, Germany.

Rifkin, J. (1981). *Entropy. A new world view*. Toronto, ON: Bantam Books.

Rogers, D. (2013, February). Shale and Wall Street: Was the decline in natural gas prices orchestrated? *Energy Policy Forum*.

Speth, G. (1980). The global 2000 report to the President. *Boston College Environmental Affairs Law Review, 8*(4), 695–703.

Stoddard, E. (2013). Reconsidering the Ontological Foundations of International Energy Affairs: Realist geopolitics, market liberalism and a politico-economic alternative. *European Security, 22*(4), 437–463.

Tilton, J., & Lagos, G. (2007). Assessing the long-run availability of copper. *Resources Policy, 32* (1), 19–23.

Wald, M. L. (2013, November 12). Shale's effect on oil supply is forecast to be brief. *New York Times*.

Waxman, H., & Markey, E. (2010). Memorandum to members of the subcommittee on energy and environment of the house of congress examining the potential impact of hydraulic fracturing. *Memorandum* 18th February. Accessed January 30, 2014, http://democrats.energycommerce.house.gov/Press_111/20100218/hydraulic_fracturing_memo.pdf

The Socio-Technical Construction of Technology in German-Argentine ICT Cooperation

Alejandra S. Kern and Hernán E. Thomas

Abstract The purpose of this chapter is to present and apply an analytical framework to study the relationship between science, technology, and politics in international cooperation processes. An interdisciplinary theoretical and methodological framework is elaborated, with an analytical focus on the complexity of the process captured by the notion of "socio-technical". Allowing for a broad understanding, the model is based on a combination of concepts from different fields of social science, particularly International Relations Theory and the Social Studies of Technology. To understand how international cooperation comes about, two dimensions are identified: (a) the of technologies, both at the international and national level; (b) the dynamics of interaction between human and non-human elements (research teams, technologies, governmental institutions, forms, texts, etc.) explained as different configurations of techno-political-economic networks. This framework is applied to the analysis of a cooperation case between Argentina and Germany in the field of Information and Communication Technologies during the 1990s (A more complete description and analysis of this case was developed in Kern 2008). This case study serves to shed light on relevant features of cooperation between developed and developing countries.

A.S. Kern (✉)
Universidad Nacional de San Martin, San Martin, Buenos Aires, Argentina
e-mail: akern@unsam.edu.ar; Alejandra.kern@gmail.com

H.E. Thomas
Instituto de Estudios sobre la Ciencia y la Tecnología, Universidad Nacional de Quilmes – CONICET, Bernal, Buenos Aires, Argentina
e-mail: thomas@unq.edu.ar

M. Mayer et al. (eds.), *The Global Politics of Science and Technology - Vol. 2*, Global Power Shift, DOI 10.1007/978-3-642-55010-2_7,

1 Introduction

From an International Relations perspective, the relationship between politics, science and technology has been mainly examined through two sources of explanation[1]: "structure"—the effects of technology on power relationships[2]—and "interaction capacity"—technology as a means that facilitates interactions (see Keohane and Joseph 1998; Rosenau and Singh 2002). Although we consider these sources to be important factors in explaining this relationship, we concentrate here on a third source of explanation, "process," which refers to the dynamics of interactions on different levels of analysis (the international system, the State and sub-national units). In order to fully understand international processes dealing with technologies, we propose a complex perspective, which goes beyond pure political analysis and includes a constructivist view of the Social Studies of Technology (SCOT). SCOT offer a wide spectrum of analytical concepts to understand the way in which technical, social, economic and political considerations shape technological development: dynamics of researcher groups (Shinn 1982; Latour 1989); techno-economic networks (Callon 1992); relevant social groups and interpretative flexibility (Collins 1985; Pinch and Bijker 1987); technological frameworks, socio-technical ensembles (Bijker 1997); social construction of working, socio-technical dynamics and trajectories (Thomas 1999, 2001); processes of co-construction (Vercelli and Thomas 2007), and socio-technical adequacy (Dagnino et al. 2004; Thomas 1999, 2008a, b). According to SCOT, it is not tenable to regard science, technology and society as strictly separate spheres. On the contrary, the three are closely linked as a complex whole, constituting a "seamless Web" (Hughes 1986; Pinch and Bijker 1987). In this vein, our approach intends to "open the black box of technology" and to deepen an understanding of the constituting role of technology in international cooperation processes.

This chapter concentrates on a specific area of international relations where politics, science and technology interrelate: scientific and technological international cooperation. Cooperation is understood as a public policy involving not only competing interests and preferences, but also definitions regarding the usefulness of knowledge in general and of specific technologies in particular. In terms of process, cooperation also involves evolving interactions between different kinds of humans—mainly policy makers and scientists—and technologies.

In the following section, we present an analytical framework for capturing the complexity of this process, which we then apply to explain a cooperation case between Germany and Argentina in the field of ICT (Kern 2008, 2009).

The case of Argentina and Germany is well suited to illustrate the complex dynamics of scientific and technological cooperation in the context of North-South

---

[1] "Sources of explanation" is taken here from Buzan and Little (2000).

[2] The work of Kenneth Waltz (1981) is here paradigmatic. From the field of International Political Economy Susan Strange (1988) offers an alternative perspective on the concept of "structural power".

relations. In this sense, the asymmetries between Germany and Argentina in terms of their material resources and differences in the politicization of ICTs are crucial. However, these characteristics did not result in an automatic subordination of the Argentineans to the agenda and the procedures determined by the Germans. The evolution of this cooperation process is much better explained by the way diverse social and technological factors interrelate throughout time.

2 International Cooperation in Science and Technology: An Analytical Framework

For analytical purposes, we present a two-dimensional framework. The first dimension refers to politics and policies and focuses on the concept of politicization of technology—framed in a historical structure and based on discourses. The second dimension refers to the interaction dynamics that occur throughout the entire process of cooperation. In this dimension, cooperation is considered a socio-technical trajectory constituted by changing configurations of techno-political-economic networks.

Regarding the political dimension, we understand international scientific and technological cooperation as a political process supported by certain ideas and interests forming the political and institutional framework in which research groups interact. This cooperation is framed by a process of *politicization of technology*, namely a process through which certain technologies enter political debate as subjects of resource allocation.[3] Politicization occurs at both, the international and the national level in the context of a particular historical structure.[4] Following Robert Cox (1987), although structure is not a mechanical determinant for action, it creates the context for habits, pressures, expectations and limitations within which action takes place. This means that the main features of a particular historical structure influence the way a specific technology and its utility are understood, as well as how it is disseminated. They also influence the policies that are put into practice. Even when cooperation policy is defined at a national level, it is embedded in a structure of ideas, material capabilities and institutions that characterize a particular international order (Cox 1987) and that therefore influence its formulation. In consequence, both the national and international levels have an important role in defining the political and institutional cooperation framework within which politicians, scientists and technologies interact.

---

[3] "Politicization" is closely related to the interplay between "scientific policy" and "politics of science" (Elzinaga and Jamison 1995). Herrera (1995) also explores the relation between power, politics, and science in Latin America and his work is also relevant to understand the idea of politicization applied here.

[4] Cox defines an historical structure as a "...particular combination of thought patterns, material conditions and human institutions that have some coherence between its elements" (Cox 1986: 141).

According to this perspective, the production and dissemination of technology should be understood as result of a social process involving power relations, which constitute a core-periphery divide in the world order. Ideas in the form of discourses support this divide and influence how politicization of technology occurs in developing countries, shaping their scientific and technological policies. Therefore, discourses supporting different ways of politicization influence the content of the international cooperation framework that results from the intersection of scientific and foreign policies.

The second dimension of this framework, the interaction dynamics of international cooperation, is clearly embodied in the above described political dimension. To analyze these dynamics, we define cooperation processes as *socio-technical trajectories. Socio-technical trajectories* are analytical constructs that involve a mutually constitutive relationship between several elements: "…products, production processes and organizations, institutions, user-producer relationships, learning processes, problem-solving relationships, processes of construction of "working/not-working" technology, rationales, policies and strategies of actors (firms, R&D institutions, universities, etc.), or more complex concepts such as *technological frames* (Bijker 1997) or *socio-technical constituencies* (Molina 1989)" (Thomas et al. 2005). This definition includes the evolution of cumulative and self-generated technological activities that reflect both the impact of previous material conditioning and socio-technical dynamics (Thomas 1999). Despite the fact that these trajectories may be partially explained by the intentions of some actors, such as governments, corporations, institutions, technicians and scientists, they are in fact self-organized processes (Thomas et al. 2005). This concept is applied in this paper in two interrelated aspects. First, we take research groups as the unit of analysis, and find that socio-technical trajectories result in specific "*socio-technical styles*": relatively stable forms of producing technology and developing definitions of its functioning and utility (Thomas et al. 2005). Secondly, when we look at a cooperation process as the unit of analysis, we find that it is a *socio-technical trajectory* itself, which is influenced by the interaction of different socio-technical styles. In order to analyze this last aspect, we understand the cooperation trajectory as the evolution of "Techno-Political-Economic Networks" (Kern 2008). This concept derives from the definition of "techno-economic networks" (TEN) proposed by Michel Callon. A TEN is defined as a coordinated set of disparate actors—such as government laboratories, technical research centers, companies, funding organizations, users and public authorities—collectively involved in the design, development, and production/distribution of production processes, goods and services, some of which may lead to a commercial transaction (Bell and Callon 1994: 68). According to this concept, scientific findings are not the result of a rational choice made by people based exclusively on their experiments and isolated from other social factors. In the same way, cooperation outcomes result from a set of factors that mobilize human and non-human elements, constituting networks. For analytical purposes, Callon, together with Gerard Bell (1994), identifies three poles around which these networks can be organized according to the type of products or intermediaries they produce: a *scientific pole* (S) includes the "production of certified knowledge" and skills; a *technological pole* (T) refers to the development of devices, involving human and non-human resources; and finally the *market* pole (M) is composed primarily by firms organizing and demand. Two additional poles act as interfaces: a transfer pole (ST)—between the scientific and technological poles—and a development pole—between the technological pole and the Market (TM). Departing from these concepts, we notice that politics and policies play a central role in the dynamics of international cooperation. Not only do they define the institutional framework where collaboration takes place, they also are key factors for the inclusion or exclusion of "local" and "foreign" elements from the network. For this reason, TPENs also include a "political pole" (P) and an interface pole between the political and scientific poles (PS). Including this political pole also offers a way of linking the politicization of technology with the explanation of interactions.

Like Callonian networks, TPENs also involve two types of elements: *actors* and *intermediaries*. An actor is any entity that defines and builds—with varying degrees of success—a world populated by others, giving them a story and an identity, and describing the relationships that bind them together (Callon 2001: 97). Intermediaries encompass everything that passes from one actor to another, constituting the form and the substance of the relations that actors establish among themselves (Callon 2001: 88). Intermediaries can, for example, be written documents (reports, articles, notes, forms, etc.), technical artifacts (scientific instruments, machines), skills and money. Any intermediary can also be an actor, if it is able to circulate other intermediaries. Following the ideas of Bell and Callon (1994), we also apply the idea of "convergence" and "divergence" to characterize TPENs. *Convergent networks* are characterized by agreement and understanding among the participants involved, while *divergent networks* involve actors who do not share their activities with others and whose results are not easily transferable to other members of the network, so that formal coordination is limited and infrequent. Convergence and divergence are applied here in particular to observe the relation between local and foreign components of a TPEN.

Finally, definitions and meanings are also important to understand the interaction dynamics dimension because they could partly explain the composition and variations of a TPEN. To this end, Callon introduces the concept of *translation*, a process through which actors impute certain interests, projects, desires, strategies, thoughts or ideas (always inscribed in intermediaries) to other entities. *Translation* is in part possible because of the *interpretive flexibility* of facts and artifacts, conceptualized by Bijker (1997) in the field of Social Studies of Technology. As he concludes, the functionality of an artifact does not result from its intrinsic properties, but from a social assessment. From his constructivist perspective, interpretive flexibility means that the meaning assigned by a relevant social group is what constitutes an artifact. In the same way, actors differ in how they understand results of cooperation (papers, artifacts, etc.) as well as technologies applied or developed in the context of cooperation. In this sense, meanings constitute translations and explain TPEN configurations and variations.

Applying this framework to the analysis of the Argentine-German case elucidates that technology matters both on the level of ideas and on the level of interactions and is a constitutive element that helps to explain the evolution of a cooperation process.

3 Explaining ICT Cooperation Between Argentina and Germany

In this case study, we examine two cooperation projects between a research team at LIFIA—an Information Technology laboratory at La Plata University—from Argentina and another at the Konstanz University of Applied Science (FH-Konstanz) from Germany. These projects dealt with software development and followed two consecutive research lines. The first one, called the "Business Process Component Framework," involved the design of a "framework[5] for business processes" (such as buying an airline ticket or a book) and was mainly based on the previous experience of the FH-Konstanz. The second line was related to the LIFIA research agenda and was aimed at enlarging a software design model called the "Object Oriented Hypermedia Design Model" (OOHDM) that was previously developed by part of this team. The objective in the second project was to introduce previous learning about business processes in Web applications design. Results consist mainly in papers and publications, but they also include some programs and software applications. Both projects were supported by public funds generated from a bilateral cooperation agreement that financed scientific exchange.

Applying our framework to this case, we first examine the discourses around the politicization of ICTs at the international and the national level. Then, we explain the interaction dynamics considering the features of each research team and the different configurations of TPENs that took place during the cooperation process.

3.1 Politicization of ICTs in the 1990s

During the 1990s, two different discourses emerged regarding the features and utility of ICTs, supporting divergent paths of politicization in the international arena. These discourses led to the institutionalization of different cooperation policies among countries. At the same time, based on power relations, both discourses complemented each other in sustaining a hegemonic order: while the first

[5] In the Information Technology field, a "framework" can be defined as a support structure that organizes various elements of software and over which different applications can be programmed. These structures are implemented as programs containing some actions that specific applications do not need to reproduce.

one supported the production of innovations in core countries, the second constituted a means for their diffusion in peripheral countries.

The first "discursive swarm" (Jäger 2003) the "innovation discourse", was built on the experience of core countries. This discourse attributed an economic value to ICTs and linked them to the notion of the accumulation and diffusion of knowledge. In the broader context of the "knowledge economy", ICTs were seen as factors for economic growth and competitiveness. Based on this discourse, the politicization of ICTs among core countries resulted in a mix of competition and cooperation mainly oriented towards innovation.

The second discursive swarm, "the development cooperation discourse" was built from a core-periphery standpoint. ICTs were understood in social terms, as means that could help to expand citizens' freedom and participation,[6] but not necessarily as economic or competitive factors. Emphasis rested on the communication aspects of ICTs and on popular "use", not on their technological development and production. In this way, the discourse reflected a "North-South" relation. The idea of a "digital divide" fits well with the asymmetrical diffusion of ICTs between the two groups of countries. Based on this discourse, the politicization of ICTs became part of development cooperation programs, aimed at providing "access" and promoting "use" of ICTs. In this way, ICTs were understood as instrumental means to achieve development goals mainly defined by the "donor" community.

Both discourses influenced the way in which ICTs were politicized at the national level. In the case of Argentina, ICTs were understood as instrumental means for communication. Policies for ICT promotion were focused on the communication potential of these technologies, but little attention was paid to ICTs as an industrial sector. As a result, the main responsibilities in defining local policies fell on the Communications Secretariat and the country's strategy was oriented towards facilitating the population's "access" to the Internet. Therefore, policies were focused on the provision of the needed infrastructure—mostly produced in industrialized countries—and the development of an open regulatory framework. The promotion of the software industry as well as the support for ICT research and development were areas of "non-decision making".

The case of Germany was completely different. ICTs were defined as means of improving German economic competitiveness. Responsibilities fell on the Ministry of Economy and the Ministry of Education and Research. In these contexts, policies were oriented towards promoting innovation and the application of ICTs—particularly software—in other traditional industrial branches.

Both politicization processes at the state level explain in part how the utility of international cooperation was understood in each country, with the other part explained by the relationship between scientific and foreign policy, considering identities and interests. During the 1990s Argentina was described as a "middle

[6] This concept was clearly expressed in the Human Development Reports and the works of Amartya Sen (1999).

income country" whose socioeconomic indicators suggested an intermediate level of development according to the Human Development Index. During the first part of this decade, the government constructed Argentina's identity as "a country in positive transition" (Kern 2008), while it complied with Washington Consensus recommendations and achieved high growth levels. The government qualified its foreign policy as "realistic", "normal" and "pragmatic", and sought to avoid confrontation with core countries and to consolidate a new strong network of international linkages with the developed world. Scientific cooperation was a means of increasing international linkages and building positive agendas with developed countries. As a result, foreign policy interests prevailed over scientific and technological aims. In the particular case of ICTs, this agenda subordination was easier because the way these technologies were politicized left them out of the priorities of the scientific policy realm.

German identity was constituted after the country's recovery from the Second World War, when it consolidated its position as a core country and constructed its identity as an "industrial and technological leader". Maintaining this leadership was a crucial interest of German foreign policy in the 1990s, and therefore foreign policy interests were linked to the scientific and technological policy in a mutually constitutive way. The science and technology policy promoted by the German Ministry of Education and Research (BMBF) had the objective of strengthening Germany's reputation in the context of new technologies, especially ICTs and biotechnology. In those areas, Germany perceived itself at disadvantage compared to the U.S. or Japan and thus ICTs constituted a key area of the country's scientific and technological policies. International cooperation was defined as a complement to this policy, intended to link local research groups with internationally recognized groups and to "attract the best minds worldwide" (BMBF 1997: 10). Hence, cooperation policy was a deliberate strategy from the scientific and technological policy realms oriented towards promoting local technological development.

3.2 Interaction Dynamics: The Socio-technical Trajectory of Cooperation

Politicization processes constitute the context in which interactions take place. To understand them, we should take into account the features of the research teams involved, which are here described according to their particular socio-technical styles. As we could see in Table 1, both research teams were completely different. However, they complemented each other in different stages of the cooperation process through a division of labor.

Some features here are especially relevant to understand the cooperation process. First, each research team built its agenda in a different way: while the German team was driven by local demands—and also local networks—the Argentine team was driven by innovations that came from the international market—like

Table 1 LIFIA and FH-Konstanz socio-technical styles

Category	FH-Konstanz team	LIFIA team
Research agenda	Driven by local firms' demands	Driven by exogenous innovations
Learning forms	Interactions with artifacts and users	Abstraction and generalization based on specific innovations Exclusion of user-producer relations
Work style	Project oriented	Event oriented
Research outputs	Software programs	Scientific texts (papers, publications)
Linkage types	Local networks (public and private organizations)	International networks linked by Object Oriented Design technological frame

hypermedia or Web applications—and attempted to understand their logics and design. Second, the linkage types influenced the learning forms and also the research outputs. Consequently, the LIFIA team, that had weak local linkages and excluded user-producer relations as a learning form, produced mainly scientific texts that reflected "creative" or "good" designs. This style contrasted with the FH-Konstanz team that produced concrete artifacts (software) or "applicable" technologies based mainly on user-producer interactions. These variations affected the sustainability of the cooperation partly because the FH-Konstanz team style implied some degree of confidentiality in working with local firms, while the LIFIA team style implied knowledge diffusion and an aim to enhance the international prestige of its members. However, cooperation was possible because both styles complemented each other in two consecutive stages of software development: LIFIA concentrated on design and the FH-Konstanz on implementation. Considering these particularities, we now examine the socio-technical trajectory of the cooperation between both teams.

The main feature of the socio-technical trajectory was its alternation between convergent and divergent TPENs. In this case study, convergence or divergence was mainly a consequence of the alignment or non-alignment between German and Argentine intermediaries. Convergent TPENs were created when texts (project proposals, articles, papers) produced together by both research groups served as "actors" aligning the TPEN. As we can see in Fig. 1, in a first moment, when the project proposal (the document which both research groups created based on a standardized form) was the actor, it success fully *translated* diverse intermediaries, producing a convergence of TPENs.

The figure illustrates how all intermediaries were defined and interrelated in the project. The requirements of the project proposal form *translated* the way previous research agendas of each group—"frameworks" and "Object Oriented Design"—complemented each other in order to fulfill a common objective: the development of a "business process framework" (T). According to their socio-technical styles, each research team represented a different pole–TM in the case of the Germans and S in the case of the Argentines. Both teams were aligned by the demands of the project proposal by fulfilling different roles and tasks. The proposal also included

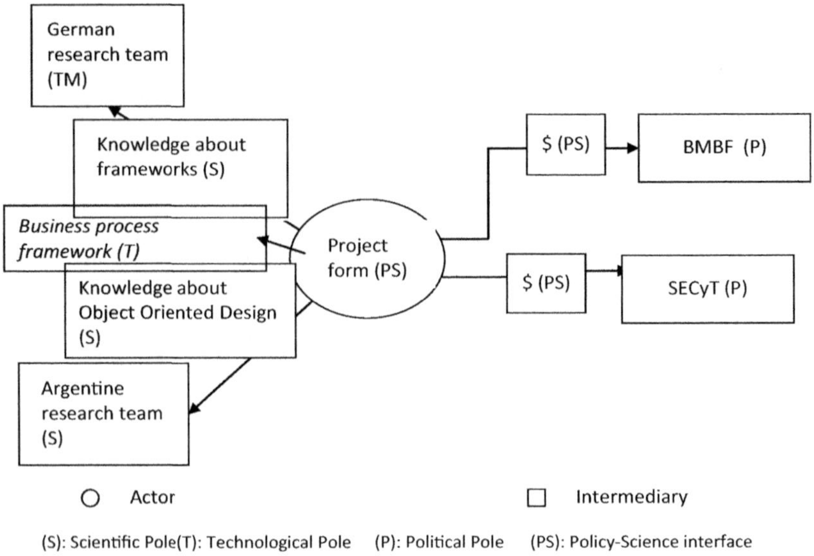

Fig. 1 Convergent TPEN

political criteria like relevance or technological transfer potential—goals to be met in order to receive public funding ($) from the Argentine and German governments (SECyT and BMBF). In this way, the proposal *translated* this funding in the form of detailed scientific visits. The project form and the funding were relevant interface poles (PS) between the scientific and political poles, as well as a source of convergence for the German and Argentine components of the TPEN.

During the cooperation process studied, convergent TPENs were circumstantial and were pushed by formal mechanisms. They were constructed when forms and papers *translated* the research agendas of each group, their socio-technical styles and the meanings assigned to the technologies involved in a complementary way. This *translation* was accomplished in two ways: through the subordination of research agendas and through the division of labor between both research teams.

The case study showed that the subordination of research agendas was neither a direct consequence of a core-periphery interstate relation, nor a pure effect of material asymmetries. Even though these factors should be considered, subordination resulted from micro-political relations and was determined by the way in which various factors came together at different times to increase the power of some agents over others. These factors included, among others, the international prestige of a scientist, the local private and public linkages that supported the research agenda of a team and the international relevance of certain research issues. In the first and second projects, convergence between FH-Konstanz and LIFIA resulted from the subordination of LIFIA's research agenda: OODM was translated as a means to design a business process framework. This subordination occurred mainly because the local linkages of the German group—firms and government funding—

pushed and supported this agenda. In a second project, things occurred the other way around. The objective was to work on a "business process for Web applications" at a time when Web applications were the main research issue of LIFIA's agenda[7]—which was working on an "Object Oriented Hypermedia Design Model"—as well as an important issue for the international scientific community dedicated to software development. The subordination of the German agenda occurred because of the Argentine scientists' great prestige in a research area that was increasing in relevance worldwide.

Considering that both research teams had different socio-technical styles, the *division of labor* was another means that explained the convergent TPEN. The Argentine team was theory-oriented, its research agenda was pushed by exogenous innovations that heavily impacted the international market and it had weak local linkages. The German team was application-oriented, its research agenda was pushed by specific firms' demands and it had stronger local linkages. Socio-technical styles complemented themselves through a division of labor in different phases of "framework" development; the Argentine team was responsible for the abstract design and programming, while the Germans measured its effectiveness and implemented it in concrete applications. However, as we shall see later, this division of labor reflected circumstantial commitments.

A divergent TPEN emerged in a second instance when each research team presented their annual reports to the science and technology governmental bureaucracies (Fig. 2).

This figure shows two overlapping TPENs when the annual reports were presented, aligning different networks linked only by the papers jointly written. The reports were important intermediaries between the research groups and BMBF and SECYT, but they became actors as soon as they *translated* other intermediaries as project "results". In the Argentine case, joint papers and publications were *translated* as "project results", while the research visits to Germany financed by SECYT were *translated* as the means for producing those results.

Figure 2 also shows a larger German network made visible by the corresponding annual report, which included three main results: a business process framework, a Web component framework and the participation of a German firm. In this way, the German report *translated* concrete artifacts and "relations with industry"—not just papers—as "results". The funding provided from BMBF to the German team was then *translated* as a means for achieving these broader results.

The formal evaluation process created a context for divergence, as the research teams filled out different forms in their own mother tongues and there was no opportunity to share and discuss them in the political arena. So the divergence between the German and Argentine components of the TPEN can be partly explained by the characteristics of the interface between the political and scientific

[7] This research issue was driven by the appearance of Amazon on the Internet. Buying on Amazon was a business process that the LIFIA team wanted to emulate explicitly in their design method. This aim was part of a wider research agenda regarding hypermedia design.

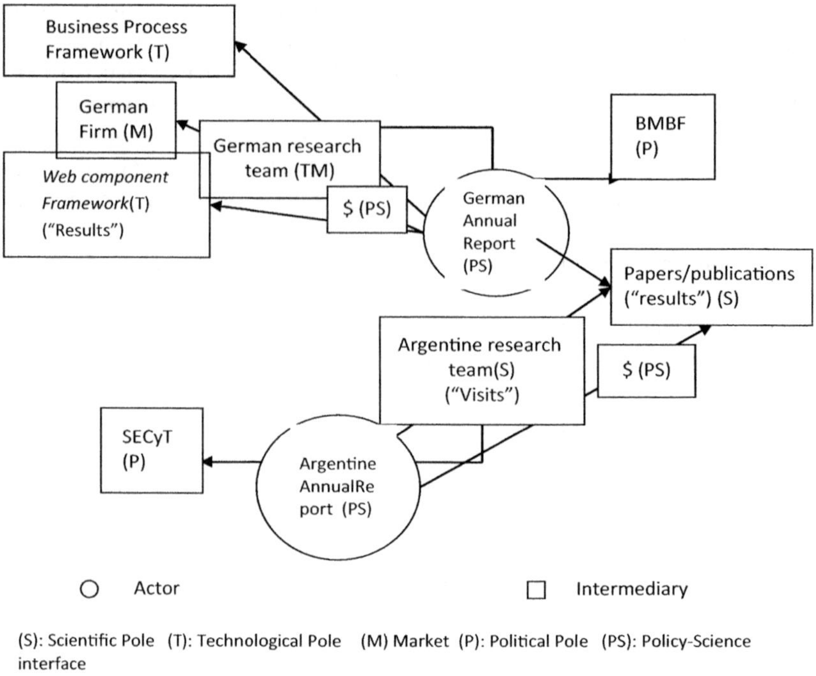

Fig. 2 Divergent TPEN

poles in the evaluation phase. A second source of explanation is related to the different meanings attributed to the results achieved by each team. As we noted previously, these variations are partly explained by the *socio-technical styles* of each team, but they are also explained by the *interpretative flexibility* of the technologies developed. A shared idea regarding software design based on "objects" created a minimum consensus about general problems and proper solutions in the area of software development.[8] However, different interpretations about what a "framework" was—and thus a "business process framework" and a "component framework for Web applications"—coexisted and generated divergence in the long term. For the LIFIA team, a "framework" was a design, an explicit model that described how to place and link the elements that constituted a software program. As a consequence, for this team a "framework" worked when its elements were "properly" and "creatively" described in a text. On the other hand, for the FH-Konstanz team, a "framework" was a program that provided a base of components which could serve to sustain new applications. A "framework" worked when it was programmed successfully and fulfilled this criterion.

[8] This consensus was based on the idea that software must be designed by programming independent and re-usable "objects". Embodied in the innovation discourse, this idea was oriented towards a more efficient process for software development.

4 Conclusions

In this chapter, we presented and applied an original analytical framework that combines concepts stemming from the fields of International Relations Theory and the Social Studies of Science and Technology. This framework was aimed at integrating technological, social, economic and political factors in the analysis of international cooperation. One of the advantages of this framework is that knowledge and technologies are treated as constitutive elements of the process. Therefore, the direction and results of scientific and technological cooperation are not studied as elements which are purely determined by foreign policy interests, or by the configuration of a particular world order. Although these factors influence the process and are considered particularly important in the politicization of technology, cooperation evolves on a self-organized trajectory that involves social and technological interactions. To understand this trajectory, we built on the idea of *techno-political-economic networks* that evolve and change during the process. TPEN configurations enabled us to identify convergence and divergence among foreign and local components in a cooperation process. *Socio-technical styles*, *translations* and the *interpretive flexibility* of technology allowed us to explain different TPEN configurations and therefore the course of a cooperation process.

Some noteworthy findings arise from applying this framework to the cooperation between Argentina and Germany.

Firstly, the case study illustrates how the structure of international order influenced the way ICTs were politicized in the 1990s. At the international level, politicization was sustained by two discourses with complementary goals—"innovation" in the core and "access" in the periphery—reproducing an unequal order. These discourses also influenced the way ICTs were politicized in Argentina—where emphasis was on the *use* of ICTs as a means of communication—and in Germany, where their industrial application was a subject of interest. Politicization at both the international and national levels appeared as relevant processes for understanding the political and institutional framework in which cooperation took place. The lack of public support for ICT research and development in Argentina rendered international cooperation an important source of funding and recognition for local scientists. For the Germans, however, international cooperation was merely complementary to other political instruments for ICT development.

Politicization appears to also be an important factor for understanding the relationship between foreign and scientific policies. The Argentine case suggests that the subordination of scientific to foreign policy interests seems to be easier when the politicization of a specific technology does not prioritize research and development. The German case shows that this relation could be mutually constitutive, depending on how a country constructs its identity.

Secondly, the case study showed that the evolution and the results of scientific and technological cooperation can largely be explained by the way in which research teams interact in the process, articulating or failing to articulate their socio-technical styles, subordinating or not subordinating their own research agendas, prioritizing local or foreign linkages and pushing for alternative meanings attributed to technologies. These interaction dynamics vary and are the origin of convergent or divergent TPENs, particularly regarding foreign and local elements of the network. The case study also rendered visible that the forms constituting the interface pole between science and policy play an important role in strengthening or weakening the links that bind the network. As was demonstrated by means of presenting the annual reports, their design not only led to a lack of convergence between foreign and local elements of the TPEN, but also to an incomplete evaluation of results. Additionally, divergence evolves from the different meanings that each research team attributed to the technologies used and developed while cooperating. It can be assumed that these meanings are related to the socio-technical styles of the research teams. The way they build their agendas, their linkages, their research outputs, etc. consistently shapes how they understand technology and its utility. While agency is important here, structure also has to be taken into account. Socio-technical styles are developed against a background of structural differences between developed and developing countries. Politicization makes this clear, but we should also look at the different history of R&D institutions and their linkages with production and society in core and periphery states. And finally, this case study served to demonstrate that structure is not determinative all the way down. Agenda subordination may vary along diverse power resources, e.g. international prestige, knowledge expertise, or research funding. In sum, cooperation paths may vary depending on the respective technological field, on the manner of politicization and the micro-level dynamics between social and technological factors.

References

Bell, G., & Callon, M. (1994). Techno-economic networks and science and technology policy. *Science Technology Industry, 14*, 59–117.

Bijker, W. E. (1997). *Of bicycles, bakelites and bulbs. Towards a theory of sociotechnical change*. Cambridge: MIT Press.

BMBF. (1997). *Bundesbericht Forschung 1996*. Bonn: BMBF.

Buzan, B., & Little, R. (2000). *International systems in world history. Remaking the study of international relations*. Oxford: Oxford University Press.

Callon, M. (1992). The dynamics of techno-economic networks. In P. Saviotti, R. Coombs, & V. Walsh (Eds.), *Technological change and company strategies* (pp. 72–102). London: Academic Press.

Callon, M. (1994). El proceso de construcción de la sociedad. El estudio de la tecnología como herramienta para el análisis sociológico. In M. Doménech & F. Javier Tirado (Eds.), *Sociología Simétrica. Ensayos sobre ciencia, tecnología y sociedad* (pp. 143–170). Barcelona: Gedisa.

Callon, M. (2001). Redes tecno-económicas e irreversibilidad. *Redes, 17*, 85–126.

Collins, H. (1985). *Changing order: Replication and induction in scientific practice*. London: Sage.

Cox, R. W. (1986). Social forces, states and world orders: Beyond international relations theory. In R. Keohane (Ed.), *Neorealism and its critics* (pp. 204–254). New York: Columbia University Press.

Cox, R. W. (1987). *Production, power and world order. Social forces in the making of history.* New York: Columbia University Press.

Dagnino, R., Brandão, F. C., & Novaes, H. T. (2004). Sobre o marco analítico-conceitual da tecnología social. In *Tecnología social. Uma estratégia para o desenvolvimento* (pp. 15–64). Rio de Janeiro: Fundacão Banco do Brasil.

Elzinaga, A., & Jamison, A. (1995). Changing policy agendas in science and technology. In S. Jasanoff et al. (Eds.), *Handbook of science and technology studies* (pp. 572–592). London: Sage.

Herrera, A. (1995). Los determinantes sociales de la política científica en América Latina. Política científica explícita y política científica implícita. *Redes, 2*(5), 117–131.

Hughes, T. P. (1986). The seamless web: Technology, science etcetera. *Social Studies of Science, 16*, 281–292.

Jäger, S. (2003). Discurso y conocimiento: aspectos teóricos y metodológicos de la crítica del discurso y del análisis de dispositivos. In M. de Análisis & C. del Discurso (Eds.), *Ruth Wodak and Michael Meyer* (pp. 61–100). Barcelona: Gedisa.

Keohane, R., & Joseph, N. (1998). Power and interdependence in the information age, *Foreign Affairs, 77*(5), 81–94.

Kern, A. (2008). *Las relaciones entre ciencia, tecnología y política en procesos de cooperación internacional.* La experiencia de cooperación entre Argentina y Alemania en el campo de las Tecnologías Informáticas. PhD Thesis at FLACSO Argentina.

Kern, A. (2009). Intereses, identidades, discursos e interacciones en la cooperación internacional en ciencia y tecnología. Observaciones de un estudio de caso entre Argentina y Alemania en el campo de las Tecnologías Informáticas. *Miríada* N° 2.

Latour, B. (1989). *Science in action.* Paris: La Découverte.

Molina, A. (1989). *The transputer constituency - Building up UK/European capabilities in information technology.* Edinburgh: Edinburgh University.

Pinch, T. J., & Bijker, W. E. (1987). The social construction of facts and artifacts: Or how the sociology of science and the sociology of technology might benefit each other. In W. E. Bijker et al. (Eds.), *The social construction of technological systems: New directions in the sociology and history of technology* (pp. 18–50). Cambridge, MA: The MIT Press.

Rosenau, J., & Singh, J. P. (2002). *Information technologies and global politics. The changing scope of power and governance.* New York: State University of New York Press.

Sen, A. (1999). *Development as freedom.* Oxford: Oxford University Press.

Shinn, T. (1982). Scientific disciplines and organizational specificity: The social and cognitive configuration of laboratory activities. In N. Elias, H. Martins, & R. Whitley (Eds.), *Scientific establishments and hierarchies* (pp. 239–264). Dordrecht: Reidel Publishing.

Strange, S. (1988). *States and markets. An introduction to international political economy.* London: Pinter.

Thomas, H. (1999). *Dinâmicas de inovação na Argentina (1970-1995) Abertura comercial, crise sistêmica e rearticulação.* PhD Thesis, UNICAMP, Campinas.

Thomas, H. (2001). Estilos socio-técnicos de innovación periférica. La dinámica del SNI argentino, 1970-2000, en *IX Seminario Latino-Iberoamericano de Gestión Tecnológica: Innovación Tecnológica en la Economía del Conocimiento*, CD ISBN: 9968-32-012-9, San José de Costa Rica.

Thomas, H. (2008a). Estructuras cerradas vs. Procesos dinámicos: trayectorias y estilos de innovación y cambio tecnológico. In H. Thomas & M. Fressoli (Eds.), *Actos, actores y artefactos. Sociología de la Tecnología* (pp. 217–262). Editorial de la UNQ: Buenos Aires.

Thomas, H. (2008b). *En búsqueda de una metodología para investigar Tecnologías Sociales.* Paper presented at the workshop Tecnologías para la inclusión social y políticas públicas en América Latina, Rio de Janeiro.

Thomas, H., Lalouf, A., & Versino, M. S. (2005). *Trayectorias socio-técnicas, estilos de innovación y cambio tecnológico, resignificación de tecnologías y conocimientos genéricos en países subdesarrollados*. VI Jornadas Latinoamericanas de Estudios Sociales de la Ciencia y la Tecnología (ESOCITE).

Vercelli, A., & Thomas, H. (2007). La co-construcción de tecnologías y regulaciones: análisis socio-técnico de un artefacto anti-copia de Sony- BMG. *Revista Espacios, 3*, 5–30.

Waltz, K. (1981). The spread of nuclear weapons: More may better. *Adelphi Papers, Number 171*. London: International Institute for Strategic Studies.

Hidden in Plain View: Exploring the Knowledge Power of States

Katharina C. Below, Sarah Herweg, Ruth Knoblich, and Krystin Unverzagt

Abstract Inspired by the notion of "structural power", this paper presents an approach to conceptualizing the knowledge power of states. Knowledge power derives from the occupation of a favorable position in the global knowledge-structure. Regarding states as actors in the structure, four clusters of proxy variables serve to empirically map structural effects and dynamics: grass-roots, cutting-edge artifacts, infrastructural embeddedness and knowledge regimes. Empirical findings confirm initial theoretical considerations on the asymmetric and path-dependent nature of the knowledge-structure. Due to path-dependency, the structure exerts a hampering effect on change. It provides an inert environment, in which states face restraints on their way up to leading positions. As the asymmetric distribution of knowledge capacities derives from past differences in foundational knowledge indicators, power shifts cannot appear out of a sudden, and established powers are likely to prevail.

Keywords Knowledge power • Global knowledge-structure • Asymmetry • Path-dependency • Technology clubs • Digital divide • Susan Strange • Innovation • Science and technology • Cutting-edge artifacts • Infrastructure

---

K.C. Below
University of Konstanz, Konstanz, Germany
e-mail: katharina.below@uni-bonn.de

S. Herweg (✉)
Free University Berlin, Berlin, Germany
e-mail: sarah.herweg@fu-berlin.de

R. Knoblich
Ruhr-University Bochum, Bochum, Germany

Bonn University, Bonn, Germany
e-mail: ruth.knoblich@rub.de

K. Unverzagt
Bonn University, Bonn, Germany
e-mail: krys.unverzagt@gmail.com

M. Mayer et al. (eds.), *The Global Politics of Science and Technology - Vol. 2*, Global Power Shift, DOI 10.1007/978-3-642-55010-2_8,

1 Introduction

Although political scientists have acknowledged the crucial role of knowledge for power in international politics, there is only limited understanding of its actual impact. Most theoretical approaches to knowledge have been developed in the context of economic research. Economists use indicators and macro data to analyze global shifts in know-how and technology as well as the effects of different national innovation performances (see e.g. UNESCO 2010; Schwab 2011; Dutta 2012). IR theories, on the other hand, often include the issue of knowledge implicitly only, as a constituent of either economic or military capabilities (Fritsch 2011, 2014; Krige and Barth 2006).[1] Two approaches already elaborate on the influence of knowledge and information diffusion on international power constellations: the concept of global "digital divide" and the "technology clubs" approach. Both identify large gaps between states due to different technological and innovative capacities. Similar to studies conducted by the OECD (2010) and the World Bank (2010), these approaches principally cling to a purely instrumental understanding of knowledge. For instance, when economists assume that less developed states could overcome the digital divide and other technology gaps, they often presuppose that governments and enterprises use knowledge as an instrument for development. Yet, this notion neglects the persistence of power-dimensions deriving from the asymmetric cross-country distribution of knowledge. A purely instrumental understanding fails to grasp how 'knowledge' itself unfolds a genuine impact, determining the range of choice for state agency and restraining relative opportunities for development and growth through network functionalities (cf. Singh 2002: 11).

Besides these theoretical limitations, there are further methodological shortcomings. Current measurements of innovation and knowledge provide various econometric indices and rankings that tend to picture countries in direct competition: Based on performance and intensity indicators, ranking lists suggest a more or less leveled playing field. Accordingly, numerous developing countries and emerging economies seem to catch up with the traditional industrial powers (see e.g. Lazar et al. 2009; WIPO Global Innovation Index 2013). In the same vein, Thomas Friedman (2009) famously declared that the "world is flat". However, as will become clear when concentrating less on performance indicators and flow-charts but broadening the picture to structural variables instead, large differences still remain the preeminent condition in global knowledge competition. Hence, it seems

[1] Realist scholars, for instance, examine technologies as material capacities with regard to their potential to change the security architecture and offensive of defensive dynamics (exemplary: Gilpin 1981). Liberal scholars, pointing to the transformational power of information, regard information technology as the decisive exogenous driving force behind the de- and restructuring of world order (see Rosenau 1990). Constructivists deal with technology and knowledge as socially constructed phenomena while emphasizing that technology does not determine power per se, but gains meaning through being embedded into social discourse (see Ruggie 1975: 558; Wendt 1995: 71–81).

that structural aspects of knowledge deserve to be given more weight if we want to come to terms with "knowledge power" in IR.

Against this background, the following considerations have two aims: shedding fresh conceptual light on the international power effects of knowledge, and providing empirical data to bear on its explication. With reference to Susan Strange's work on "structural power" (Strange 1988) we propose the existence of a *global knowledge-structure*. This global knowledge-structure entails a power-dimension stemming from the positioning of actors therein. Although we can neither claim to fully operationalize the concept, nor to provide a comprehensive theoretical and methodological treatment, we nonetheless move beyond Strange's conceptual considerations, assuming the actual existence of a global knowledge-structure. Thereby, we adopt an empirical approach, capturing the global knowledge-structure through a fourfold set of proxy indicators in order to disclose the structure and its dynamics. In this sense, our findings shall serve to illuminate the nature of knowledge power and to contribute to the academic debate on the role of technology and knowledge in global power shifts.

Expanding our theoretical and methodological considerations, we proceed as follows: Sect. 2 outlines the assumed appearance of the global knowledge-structure. It introduces two main features of the structure, partly derived from literature on structure-agency relationships: the structure's asymmetry and the impact of path-dependency. Section 3 presents our indicators and Sect. 4 seeks to empirically trace the global knowledge-structure. Crucially, besides measuring quantitative performances, we integrate structural indicators on knowledge creation and data collection, global information networks as well as global knowledge regulation processes. Focusing on states as main actors, we eventually disclose what otherwise remains "hidden in plain view": the structural inertia of asymmetric positions within the global knowledge-structure and the existence of structural constraints limiting states' range of choice and impeding innovational activities and technological advance.

2 The Global Knowledge-Structure

Against the impression of a head-to-head race or a leveled playing field, the technology clubs approach and the digital divide discourse emphasize the asymmetric distribution of knowledge-related capacities. From an economic point of view, the technology clubs approach points to the significant role of technological knowledge in the debate on convergences in the global economy. Implicitly, it holds that global power asymmetries arise from the uneven distribution of technological knowledge among groups of states. Cross-national differences in technological development and unequal access to knowledge stocks allow to cluster countries in three "clubs": industrial countries, emerging economies and developing countries (Castellacci 2008; Castellacci and Archibugi 2008).

The notion of digital divide instead observes that access to Information and Communication Technologies (ICTs) is unevenly distributed among countries and people. Considering various factors such as gender, age, education and skills that impact the digital divide on an intra-national level, it goes beyond the purely economic lens offered by the technology clubs approach (Van Dijk and Hacker 2003; Hargittai and Shafer 2006; Vicente and López 2010). However, most scholars still advance an economic understanding, pointing to the link between economic development and access to telecommunication infrastructure as the relevant explaining factor on a global level (Norris 2001; Guillen and Suarez 2005; Chinn and Fairlie 2007; Cruz-Jesus et al. 2012).

In short, both concepts only implicitly hint at the implications that knowledge asymmetry holds for power. Apart from a theoretical bias, this conveys a problematic political message. By substantiating global knowledge and information asymmetries with distinctions between economic progress and backwardness, it seems that global information asymmetries can be resolved directly through fostering economic development. This view, however, appears shortsighted considering the power-dimension of knowledge distribution. In order to disclose this dimension, we consult Susan Strange's work on "structural power" for a theoretical groundwork. In *States and Markets* (1988), Strange suggests four mutually supporting structures which constitute structural power in the international political economy: the primary structures of security, production, finance and knowledge. The knowledge-structure determines "what knowledge is discovered, how it is stored, and who communicates it by what means to whom and on what terms" (Strange 1988: 117). Those who occupy "key decision-making positions in the knowledge structure" are entrusted with the storage and generation of knowledge. Consequently they are able to control "the channels by which knowledge, or information, is communicated" (ibid.).

Critics have claimed that, although Strange develops a cohesive concept, it is analytically misleading. For instance, Stefano Guzzini (1993) critically points at the diversity of elements proposed as parts of the knowledge-structure. Material computers, for instance, are mentioned along with "believe-systems", reminding of what Peter Haas (1992) amongst others conceptualized as epistemic communities (May 1996). Given the broad conceptual difference between elements supposedly constituting the knowledge-structure, it is hardly possible to construe a definition apt for operationalization. Nevertheless, Strange's conception of power leads us to a holistic comprehension of the relevance of knowledge in a globalized world. While acknowledging the value of "hard" capabilities for exerting influence over external actors, Strange does not reduce power to a merely instrumental use of resources vis-à-vis other agents. Instead, she stresses the distribution of power within structures and the possibility of "non-intentional" power (Guzzini 1993), deriving from the structural setting. Actors are affected by the positions and actions of others regardless of those actors' intentions.

Strange also points out the significance of historical preconditions and the inertia of change within these structures. This speaks to scholarly work on the historical impact of knowledge organized in global networks or structures. Historians have found that an increasingly asymmetric distribution of knowledge-capacities and access to expertise already characterized the emerging global communication system in the early nineteenth century. They demonstrate how limited access to the new technologies of the Industrial Age constrained huge parts of the world, while European countries and enterprises simultaneously built up and secured their dominance (Headrick 1981, 1988; Clark and Feenstra 2003). Indeed, it can be assumed that the evolution of a global knowledge-structure dates from the eighteenth and nineteenth century, when tangible communication and transportation systems started to spread around the globe, enabling truly materialized inter-connectivity and extensive exchange of knowledge (Bayly 2004; Osterhammel 2009). In this context, the emergence of infrastructural technologies such as roads, railroads, electricity as well as information and communication transfer strengthened the state's power by expanding its territorial reach and its penetration of society (Mann 1984; Soifer 2008). John Krige, examining U.S. hegemony as an outcome of the structural imbalance in scientific and technological capacity after World War II, has linked a powerful role in world politics to a favorable position within the global science-structure (Krige 2008). From a different angle, Manuel Castells argues that information technologies establish the structure of a networked global society, determining the inclusion and exclusion of different actors (Castells 2000).

These historical and sociological insights reveal important analytical suppositions for the study of the global knowledge-structure. Historically, knowledge distribution has always been *asymmetric* while *path-dependency* is a key to explain its persistent character. Path-dependency implies that change "is tied to previous decisions and existing institutions" (Wilsford 1994: 252). Path-dependency holds that each decision is built on a multitude of decisions made before. Each decision contributes to the formation of a pattern which will eventually result in a higher probability of continuing along the established road than admitting something new (ibid.). Path-dependency, hence, produces structural inertia or continuity restraining radical innovation or change. A change in position within the structure is thus difficult to achieve and decelerated. In brief, path-dependency entails a competitive advantage for established leading countries while constraining followers seeking to improve their position.

Based on these ideas, we propose the following definition of a global knowledge-structure: *The global knowledge-structure consists of the power-relevant set of knowledge-related processes, practices and interactions.* The power *effects*, which arise through the knowledge-structure, termed as "knowledge power", derive from *the occupation of a certain position within the global knowledge-structure.* In practice, we assume that the position of a state within the global knowledge-structure largely shapes its ability to innovate, its access to cutting-edge technologies, as well as its access to expertise and scientific knowledge (cf. Below et al. 2012; Mayer 2012). Four components will be employed to empirically map the global knowledge-structure: national performance in basic capacities such as education, considered as a pre-condition for scientific outcomes, as well as the possession of exclusive knowledge; production or possession of cutting-edge artifacts; embeddedness in global informational infrastructures; and participation in negotiations on global issues and standardization processes. It needs to be noticed that what constitutes knowledge power is not the possession of a single cutting-edge artifact or an active role in negotiations on a single global issue. Neither are investment and respective increases in basic capacities (such as education) by itself sufficient indicators for power. Knowledge power increases or decreases in relation to the entire structure, while components are mutually reinforcing. Leading knowledge powers thus typically display a strong performance in the whole range of components which constitute the global knowledge-structure.

Regarding changes within the global knowledge-structure, we assume a virtuous circle in operation. Strong performance in one category may even be impossible without commensurate groundwork in a certain other. This observation deepens an understanding of path-dependency: While a state's capacities with regard to science, technology and information input might change fast if fostered by suitable policies, a commensurate change in position might not happen immediately. The slow momentum transfer of an increase in capacity into good performance on complex structural levels, such as participation in international scientific negotiations, hampers positional changes within the structure. Vice versa, once in a structurally dominant position, a state is not immediately endangered to lose its position as structural inertia stabilizes the asymmetric positioning to the advantage of first movers. Consequently, our conception of the global knowledge-structure and our assessment of asymmetries differs from existing works in the sense that we regard states to derive power from the asymmetry of a global knowledge-structure. Apart from bearing economic advantages, a favorable position within the uneven knowledge-structure widens an actor's range of choice and enables active participation in the continuous development of the structure. The notion of asymmetry thereby challenges the rather neutral description of differences. It a priori holds a connotation of power, advantage and control.

The following section attempts to explore state performance in the different constituents of the knowledge-structure. Based on national data sets entailing time series, we visualize a *representative* image of the global knowledge-structure. Although the structure can be assumed to comprise many different types of actors, we concentrate on states as main actors. State agencies engage on all levels of knowledge production, application and storage. Therefore, states are most suitable for our analytical purpose due to their involvement with a broad variety of knowledge-structure indicators, while other actors presumably engage in only some of them. A private company, for instance, might generate knowledge without being involved in basic education. The section of the structure opened up by the state-as-actor lens is thus the broadest. Additionally, from a technical perspective, national data are the easiest to come by, they are already packed in standardized indicators and comparable without needing further clarification. Lamentably, we cannot always count on complete data sets. In fact, our collection is characterized through a recurrent lack of comparable data for non-OECD countries.

3 The Four Structural Components

The four components used to approximate the global knowledge-structure cluster different sorts of indicators according to their structural relevance and their effects on possible structural changes. All components enable us to extrapolate the asymmetry of the structure as well as its inertia and path-dependency.

The first component ("grass-roots") comprises indicators measuring national performances in basic knowledge capacities, such as human capital and the ability to generate scientific expertise and technological innovation, hence indicators traditionally listed in innovation indices and various cross-country studies providing reference data for economic growth and national innovation activity (Archibugi and Coco 2005: 178). Furthermore, the literature on technology clubs also employs this kind of data in order to point at the asymmetric global distribution of knowledge capacities, clustering countries along their ability to imitate, to innovate or to do neither (Castellacci 2008; Castellacci and Archibugi 2008). Yet, given the incongruity of performance levels in input and output indicators, this component points at the retarding effects of path-dependency on structural change. This non-equivalence is representative for large-scale and time-spanning structural differentiations and asymmetries.

The second component clusters indicators measuring the cross-country distribution of "cutting-edge artifacts" such as particle accelerators, satellite ranging systems or supercomputers. So far, the ability to create or to sustain cutting-edge artifacts has not been considered by conventional indices. Yet, they are central to technological development and knowledge prosperity. Due to the fact that material artifacts are used mainly to gather scientific insights, their relevance for the global knowledge-structure is twofold: Firstly, they represent the latest advances in the development of sophisticated technology. Secondly, observatories as well as particle accelerators, satellites and supercomputers do not only serve to store obtained data and information, but also to generate new knowledge from huge amounts of data. From seventeenth century astronomic observatories to today's supercomputers, cutting-edge artifacts have served people to collect knowledge and information on their environment in order to make sense of the world. Historically, the creation of cutting-edge artifacts in the Industrial Revolution can be considered a key moment for the evolution of international knowledge-linkages forming today's global knowledge-structure (Mokyr 2002). Nowadays, the significance of material artifacts for the global knowledge-structure is no less elementary: Due to the fact that specific artifacts shape respective practices of knowledge creation and data collection, they evolve into crucial and powerful tools of knowledge, granting great advantage to their users while excluding others or holding others in dependency. Thus, this component elucidates asymmetries within the global knowledge-structure with a strong element of structural path-dependency, displaying how latecomers are kept in dependencies. However, it also illustrates that distance can be leapfrogged through investment in cutting-edge artifacts.

Alike, "infrastructural embeddedness", the third component of the global knowledge-structure, consists of structural indicators that do not form part of conventional knowledge indices but add up to vital parts of the overall knowledge-structure: Besides the resources and conditions mentioned above, the extent to which a state is embedded in the global network of information-flows and knowledge-distribution is another decisive factor for its position in the structure. Degrees of embeddedness are contingent upon human-built infrastructures, such as harbors and deep-sea cables that, naturally, are conditioned by advantageous and disadvantageous geographical realities. These mediating networks, their knots or hubs in particular, constitute a cluster of infrastructural indicators that, again, illustrate path-dependency. In the context of communication infrastructure, globalization in the nineteenth century did not only lead to increasing world trade, but was facilitated and institutionalized by flows of knowledge (Mokyr 2002: 29–30). Thereby, a global "closed political system" was established in which every action is "sharply re-echoed from the far side of the globe" (Mackinder 1904: 422). New pathways of information and communication exchange—telegraph cables, new technological means and new passages for transport and traveling, etc.—created new sceneries of international competition, influence and interdependence (Headrick 1991; Hugill 1999; Wenzelhuemer 2013). In addition, harbors, which have always been vital to a favorable position in trade and military issues, historically served as important nodes in the global transfer of knowledge-intensive artifacts. In the same vein, recently, big international airports have developed into hubs for trade and trade-related knowledge (c.f. "aerotropolis" or "airport metropolis", Keast et al. 2012). Saskia Sassen (2001) assumes the global web of airports to support the continuous development of a social infrastructure in which the interlinkages of "global cities" are built on the maximization of connectivity and information flows.

The fourth component of the global knowledge-structure centers on "knowledge regimes". It comprises indicators representing national influence on different sorts of global knowledge regulation. The idea of knowledge regimes builds on regime theory.[2] Different approaches concentrate on the importance of the functional co-operation of states and regard regimes as primary regulatory platforms in the international system (Krasner 1983: 2). By examining the composition of knowledge regimes we pay regard to the shift of state activity into a new institutional and organizational context in which influence is henceforth exerted through participation. Although somewhat simplistic, we deduce a state's influence on global knowledge and information regulation from its participation in certain knowledge regimes by looking at membership data. The International Bureau of Weights and Measures (BIPM) as well as the International Organization for Standardization (ISO) and the Antarctic Treaty System (ATS) can be considered "knowledge

---

[2] The best known definition of regimes has been proposed by Stephen Krasner. He defines regimes as "set(s) of implicit or explicit principles, norms, rules and decision-making procedures" (Krasner 1983: 2).

regimes". These regimes have been created in order to organize and coordinate state behaviors and global functionalities as far as issues of knowledge are concerned (Adler 1992: 104; Adler and Haas 1992: 389). Standardization, the artificial construction of units, the fundamental condition for a global system of science, goes back to the ur-metre and the time convention. Both were installed when Edward Newton and his European contemporaries laid the groundwork for the integration of physics and mechanics into the classical canon of scientific research. Nowadays, trade, communication and information networks are largely reliant on globally agreed standards and norms. It is evident that participation in global knowledge regulation and influence on decisions taken in standardization processes again illustrate the knowledge-structure's inertia and path-dependency. A change of positions within the asymmetric structure is tied to already existing behaviors, decisions and institutions.

4 Empirical Evidence for Structural Asymmetries

In the following, empirical evidence is presented by means of the four sets of proxy variables. The data sheds light on the dynamics of asymmetry and shifting positions in the global knowledge-structure, thus delivering punctual insight into state knowledge power.

4.1 Grass-roots

Regarding cross-country differences in the distribution of technological capabilities and innovation activity, there is a major block of variables around the national human capital base. The formation of well-skilled human capital is considered a pre-condition for advanced national innovative activity. Public spending on education is an essential content of national policies pursuing economic growth and welfare (Fig. 1). To be sure, taking into account the share of public spending in overall government expenditures, as we do here, does not lead to assertions on the overall quantity of education investments. However, it hints at the importance that governments ascribe to public spendings on education in relation to overall government spendings. Thus, taking a look at the data set, both, least developed countries and emerging countries such as Brazil, Russia and South Africa invest large parts of their government budget in education. In Brazil and South Africa, the respective shares of public spending even surpass those of highly industrialized nations such as the United States, the Scandinavian countries and the average of the European Union member states.

On first sight, this data set shows something simple and evident: Countries seeking to foster their technological catch-up and to improve their position in the global knowledge-structure expand their investment in education. However, taking into account further data on the condition of basic knowledge capacities, it appears that it needs time to translate investment in education into an advanced human capital base. Data on the number of researchers per thousand employess and university

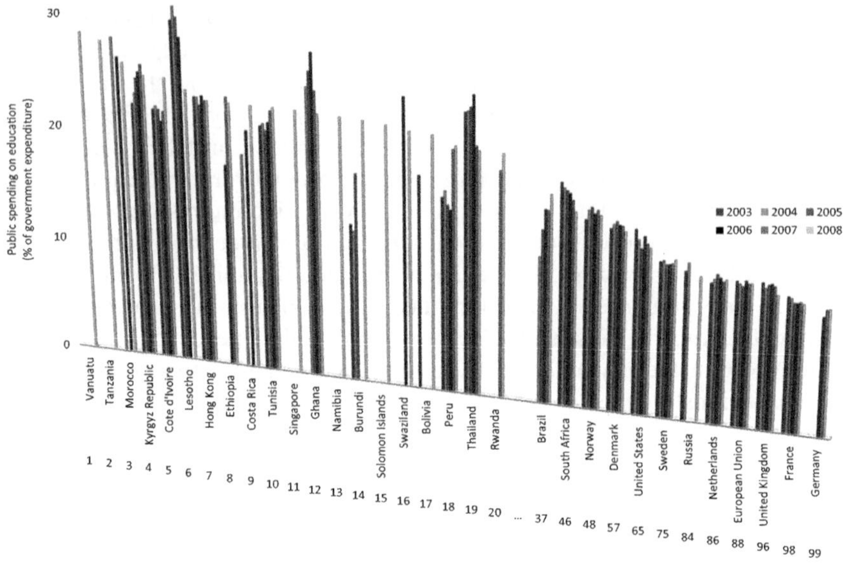

Fig. 1 Public spending on education (% of government expenditure) (*Source* World Bank Indicators 2014, http://data.worldbank.org/indicator/SE.XPD.TOTL.GB.ZS. Note: Complete data set for 2008. For years 2003–2008, data sets are incomplete. Countries ranked according to 2008.)

rankings, amongst others, compellingly illustrate that countries looking back on a long history of educational and academic tradition, especially Europe and North America, are still better off regarding their position in the global knowledge-structure. Despite high investment rates in education over more than a decade, China, Brazil and South Africa—like all developing countries—still considerably lack behind in terms of R&D personnel (Fig. 2).

The same asymmetry reflects in university rankings (Fig. 3). Only China, which is ranked ninth, seems about to close ranks here as a single exception. Future developments will show whether current investment on the input side of the grass-roots equation will further enhance the state's position with regard to university quality.

Likewise, grass-root indicators measuring national innovation performances in terms of knowledge-output (patents, scientific and technical journal articles, etc.) clearly point to the persisting asymmetry and path-dependency of the global knowledge-structure. Again they mirror the structural advantage of countries with long-standing educational and academic tradition. Among all states that, for instance, appear in data on citations of scientific articles, the United States and European countries such as UK, France, the Netherlands and Germany show an outstanding performance (Fig. 4). Again, only China has managed to establish itself among the top ten performers.

For sure, the high ranks of Anglophone states are partly due to English being the main scientific language—thus, it becomes obvious that English as official language can be considered a structural advantage in itself. However, the United

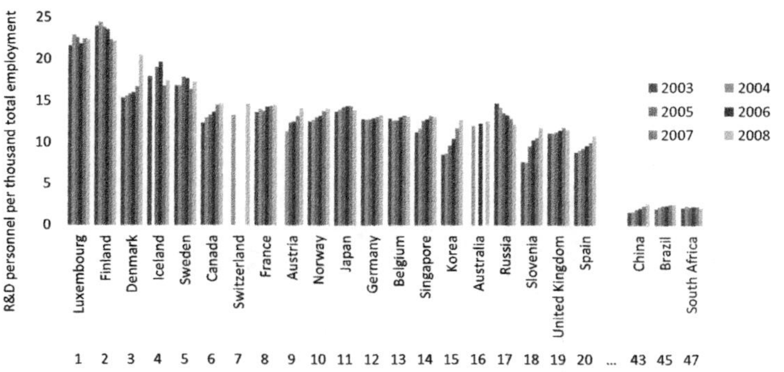

Fig. 2 Total R&D personnel per thousand total employment (*Source* UNESCO 2014, http://data.uis.unesco.org/Index.aspx?queryid=61#. Note: Full time employment. Complete data set for 2008. For years 2003–2007, data sets are incomplete. Countries ranked according to 2008.)

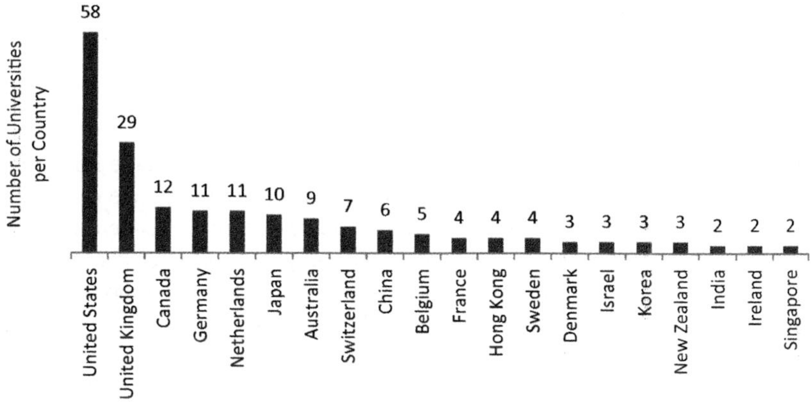

Fig. 3 Top20 performer in the Times Higher Education Wolrd University Ranking 2008 (*Source* Times Higher Education Ranking 2008, http://www.timeshighereducation.co.uk/. Note: Times Higher Education ranks the world's top 200 universities.)

States' high scores—compared to the lower performances of other English speaking nations such as Australia and Singapore—also point to its extraordinary scientific performance. In contrast, emerging countries are rarely present, with the exception of China and Korea.

Overall, these empirical findings entail two main observations: First, it has been demonstrated that states' investment in knowledge-input needs time to eventually affect knowledge-output. Thus, if Tanzania aims at improving its position in the global knowledge-structure by investing almost 30 % of government spending in education, it is unclear how much time it will actually need to reach high ranks in indicators measuring advanced human skills, scientific expertise or innovation. Consequently, overcoming the structure's asymmetry in the grass-roots component already presents itself as a large-scale and long-time project.

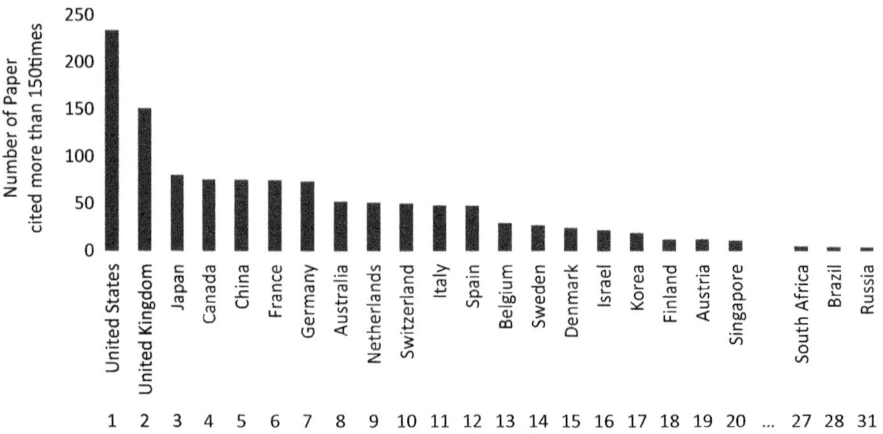

Fig. 4 Scientific papers cited more than 150 times in 2008 [*Source* Reuters Web of Knowledge, Accessed via http://wokinfo.com/ 1 May 2013, Citation Database/counting procedure developed by research group "Rising Knowledge Powers". Note: Number of citations includes only papers cited 150 times or more. Each author translates into 1 point, whereas co-authors count with the reciprocal value of co-authors in a paper (33 co-authors = adds 1/33)]

Secondly, there is a persistent and channeling structural advantage for European and Western countries, which goes back to the historical emergence of the global knowledge-structure itself. Countries with a good position in academic and educational structures have not only been investing in grass-roots for centuries. They have furthermore shaped appropriate structures: The roots of today's university institutions can be traced back to medieval Europe. Over time, those institutions grew and widened their transatlantic scope. With regard to scientific journal articles and citations, Derek J. de Solla Price shows that the introduction of the scientific journal as a "cumulating device" in the mid-seventeenth century induced the explosion of the exponential growth of science (Price 1965: 562). Scholars generally agree that modern science essentially stems from European tradition, which gradually established itself as universal code of conduct (e.g. Whitehead 1926; Butterfield 1951; Crombie 1961; Hall 1983). Through "chance meeting[s]" on the European continent (Inkster 2003: 847) a network of scientific exchange and mutual interaction was established, an early form of knowledge-structure with limited (European) scope. Some scholars assume that lacking universities as neutral spaces of autonomous research which could serve to resolve conflicts between religious and political points of view, other regions like China, India or the Islamic world did not develop the same scientific mentality (Whitehead 1926: 7; Huff 2003: 319). The global science-structure thus evolved when knowledge systems of European background were globalized in response to modern discoveries and insights (Renn and Hyman 2012). Thereafter, the European division of academic science along disciplines spread to other regions around the world founding a European-style uniform system of science. This university system, which originated in "Bologna, Oxford, and Paris" and then spread throughout Europe and the newly discovered American continent, has nowadays reached globality (Olesko 2003).

4.2 *Cutting-Edge Artifacts*

The distribution of cutting-edge artifacts also sheds light on the asymmetric positioning of states in the global knowledge-structure. Surely, an interpretation of their distribution should not take place without considering the financial resources they presuppose, but due to their influence on practices of knowledge creation and data collection they also represent material artifacts favoring their users while excluding others from the information they generate.

Looking at the current number of astronomic observatories per country, there is an obvious gap between highly industrialized countries and less developed countries. With 148 astronomic observatories, the United States fully outraces all other countries. Germany and Japan, each counting 25, rank second, Italy ranks third (21 astronomic observatories). Hence, mainly European and industrialized countries such as France, UK, Spain, Switzerland, Canada and Australia possess astronomic observatories, which again points to the path-dependency of the structure. Astronomic observatories emerged in line with the Enlightenment in Central and Northern Europe. Besides European and industrialized countries, the BRICS-countries China, Russia, India, Brazil and South Africa also possess astronomic observatories, which indicates a catch-up in terms of cutting-edge artifacts (Fig. 5). Additionally, Chile and Argentina rank relatively high, as their geography offers perfect conditions for the establishment of large telescopes like the Atacama Large Millimeter-Submillimeter Array (ALMA).

Similarly, regarding the number of particle accelerators, the United States (32) and Germany (17) rank highest. In fact, there is a huge gap between these two and all other countries, which mostly possess only three particle accelerators or less. All countries possessing at least one particle accelerator belong to the group of industrialized countries or newly emerging economies like China, India, Brazil, Korea and South Africa. Indeed, the actual distribution of particle accelerators also mirrors past developments: After World War II, the United States, Great Britain and continental Western Europe emerged as leading countries in the global particle physics community (Mersits 1987: 13). In short, the uneven distribution of particle accelerators points to the asymmetric character of the global knowledge-structure (Fig. 6). Its path-dependency is deducible from no less developed countries being listed. Although nowadays most particle accelerator experiments are transnational research projects involving different states and researchers from all over the globe, the countries depicted here will supposedly remain capable of partly controlling knowledge production, storage and dissemination through possession of particle accelerators.

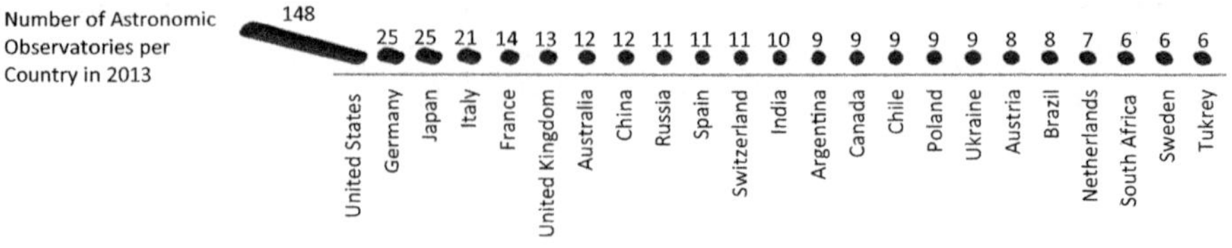

Fig. 5 Astronomic observatories (*Source* The Astronomical Almanac 2013, http://asa.usno.navy.mil/SecJ/Observatories.html. Note: Countries with more than five observatories are listed)

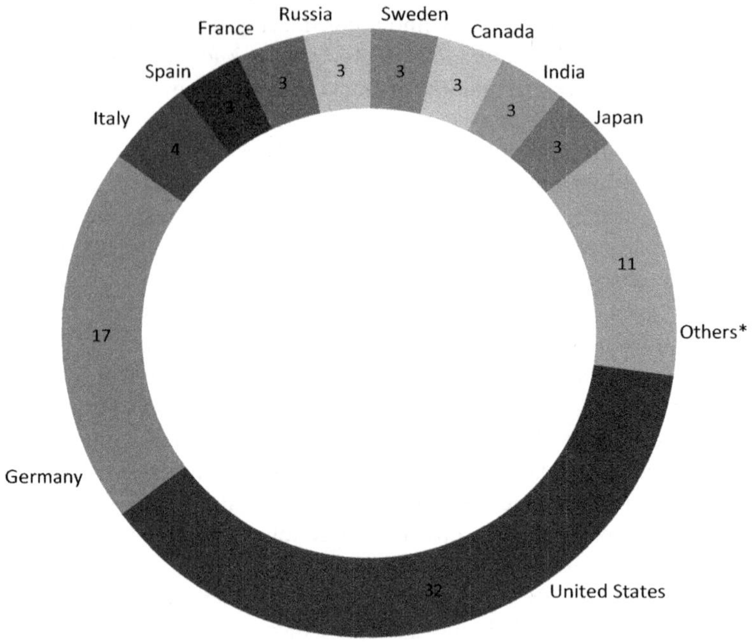

Fig. 6 Number of particle accelerators per country (Source ELSA University of Bonn (2013), http://www-elsa.physik.unibonn.de/accelerator_list.html, 2013. *Note*: *Others** include countries with one particle accelerator facility (Netherlands, Belgium, Denmark, Finland, Slovenia, Brazil, Korea, Jordan, Taiwan, South Africa, Australia)

Supercomputers and satellite systems are other types of cutting-edge artifacts. Prominently, the only ubiquitously deployed Global Satellite Navigation System (GNSS) is the Global Positioning System (GPS) owned by the United States. The United States, which had seized important allocation frequencies, impeded European aims to deploy its own GNSS called Galileo until both parties reached an agreement in 2004 (Beidleman 2006; Giegerich 2007). However, European plans to completely deploy Galileo apparently stand still due to technical and economic problems. Besides the European Union, Russia, China and India aim at implementing satellite navigation systems with global reach (Someswar

	2004	2008	2012
United States	266	290	251
China	17	15	72
Japan	30	17	32
United Kingdom	42	46	24
France	15	25	21
Germany	33	25	19
India	8	8	8
Italy	15	11	7
South Korea	11		
Brazil	6		
Russia		8	8
Canada	8		11
Sweden		7	6
Australia			7

Fig. 7 Number of supercomputers per country (*Source* http://www.top500.org. 2012 Note 1: The countries displayed here belong to the top10 performers in 2004, 2008 and/or 2012. Not all countriees belong to the top10 performers in each year, more than 10 countries are therefore displayed in total. Same number of supercomputers means same rank, more than 10 countries may therefore be displayed in one year. Note 2: In order to classify and rank the computers, the Linpack benchmark system is used. The computers are therefore ranked by their "ability to solve a set of linear equations, A x = b, using a dense random matrix")

et al. 2013). At present, however, the GPS is the only GNSS with global reach and may prevail for another couple of years.

Supercomputers greatly matter for international knowledge and information politics due to their capacity for data-driven research and data storage, which is crucial for current Big Science projects (Fig. 7). Most of today's cutting-edge artifacts emerged from Big Science projects initiated after World War II (Galison and Hevly 1992). Big Science is relevant for the constitution of international order as research is always conducted in reaction to a cultural and social context and to what politics and culture present as necessary or desirable (Latour and Woolgar 1986; Everitt 1992: 213; Potthast 2010; Roelke 2010). More than half a century after the launch of Sputnik 1, Russia seems to have lost much of its cutting-edge capacities; it considerably falls short of the United States, which leads with a total number of 251 supercomputers in 2012. China as only non-Western country has caught up significantly since 2004. Ranking second with a total number of 72, China has diminished the gap to industrialized countries.

As predicted, empirical data in the field of cutting-edge artifacts also corresponds to the theoretical conceptualization of the global knowledge-structure: European and industrialized countries dominate the asymmetric distribution of material artifacts while emerging countries only gradually develop into players in the field. Less developed countries do not appear in any of the rankings, which does not only underline their economic and financial backwardness, but demonstrates the path-dependent character of the distribution of cutting-edge artifacts.

4.3 *Infrastructural Embeddedness*

Until today, deep-sea cables are the backbone of modern communication. It is an emerging trend for states to treat fibre-optic communication cables as a critical element of infrastructure. This seems only natural considering that the United States, for instance, relies on cables for over 95 % of its international voice and data traffic. By contrast, satellites would only be capable of carrying 7 % of this traffic (Carter et al. 2009: 28). Similarly, there are three international cables carrying 99 % of Australia's voice and data traffic (Carter et al. 2009: 28). In that sense, the value of deep-sea cables is going to increase, as most nations depend on cables not only for participating in the global economy but also for being connected to global information- and knowledge-flows. Moreover, being embedded in the deep-sea cables network interrelates with national security. Recent revelations sustain this perception, reporting on the activities of Britain's intelligence agency (GCHQ), which has tapped into the deep-sea net of fibre-optic cables and shared the targeted communication with the National Security Agency (NSA) (MacAskill et al. 2013).

Looking at world maps that display the position of deep-sea cables, it becomes apparent that regions embedded in the early networks of the Modern World still remain at the center of the world's communication and information web.[3] Internet exchange points (IXPs) serviced by deep-sea fibre optic cables (Carter et al 2009:

[3] See, for instance, the interactive Submarine Cable Map, provided as a free resource by TeleGeography: http://www.telegeography.com/telecom-resources/submarine-cable-map/index.html (accessed Feb. 2014). As part of the Cluster of Excellence "Asia and Europe in a Global Context", University of Heidelberg, Germany, Roland Wenzelhuemer and colleagues are currently working on the question how far current digital divide appears similar to the asymmetrical pattern of the ninteenth century telegraph network and appropriate information flow. For further information see: http://www.asia-europe.uni-heidelberg.de/de/forschung/b-public-spheres/b9/research-field.html (17.12.13).

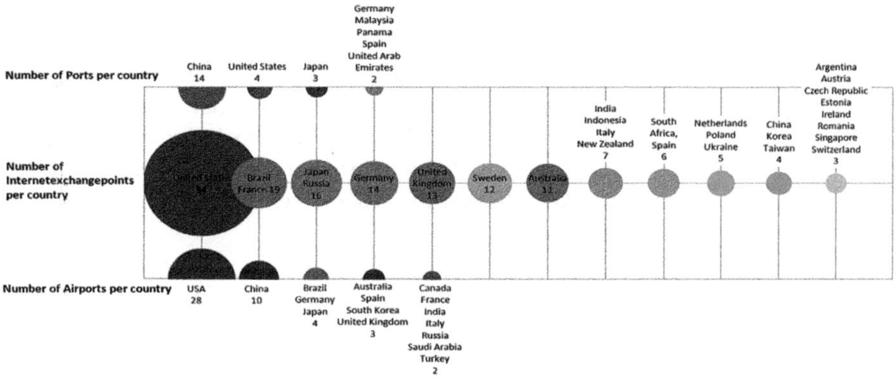

Fig. 8 Ports, airports and Internet exchange points per country (Source "*ports*" World shipping Council 2012. Source "*IXPs*" Packet Clearing IIousc 2013. Source "*airports*" Airports Council International 2011. *Note*: Countries that have more than 1 port among the top 50 ports, and/or more than 2 internet exchange points, and/or more than 1 airport among the top100 airports [in total passengers enplaned and deplaned, passengers in transit counted once])

16) also provide data representative of infrastructural embeddedness (Fig. 8). An IXP is a physical element of infrastructure that allows Internet Service Providers (ISPs) to peer with each other and exchange trafficking data. Through interconnection, IXPs can reduce per-bit delivery cost and improve the quality of service. Furthermore, direct interconnection leads to distance reduction between networks and thus speeds up data exchange, increases data traffic and encourages the development of local host content and services (Weller and Woodcock 2013: 9 f.). In the long run, a good IXP infrastructure helps to reduce Internet access tariffs and results in increased Internet proliferation and usage, as has been observed in Kenya and Nigeria of late (Kende and Hurpy 2012). The possession of IXPs does not only give economic strength, as it consolidates data traffic and possibly leads to a "halo effect" for industries outside the telecommunications sector (Ryan and Gerson 2012).

In terms of IXPs, the United States outperforms all other countries. Accounting for 84 IXPs, the USA possesses four times more than France and Brazil, which rank second. It is mostly industrialized countries that hold a high number of IXPs. As IXPs are often installed close to oil or gas pipelines, Russia also possesses a surprisingly high amount. Brazil has installed 12 of its 19 IXPs in the last decade.[4] China, instead, ranks very low. The Chinese strategy on Internet telecommunication might ensure control, but at the same time it results in limited opportunities to intervene in the networks of others, a structural disadvantage in the global knowledge-structure. The number of harbors and airports, on the other hand, mirrors China's intensive trade links, but also underlines its embeddedness in the global transfer of knowledge-intensive artifacts. Expectably, the United States has

[4] Center of Study and Research in Network Technology and Operations: http://www.ceptro.br/CEPTRO/EnglishHome (17.12.13).

the highest amount of globally significant airports (28 airports) while the distance to China, ranking second (10 airports), and Brazil, Germany and Japan as third ranked countries (4 airports) is vast. What this infrastructural data makes clear besides asymmetry and path-dependency, is the outranging effort that is needed for a country to overcome its infrastructural disadvantages and to improve its position in the global knowledge-structure. Frankly speaking: For a country that historically has not been part of the global knowledge infrastructure, chances to quickly rise to a significant player in innovation are rather low.

4.4 Knowledge Regimes

The International Bureau of Weights and Measures (BIPM) was founded in France in 1875 as the first institutionalization of the System of Units (SI). In the same year, the BIPM defined the first prototype of the meter. Up to the present, it sets global time standards and defines weights and lengths, thus promoting the globality of science by enabling worldwide scientific reproducibility and comparability (Kühne 2013: 93, see also BIPM 2013). Experts work in consultative committees that prepare recommendations for discussion within the Comité International des Poids et Mesures (CIPM), the supervising organization of the BIPM (Fig. 9). Today, there are 12 committees with different thematic focuses. Standards and norms are produced collaboratively (Saltzman et al. 2008: 286).

Empirical data strikingly illustrates that only a few states dominate throughout all the different committees: China, France, Germany, Russia, the United Kingdom and the United States participate in all 12 consultative committees. All other countries are represented in fewer committees. Norway, Ireland, Romania, Portugal, Belgium and Israel are only represented in a single one. It is also striking that while South Africa, Brazil and Mexico participate in 9 and 7 committees respectively, other countries from the African and Latin-American continents are completely absent. Hence, seats in BIPM committees are distributed asymmetrically with industrialized countries dominating.

A similar picture emerges regarding the composition of the International Organization for Standardization (ISO). The ISO is the main international body for global standard-setting. Founded in 1947, it develops international standards in various technical committees by consensus among member states. These standards give "state of the art specifications for products, services and good practice" (ISO Homepage 2013). Mirroring their influential role within the BIPM, China, France, Germany, Russia, the United Kingdom and the United States also dominate the technical committees of the ISO (FIPRA 2010: 5). Generally, standard-setting institutions are often associated with barrier reduction in the global economy. However, interestingly, standards and norms can also be considered new forms of trade barriers if interpreted as technical tariffs. As such, they matter for international power constellations in bi- or multinational politics. Standards might force enterprises to invest in the adaption of technologies in order to gain access to new markets.

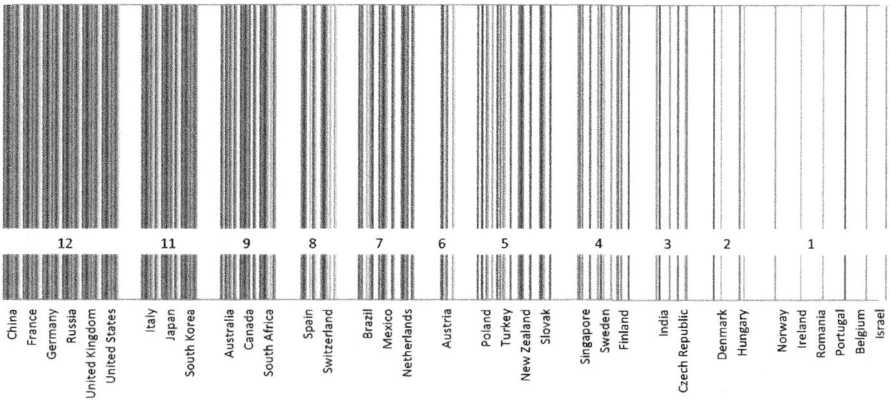

Fig. 9 Participation in the consultative commitees of the BIPM (International Bureau of Weights and Measures) (*Source* International Bureau of Weights and Measures, 2013, http://www.bipm.org/. Note: Each bar stands for participation in one of the 12 existing consultative committees. Consequently, the more bars, the more participation)

The Antarctic Treaty System, another knowledge regime, dates from a conference initiated by the United States (Triggs 2011: 42). The ATS, founded in the 1960s, is of interest for global knowledge regulation as it transformed the Antarctic continent into a nonnuclear zone and a global joint research laboratory, symbolically dividing the Antarctic territory among its members for the purpose of scientific exploration (Triggs 2011: 40). In order to become a member of the Antarctic Treaty, states have to demonstrate their ability to carry out substantial scientific research in the region. Members attend yearly meetings and are allowed to vote on binding measures. Nowadays, several additional member states have joined the 12 founding member states of the Antarctic Treaty. Non-founding members, such as Japan or Germany, are also allowed to send national experts to the Antarctic Treaty Consultative Meetings (ATCM) in order to communicate their national scientific interest in the Antarctic region (Scott 2003: 475).

It is not surprising that states such as Australia and Chile, which have made early claims on Antarctic territories due to geographical proximity, are among the highly represented Consultative Parties of the ATS, measured in the total number of delegates at the Antarctic Treaty Consultative Meetings (ATCM) (Fig. 10). More interesting is the participation of countries that are not geographically related to the Antarctic region such as Germany, Japan, Russia and the United States. Despite geographical distance, the United States has been the driving force behind the Antarctic Treaty System. Accordingly, it has been among the most represented countries ever since the first consultative meetings.

The cases of BIPM and ISO compellingly illustrate that the influence of a few states within standard finding processes is crucial to the global state-of-the-art of standardization. Correspondingly, the relatively small group of member states of the ATS "has been criticized in the past as comprising a hegemonic consortium of world powers" (Scott 2003: 475), which regulates large parts of the planet. Thus,

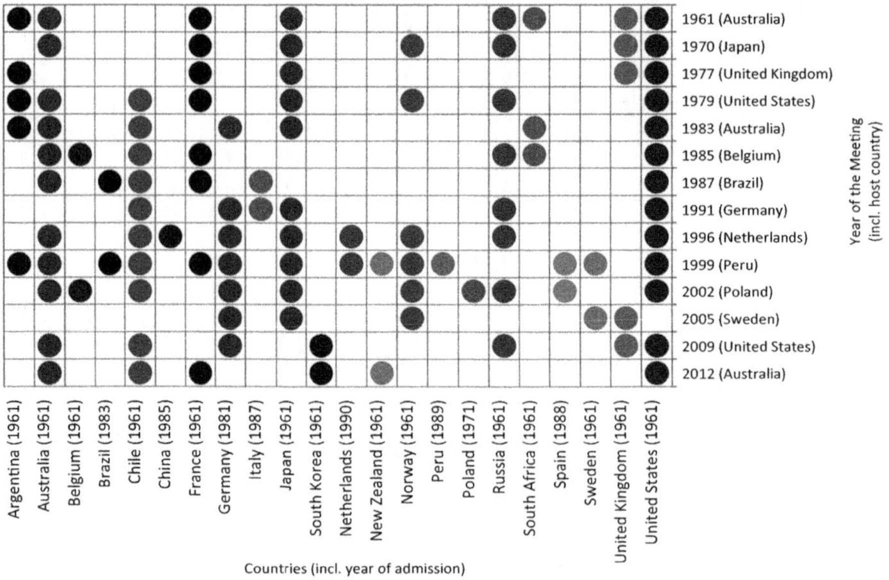

Fig. 10 Participation in the Antarctic treaty consultative meetings (ATCM) (*Source* Secretariat of the Antarctic Treaty, 2013, http://www.ats.aq/index_e.htm. Note: Each row marks the top five countries with most national delegates. *Dots* on more than five countries occur if several share a rank)

the composition of global knowledge regimes distinctly illuminates the asymmetry of the global knowledge-structure: A small consortium of highly industrialized countries dominates global knowledge forums with decisive impact on global knowledge production and dissemination. Only China serves as a compelling example of how to break the path-dependent manifestation of this structural component. It participates actively in all three evaluated knowledge regimes. On the contrary, less developed countries are almost completely missing unless they are geographically or otherwise directly linked to the issues in question.

5 Conclusion

This chapter has presented various proxy indicators to trace the global knowledge-structure. Empirical data on four components, grass-roots, cutting-edge artifacts, infrastructural embeddedness and knowledge regimes, served to shed new theoretical light on the global power effects of knowledge, which according to our conceptualization manifest in structural advantages. The consideration of structural indicators and innovative variables such as cutting-edge artifacts and embeddedness in global infrastructures served to overcome an important theoretical lack in concepts solely relying on flow and intensity indicators such as digital divide or technology clubs, which neglect the power-dimension of knowledge itself.

The empirical insights presented above strongly support our initial notion of the global knowledge-structure being composed *asymmetrically* and characterized by *path-dependency*. It entails a structural advantage for Western industrialized countries, which can be traced back to the historical emergence of the global knowledge-structure, the proliferation of the European university system, geographic discoveries following the Age of Enlightenment, and new technologies enabling global communication and trade, which required global rules and standards. Although emerging economies like India, Brazil and China in particular are catching up in some respects, the United States, Japan and the European countries still outperform all other countries in all four components of the global knowledge-structure. Path-dependency is the key to explaining this persisting asymmetry. The structure entails a competitive advantage for established leading countries while constraining followers in their effort to rise. Consequently, our empirical findings also lead to the conclusion that shifts in capacity only gradually develop into shifts in knowledge power, defined as the occupation of a favorable position in the global knowledge-structure. This becomes obvious, for example, with regard to countries which invest heavily in education but do not occupy strong positions in any other variable. Hence, actors do not depend solely on ongoing investment in order to rise. Due to structural dynamics, initial positions strongly influence ascendency. The United States, Japan and the European countries as established powers require less effort to prevail than ascending powers like the BRICS countries need to rise, and states outside of the structure can be assumed to be coerced into an even bigger effort in order to get in.

Our preliminary findings on the structure thus enable us to contribute to current policy-debates: Our exemplary data on the global knowledge-structure neither back up the notion of an even global arena (Friedman 2009), nor do they support claims that emerging countries are about to become innovation leaders (Altenburg et al. 2008; Frietsch and Schüller 2010), or affirm alarming voices pointing at the eventual end of European and US hegemony (Sainsbury of Turville 2007). Our findings also permit to face the problematic political message that implicitly inheres in the aforementioned current econometric approaches to knowledge and innovation: that by fostering economic development in less developed countries, global information and knowledge asymmetries can be more or less easily overcome. Contrary to that assumption, our data show that improvements in foundational variables such as investment in education do not automatically translate into increased performance. Rather, knowledge power increases or decreases in relation to the entire structure. As its components are mutually reinforcing, leading knowledge powers typically display a strong performance in the whole range of components while follower countries face structural restraints.

Undoubtedly, further systematic research has to be carried out in order to grasp the power effects of the global knowledge-structure. Proxy indicators presented in time series, as well as the focus on states as actors, enable a mapping of changes and persisting asymmetries, but cannot visualize the actual shape of the structure as a totality. Yet, we have been able to exemplify the structure's inert logic and asymmetric form.

References

Adler, E. (1992). The emergence of cooperation: National epistemic communities and the international evolution of the idea of nuclear arms control. *International Organization, Special Issue: Knowledge, Power, and International Policy Coordination, 46*(1), 101–145.

Adler, E., & Haas, P. M. (1992). Conclusion: Epistemic communities, world order, and the creation of a reflective research program. *International Organization (Special Issue: Knowledge, Power, and International Policy Coordination), 46*(1), 367–390.

Airports Council International. (2011). *ACI Annual World Airport Traffic Report*. Accessed 1 October, 2012, from http://www.aci.aero/Data-Centre/Annual-Traffic-Data/Passengers/2011-final.

Altenburg, T., Schmitz, H., & Stamm, A. A. (2008). Breakthrough? China's and India's transition from production to innovation. *World Development, 36*(2), 325–344.

Archibugi, D., & Coco, A. (2005). Measuring technological capabilities at the country level: A survey and a menu for choice. *Research Policy, 34*, 175–194.

Bayly, C. A. (2004). *The birth of the modern world, 1780-1914: Global connections and comparisons*. Oxford: Blackwell.

Beidleman, S. W. (2006). *GPS versus Galileo: Balancing for position in space* (CADRE paper 23). Maxwell: Air University Press.

Below, K. C., Bösl, R., Franken, J.-P., Herweg, S., Knoblich, R., Krupp, M. (2012). Der Aufstieg der BIC-Staaten als Wissensmächte? *Zeitschrift für Politik, 59*(1), 3–23.

Bureau International des Poids et Mesures. (2013). *Comités consultatifs*. Accessed March 3, 2013, from http://www.bipm.org/fr/committees/cc/.

Butterfield, H. (1951). *The origins of modern science. 1300-1800*. London: G. Bell.

Carter, L., Burnett, D., Drew, S., Marle, G., Hagadorn, L., Bartlett-McNeil, D., et al. (2009). *Submarine cables and the oceans – connecting the world* (UNEP-WCMC Biodiversity Series 31). Accessed March 1, 2013, from http://www.iscpc.org/publications/ICPC-UNEP_Report.pdf.

Castellacci, F. (2008). Technology clubs, technology gaps and growth trajectories (MPRA Paper 27595). Accessed March 1, 2013, from http://mpra.ub.uni-muenchen.de/27595.

Castellacci, F., & Archibugi, D. (2008). The technology clubs: The distribution of knowledge across nations. *Research Policy, 37*, 1659–1673.

Castells, M. (2000). *The rise of the network society: The information age: Economy, society and culture*. Cambridge: Blackwell.

Chinn, M. D., & Fairlie, R. W. (2007). The determinants of the global digital divide: A cross-country analysis of computer and internet penetration. *Oxford Economic Papers, 59*(1), 16–44.

Clark, G., & Feenstra, R. (2003). Technology in the great divergence. In M. Bordo, A. M. Taylor, & J. G. Williamson (Eds.), *Globalization in historical perspective* (pp. 277–322). Chicago: University of Chicago Press.

Crombie, A. C. (1961). *Medieval and early modern science*. Cambridge: Harvard University Press.

Cruz-Jesus, F., Oliveira, T., & Bacao, F. (2012). Digital divide across the European Union. *Information and Management, 49*(6), 278–291.

De Solla Price, D. J. (1965). Is technology historically independent of science? A study in statistical historiography. *Technology and Culture, 6*(4), 553–568.

Dutta, S. (2012). *The global innovation index 2011* (Accelerating growth and development). Fontainebleau: WIPO.

ELSA, University of Bonn. (2013). *List of particle accelerators*. Accessed 1 May 2013, from http://www-elsa.physik.unibonn.de/accelerator_list.html.

Everitt, C. W. F. (1992). Background to history: The transition from little physics to big physics in the gravity probe B relativity gyroscope program. In P. Galison & B. Hevly (Eds.), *Big science: The growth of large-scale research* (pp. 212–234). Stanford: Stanford University Press.

FIPRA International. (2010). *Standard setting in a changing global landscape*. Report to the European Round Table of Industrialists (ERT). Accessed August 20, 2013, from http://www.ert.eu/sites/default/files/Standard%20setting%20in%20a%20changing%20global%20landscape%20Final%20Report_0.pdf.

Friedman, T. (2009). *The world is flat: The globalized world in the twenty-first century*. London: Penguin.
Frietsch, R., & Schüller, M. (Eds.). (2010). *Competing for global innovation leadership: Innovation systems and policies in the USA, Europe and Asia*. Stuttgart: Fraunhofer.
Fritsch, S. (2011). Technology and global affairs. *International Studies Perspectives, 12*(1), 27–45.
Fritsch, S. (2014). Conceptualizing the ambivalent role of technology in international relations: Between systemic change and continuity. In: M. Mayer et al. (Eds.), *The global politics of science and technology* (Vol. 1. Concepts from international relations and other disciplines). Heidelberg: Springer.
Galison, P., & Hevly, B. (1992). *Big science: The growth of large-scale research*. Stanford: Stanford University Press.
Giegerich, B. (2007). Navigating differences: Transatlantic negotiations over Galileo. *Cambridge Review of International Affairs, 20*(3), 491–508.
Gilpin, R. (1981). *War and change in international politics*. Cambridge: Cambridge University Press.
Guillen, M. F., & Suarez, S. L. (2005). Explaining the global digital divide: Economic, political and sociological drivers of cross-national internet use. *Social Forces, 84*(2), 681–708.
Guzzini, S. (1993). Structural power: The limits of neorealist analysis. *International Organization, 47*(3), 443–478.
Haas, P. M. (1992). Introduction: Epistemic communities and international policy coordination. *International Organization, 46*(1), 1–35 (Special Issue: *Knowledge, Power, and International Policy Coordination*).
Hall, R. A. (1983). *The revolution in science 1500–1750*. Harlow: Longman.
Hargittai, E., & Shafer, S. (2006). Differences in actual and perceived online skills: The role of gender. *Social Science Quarterly, 87*, 432–448.
Headrick, D. (1981). *The tools of empire: Technology and European imperialism in the nineteenth century*. New York: Oxford University Press.
Headrick, D. (1988). *The tentacles of progress: Technology transfer in the age of imperialism, 1850-1940*. New York: Oxford University Press.
Headrick, D. R. (1991). *The invisible weapon: Telecommunications and international politics 1851-1945*. New York: Oxford University Press.
Huff, T. (2003). *The rise of early modern science: Islam, China and the West*. Cambridge: Cambridge University Press.
Hugill, P. J. (1999). *Global communications since 1844: Geopolitics and technology*. Baltimore: The Johns Hopkins University Press.
Inkster, I. (2003). Technological and industrial change: A comparative essay. In R. Porter (Ed.), *The Cambridge history of science* (Vol. 4, pp. 845–882). Cambridge: Cambridge University Press.
ISO Homepage. (2013). Accessed September 30, 2013, from http://www.iso.org/iso/home/about.htm
Renn, J., & Hyman, M. D. (2012). Survey: The globalization of modern science. In J. Renn (Ed.), *The globalization of knowledge in history* (Max Planck Research Library for the History and Development of Knowledge. Studies II, pp. 561–604). Accessed August 10, 2013, from http://www.edition-open-access.de/studies/1/index.html.
Keast, R. L., Baker, D. C., & Brown, K. (2012). Sustainable Airport infrastructure: Balancing infrastructures for the Airport Metropolis. In *Regional development: Concepts, methodologies, tools, and applications* (pp. 1505–1515). Hershey, PA: Information Resources Management Association.
Kende, M., & Hurpy, C. (2012). Assessment of the impact of internet exchange points - Empirical study of Kenya and Nigeria. *Report for the Internet Society*. Accessed January 3, 2013, from http://www.internetsociety.org/assessment-impact-internet-exchange-points-empirical-study-kenya-and-nigeria.
Krasner, S. (Ed.). (1983). *International regimes*. Ithaca, NY: Cornell University Press.

Krige, J. (2008). *American hegemony and the postwar reconstruction of science in Europe*. Cambridge: MIT Press.

Krige, J., & Barth, K-H. (2006). Introduction: Science and technology in international affairs. In K. John & Barth, K.-H. (Eds.), *Global power knowledge: Science and technology in international affairs* (Osiris, pp. 1–21). Chicago: University of Chicago Press.

Kühne, M. (2013). News from the BIPM - 2012. *Metrologia, 50*, 93–105.

Latour, B., & Woolgar, S. (1986). *Laboratory life: The social construction of scientific facts*. Princeton: Princeton University Press.

Lazar, J., Kekic, L., Dorr Abreu, K., & Kenny, M. (2009). *A new ranking of the world's most innovative countries* (An economics intelligence unit report). Sponsored by Cisco. Accessed January 4, 2014, from http://graphics.eiu.com/PDF/Cisco_Innovation_Complete.pdf.

MacAskill, E., Borger, J., Hopkins, N., & Ball, J. (2013, June 21). GCHQ taps fibre-optic cables for secret access to world's communication. *The Guardian*. Accessed August 18, 2013, from www.theguardian.com/uk/2013/jun/21/gchq-cables-secret-world-communications-nsa.

Mackinder, H. J. (1904). The geographical pivot of history. *The Geographical Journal, 23*(4), 421–437.

Mann, M. (1984). The autonomous power of the state: its origins, mechanisms and results. *European Journal of Sociology/Archives Européennes de Sociologie, 25*(02), 185–213.

May, C. (1996). Strange fruit: Susan strange's theory of structural power in the international political economy. *Global Society, 10*(2), 167–189.

Mayer, M. (2012). Exploring China's rise as knowledge power. In E. Fels, K. Harmat, & J.-F. Kremer (Eds.), *Power in the 21st century: International security and international political economy in a changing world* (pp. 287–311). Berlin: Springer.

Mersits, U. (1987). From cosmic-ray and nuclear physics to high-energy physics. In A. Hermann, J. Krige, U. Mersits, D. Pestre, & L. Belloni (Eds.), *History of CERN* (Launching the European organization for nuclear research, Vol. 1, pp. 3–61). Amsterdam: North Holland.

Mokyr, J. (2002). *The gifts of Athena: Historical origins of the knowledge economy*. Princeton: Princeton University Press.

Norris, P. (2001). *Digital divide: Civic engagement, information poverty, and the internet worldwide*. Cambridge: Cambridge University Press.

OECD. (2010). *Measuring innovation. A new perspective. OECD innovation strategy*. Paris: OECD Publishing.

Olesko, K. (2003). University. In J. L. Heilbron (Ed.), *The Oxford companion to the history of modern science* (pp. 813–815). Oxford: Oxford University Press.

Osterhammel, J. (2009). *The transformation of the world: A history of the 19th century*. Princeton: Princeton University Press.

Packet Clearing House. (2013). *Internet exchange directory*. Accessed June 27, 2014, from https://prefix.pch.net/applications/ixpdir/.

Potthast, J. (2010). Science and technology studies. In D. Simon, A. Knie, & S. Hornbostel (Eds.), *Handbuch Wissenschaftspolitik* (pp. 91–105). Wiesbaden: VS.

Roelke, V. (2010). Auf der Suche nach der Politik in der Wissensproduktion: Plädoyer für eine historisch-politische Epistemologie. *Berichte zur Wissenschaftsgeschichte, 33*(2), 176–192.

Rosenau, J. (1990). *Turbulence in world politics: A theory of change and continuity*. Princeton: Princeton University Press.

Ruggie, J. G. (1975). International responses to technology: Concepts and trends. *International Organization, 29*(3), 557–583.

Ryan, P. S., & Gerson, J. (2012). *A primer on internet exchange points for policymakers and non-engineers* (SSRN paper). Accessed March 1, 2013, from http://ssrn.com/abstract=2128103.

Sainsbury of Turville, L. (2007). *The race to the top: A review of government's science and innovation policies*. HM Treasury. Accessed March 11, 2013, from http://www.rsc.org/images/sainsbury_review051007_tcm18-103116.pdf.

Saltzman, J., Chatterjee, S., & Raman, M. (2008). A framework for ICT standards creation: The case of ITU-T standard H.350. *Information Systems, 33*, 285–299.

Sassen, S. (2001). *The global city: New York, London, Tokyo*. Princeton: Princeton University Press.

Schwab, K. (Ed.), (2011). *The global competitiveness report 2011-2012*. Accessed March 1, 2013, from http://www3.weforum.org/docs/WEF_GCR_Report_2011-12.pdf.

Scott, K. (2003). Institutional developments within the Antarctic treaty system. *The International and Comparative Law Quarterly, 52*(2), 473–487.

Secretariat of the Antarctic Treaty. (2013). *Final Reports of the Antarctic Treaty Consultative Meetings*. Accessed June 27, 2014, from http://www.ats.aq/devAS/info_finalrep.aspx?lang=e&menu=2.

Singh, JP. (2002). Introduction: Information technologies and the changing scope of global power and governance. In J. N. Rosenau & J. P. Singh (Eds.), *Information technologies and global politics: The changing scope of power and governance* (pp. 1–38). Albany: State University of New York Press.

Soifer, H. (2008). State infrastructural power: Approaches to conceptualization and measurement. *Studies in Comparative International Development (SCID), 43*(3), 231–251.

Someswar, G., Manoj, T. P. S., Rao, C., & Chigurukota, D. R. (2013). Global navigation satellite systems and their applications. *International Journal of Software and Web Sciences, 3*(1), 17–23.

Strange, S. (1988). *States and markets*. London: Pinter.

Times Higher Education. (2008). *Times Higher Education Ranking 2008*. Accessed January 1, 2012, from http://www.timeshighereducation.co.uk/.

Triggs, G. (2011). The Antarctic treaty system: A model of legal creativity and cooperation. In P. A. Berkman, M. A. Lang, D. W. H. Walton, & O. R. Young (Eds.), *Science diplomacy: Antarctica, science, and the governance of international spaces* (pp. 39–49). Washington, DC: Smithsonian Institution Scholarly Press.

UNESCO. (2010). *Science report*. The current status of science around the world. Accessed March 1, 2013, from http://www.unesco.org/new/en/natural-sciences/science-technology/prospective-studies/unesco-science-report/unesco-science-report-2010/.

UNESCO Institute for Statistics. (2014). *Total R&D personnel per thousand employment (FTE)*. USNO & HMNAO. 2013. The Astronomical Almanac 2013. Accessed June 24, 2014, from http://data.uis.unesco.org/Index.aspx?queryid=61

Van Dijk, J., & Hacker, K. (2003). The digital divide as a complex and dynamic phenomenon. *Information Society, 19*, 315–326.

Vicente, M. R., & López, A. J. (2010). A multidimensional analysis of the disability digital divide: Some evidence for internet use. *The Information Society: An International Journal, 26*(1), 48–64.

Weller, D., & Woodcock, B. (2013). *Internet traffic exchange: Market developments and policy challenges* (OECD digital economy papers 207). Paris: OECD.

Wendt, A. (1995). Constructing international politics. *International Security, 20*(1), 71–81.

Wenzelhuemer, R. (2013). *Connecting the nineteenth-century world: The telegraph and globalization*. Cambridge: Cambridge University Press.

Whitehead, A. N. (1926). *Science and the modern world*. Cambridge: Cambridge University Press.

Wilsford, D. (1994). Path dependency, or why history makes it difficult but not impossible to reform health care systems in a big way. *Journal of Public Policy, 14*(3), 251–283.

World Bank. (2010). *The changing wealth of nations: Measuring sustainable development in the new millennium*. Washington, DC: World Bank.

World Bank Indicators. (2014). *Public spending on education, total (% of government expenditure)*. Accessed June 24, 2014, from http://data.worldbank.org/indicator/SE.XPD.TOTL.GB.ZS.

World Shipping Council. (2012). *Top 50 World Container Ports*. Accessed March 3, 2013, from http://www.worldshipping.org/about-the-industry/global-trade/top-50-world-container-ports.

WIPO. (2013). *Global innovation index 2013*. Accessed January 4, 2014, from http://www.globalinnovationindex.org/content.aspx?page=data-analysis.

www.top500.org. (2012). Accessed 1 October, 2012, from http://www.top500.org/lists/.

Conflict and Cooperation in the Development of US–China Relations in Science and Technology: Empirical Observations and Theoretical Implications

Richard P. Suttmeier and Denis Fred Simon

Abstract Science and technology have played important roles in the development of US-China relations since the late 1970s. The mechanism of scientific and technological cooperation between the two countries has been a useful tool of diplomacy and remains so today. However, the use of that tool has become more complicated over the past three decades in the face of changing political, economic and security environments, the impact of China's growing capabilities in science and technology, a deepening of economic globalization and the growing role of global production networks, and the rise of global environmental and health issues. Ethnic identity as a basis for collaboration and the changing roles played by US-based ethnic Chinese scientists and engineers have played important roles. While the imperatives for building a long-term, sustainable cooperative science and technology relationship between the two countries are stronger than ever, the potential for conflict also has increased, pointing to the need for new approaches to governance in the bilateral relationship.

Keywords US-China relations • International scientific cooperation • Science and foreign policy • Science • Technology and international relations

R.P. Suttmeier (✉)
University of Oregon, Eugene, OR, USA
e-mail: petesutt@uoregon.edu

D.F. Simon
Office of the President, Arizona State University, 300 E. University Drive, Tempe, AR 85281, USA
e-mail: denis.simon@asu.edu

M. Mayer et al. (eds.), *The Global Politics of Science and Technology - Vol. 2*, Global Power Shift, DOI 10.1007/978-3-642-55010-2_9,

1 Introduction

Science and technology have played, and continue to play, an important role in the overall relationship between China and the United States. Not surprisingly, the S&T relationship reflects a distinct, rich historical experience of engagement and dis-engagement between China and the United States. But, this bilateral "science and technology relationship" (hereafter, S&T relationship) also provides especially rich case material for examining the ways in which science and technology are related to international relations.

In general, science and technology have not figured prominently in the conceptual apparatus of international relations in spite of their significance for understanding such central concerns as national security, economic competitiveness, foreign assistance, and sustainability. More often than not, questions of science and technology in international affairs are not seen as rising to the level of high politics and, indeed, scientific and technological relationships are often regarded, at best, as low politics or as of trivial interest. S&T relationships are typically considered derivative of political relations in spite of the growing importance of the internationalization of research and innovation and its growing implications for the wealth and power of nations.

The discussion below does not purport to offer significant theoretical breakthroughs but it does present a case rich with implications for the development of conceptual schemes to enhance understandings of international relations in the twenty-first century. In the discussion that follows, we explore the evolution of the relationship highlighting how changing asymmetries of scientific and technological capabilities alter the ways in which the two countries interact, and note the strong transnational elements of the relationship which raise interesting challenges for state to state interactions. We argue that the relationship has been strongly influenced by changes in the international political and security environments, as well as by the rise of new global issues associated with energy, environmental quality, and public health. We also explore governance mechanisms for the relationship and the approaches to managing it domestically in the two countries.

2 Background

Prior to the founding of the People's Republic of China (PRC), many of China's leading scientists were trained in the United States, American missionary educators played a key role in establishing and building up Chinese universities, and a variety of public and private US initiatives supported the founding and development of such key institutions as Tsinghua University and Peking Union Medical College. This important legacy of cooperation, however, was interrupted by the Communist victory in 1949, the failure to establish diplomatic relations between the two countries, China's "leaning" toward the former Soviet Union for a decade thereafter, and the rise of anti-communism in the United States (Chang 1995). Nevertheless, the legacy never completely died and its enduring quality became evident in the early 1970s as political relations between the two countries began to thaw.

The ways in which science and technology were related to the establishment and improvement of diplomatic relations during the 1970s is an instructive example of the uses of S&T in diplomacy. There is no doubt that S&T relations would not have developed as they did without a significant political breakthrough, driven largely by shared geopolitical interests vis-a-vis the former Soviet Union. Nevertheless, shared interests surrounding science and technology offered assets to the political negotiation process that were quite distinctive. After the debilitating effects of the Cultural Revolution on its research and education system, China quickly grasped the critical importance of restoring the effectiveness of its science and technology system and, by the late 1970s, had come to understand just how damaging the country's interlude of more than a decade of radical politics had been in setting back its scientific and technological development during a time of rapid advances internationally. In particular, China largely missed the onset and contributions of the microelectronics and information revolutions that re-shaped the growth trajectories of the Western economies and became the foundation of high technology development in the US, Western Europe, and its East Asian neighbors.

Confronted with the starkness of Chinese backwardness at the time, Chinese elites were primed to give scientific and technological development an important place on their domestic policy agenda. Key American scientists, principally those of Chinese descent, encouraged this policy orientation, and many of China's senior, US trained, scientists helped revive the spirit of the pre-1949 legacy. Together, these factors made the prospect of establishing science and technology cooperation especially appealing, a fact which was appreciated by officials of the Carter administration. These various strands of interaction and engagement came together in efficacious new ways in 1978 when, prior to normalization, the US government sent a delegation of its senior science officials to China, a mission which resulted in the signing of the first of many formal government to government agreements for educational exchanges and scientific and technological cooperation. This high level visit led directly to the reestablishment of diplomatic relations in early 1979 (Smith 1998: 114–136; Suttmeier 1998: 137–164).

3 The Evolution of the Relationship

The S&T relationship was formalized with the signing of the bilateral Agreement for Science and Technology Cooperation (hereafter, Agreement) in January 1979. The Agreement called for government to government cooperation among the technical agencies of the two countries, and quickly led to the signing of a series of protocols between those agencies. Over the years, some of these have endured with new activities added, while others saw only limited activity. The Agreement called for the establishment of a Joint Commission on Science and Technology which serves as the overall governing body of the relationship. The Commission meets every 2 years and has as co-chairs the Chinese minister for science and technology and the US president's science advisor. The Ministry of Science and Technology (MoST) provides executive support on the Chinese side, with the Department of State providing overall support on the US side. (The US does not have an exact counterpart agency to China's Ministry of Science and Technology, a fact that often adds complications in terms of the way the two countries approach bilateral S&T engagement). On alternate years when the Joint Commission is not meeting, the two sides convene a lower level meeting of these executive secretaries. Over 30 years, the number of active interagency protocols has risen to more than 30.[1] While not every agreement or initiative has been successful, taken together, they have created a network of sustained interactions and on-going relationships that, in general, have proven mutually beneficial over time.

While the Agreement represents the politically most visible sign of the relationship, and has facilitated widespread cooperation via government channels, it is also useful to think of the government to government relationship as providing a framework for a broader range of other S&T related activities outside of the Agreement. These include, academic exchanges facilitated by the government to government Agreement on the Exchange of Students and Scholars, signed in 1978, which has led to hundreds of thousands of Chinese receiving advanced education in the United States. Indeed, some officials in both the US and China believe that the exchanges of talent and the professional training that has occurred represents the most tangible and significant contribution of the S&T relationship.[2] Cooperation through academic ties has by now led to expanding research cooperation and a growing number of university to university relationships.

The S&T relationship also has helped provide a framework for the growth of cooperative activities via commercial channels, which are also facilitated by the government to government Agreement on Commerce and Trade overseen by The Joint Commission on Commerce and Trade. The development of these commercial channels has facilitated a massive transfer of technology, especially as foreign direct investment expanded as a result of FDI liberalization policies in China during the 1990s. By the end of the 1990s, we also began to see a growing commitment from US companies to the establishment of R&D centers in China. Today there are over 1,300 of these foreign R&D centers—with top US multinational firms such as GE, IBM and Microsoft being among the most active in the effort to tap into China's talent pool. Thus, over the course of some 35 years, there has developed a complex web of relationships in science and technology involving governmental, academic, and commercial channels.[3] Some of the key drivers and shapers of that development include the following set of issues.

---

[1] For recent discussions of the extent of the US-China government to government S&T relationship, see White House (2012).

[2] Interviews conducted in Beijing in May 2013.

[3] For a fuller account of the evolution of the relationship, see Suttmeier (1998, 2010).

4 Issues of Asymmetry

One of the interesting characteristics of the bilateral S&T relationship is the ways in which its asymmetries have evolved and been managed. At the outset, Chinese science and technology had been seriously set back by the effects of the Cultural Revolution, as noted above. In addition to the interruptions in research and innovation agendas in most fields, the disruption of university life also meant that China had lost a generation of new scientists and engineers. Hence, the special appeal of getting access to the US university system as an expeditious way to compensate for this loss of talent. In most ways, then, in those early years, China brought little to the relationship in scientific terms other than a pool of smart, highly motivated young students and scholars and the opportunity to access distinctive natural and social phenomena in China (seismicity and other geological phenomena, climate, disease patterns, village life and social structure, etc.), which unfortunately had long been denied to US investigators.

As a result of these asymmetries in capabilities, the balance of benefits seemed very much to favor the Chinese side, especially as large numbers of Chinese students were accepted into US graduate programs with some funding from the Chinese government but with the bulk of support coming from the US side. On the other hand, both sides from the beginning were sensitive to the balance of benefits question, and attempted to allocate costs on a "benefitting side pays" formula.

Interestingly, the approaches taken by the two respective countries to conduct their evolving S&T relations differed in many ways. The US side, in general, took the position that activities under the various protocols should be paid for out of existing agency budgets in the belief that spending on activities with China would only be justified if they supported agency missions. While guaranteeing a degree of discipline in China related activities, this approach also meant that US technical agencies were somewhat constrained on the kinds of initiatives they might take, constraints which became more evident when government budgets were tight, as they often were following the "Reagan revolution" in the early 1980s. While the US government through the White House Office of Science and Technology Policy endorsed a growing Sino-US S&T relationship, the fragmented nature of American engagement often left the Chinese side disappointed in terms of the tangible aspects of the US commitment. On the Chinese side, on the other hand, special funds were set aside by the central government for international cooperation activities including those with United States. Chinese officials clearly viewed these activities as highly strategic in nature and extremely critical to China's program of so-called "four modernizations"—agriculture, industry, national defense and science and technology.

To some extent, these alternative approaches reflected fundamental institutional differences, though over time it has become clear that the Chinese approach also can be construed as more of an investment orientation in which the relationship with the US was a critical part of a national effort to re-build and modernize scientific and technological capabilities. As the poorer, less capable partner, paying for these investments was not easy, but could be readily justified in the face of manifest benefits. With the US enjoying unchallenged leadership in most areas of science and technology when the relationship began, it is perhaps not surprising that an investment orientation was less in evidence. In the 1980s, the principal focus of American civilian technological worries was the rising competition from Japan; China seemed of little immediate concern given the serious competitive challenges posed by what seemed like the Japanese technological juggernaut.

To be sure, some members of the technical community in the US saw cooperation with China in its various forms as an investment opportunity to add to the global stock of knowledge and technical talent necessary for the continuing development of science. For foreign policy decision-makers, on the other hand, the S&T relationship provided policy tools for keeping the political relationship on track. Few on the US side might have imagined in the 1980s that Chinese science and technology would progress to the extent that it had by the beginning of the second decade of the twenty-first century. This apparent lack of vision and foresight—whether viewed from the threat or opportunity perspective—has helped create some of today's uneasiness and discomfort felt by US officials in the face of the technological foundations of China's rapid economic and military progress.

For China, as suggested above, the building of national scientific and technological capabilities in the post-Mao era became a matter of high national priority. Thus, in addition to exploiting opportunities for scientific and technological development in the international environment—the richest of which were in the relationship with the US—China also set about trying to get things right domestically. This led, in particular, to the initiation of a series of on-going reforms in its system of science and technology institutions, with many of these—such as the establishment of the National Natural Science Foundation of China (NSFC)—inspired by expanding international experience, especially with the US.

This reform experience, which continues today, has involved mixing successful policy and institutional models from abroad with Chinese political and institutional realities. While not always successful, overall, the transformation of the Chinese science and technology system of 1978 into the system of 2013 is truly remarkable.[4] Thus, in contrast to the situation in the 1980s when China's S&T system suffered from a lack of funds, a dearth of talent, and a very backward infrastructure, the Chinese S&T system today is characterized seemingly by a high level of abundance—with substantial resources committed to R&D spending, the training of more high end talent, and the establishment of a large number of well-equipped modern facilities.

During this same period, US leadership in research and innovation was maintained, but its relative position as a science and technology power was changing. Advanced scientific and technological capabilities had diffused to more parts of the world, and the US faced a growing list of problems, including budget constraints, an aging S&T workforce, and seemingly intractable problems with STEM

[4] See Suttmeier and Cao (1999), Springut et al. (2011), and OECD (2008).

education (See, for instance, National Science Board (2012), Wilsdon and Keeley (2007)).

In short, the conditions which characterized the early asymmetries were changing. Chinese scientific and technological development had progressed to the point where distinctive natural phenomena, and the promise of smart people, were not the only appeals making cooperation attractive. Gradually, albeit steadily, China became a far more important locus for research and innovation, offering in some cases distinctive facilities and research environments not found elsewhere. In addition, the supply of Chinese scientists and engineers, along with those from India and other countries, became increasingly important for sustaining the innovative performance of US research and technology-oriented enterprises.

5 New Directions

Thus, in a variety of ways, the material conditions as well as the policy environment for productive cooperation between the two countries have changed. In some ways, these changes have recast the basic premises and assumptions underlying the foundation of the bilateral S&T relationship. In some instances, questions have been raised in the US about the continued value of closer S&T cooperation, especially because of the growing uneasiness regarding alleged Chinese behaviors—whether state directed or not—concerning IPR protection, industrial espionage, and cyber hacking. On the other hand, a record of some success over the past 30 years and a series of new challenges facing the technical communities of the two countries have created opportunities for new directions in the relationship.[5] There clearly are a series of new imperatives—bilateral and multilateral—that suggest that perhaps the time has come to reframe and re-structure the agenda and nature of Sino-US S&T interactions.

5.1 Global Issues

This is increasingly true with regard to the growing importance of global issues, especially climate change, the energy-environment nexus, and global health concerns. Among the latter, concerns for international cooperation in approaching pandemics has become prominent, especially after the outbreak of SARS in 2003. The growing concern for global health issues has led to a more active involvement in China of the US Centers for Disease Control, for instance, which has cooperated with the Chinese Ministry of Public Health in establishing a Chinese CDC. The

[5] For recent discussions of the extent of the US-China government to government S&T relationship, see Department of State (2012), White House (2012).

value of these interactions has yielded a variety of meaningful payoffs as illustrated by the more transparent and more collaborative approach the Chinese government has taken to handling the recent Asian flu outbreaks in contrast to the defensive, secretive way the initial SARs epidemic was dealt with by Beijing.

While energy and environmental issues have been on the agenda for quite some time, they gained importance during the Bush administration and acquired a new focus on the US side at the beginning of the Obama administration under the leadership of Energy Secretary Steven Chu and Presidential Science Advisor John Holdren. Out of this has come an interesting new experiment in international cooperation known as the Clean Energy Research Center (CERC) which some observers believe could serve as the basis for taking Sino-US S&T cooperation to the next level.

CERC is less a physical center than a coordinated program of cooperation in which both sides have contributed equal financial resources and both have organized domestic research consortia with members drawn from industry, universities, and government research institutes in the target areas of research and development: clean coal, clean vehicles, and energy efficient buildings. CERC is notable in that costs are shared equally, and the consortia are, by design, intended to facilitate intersectoral cooperation among the three main types of institutions—industry, government, and academe—which in the past often worked separately with China. Needless to say, the establishment of the domestic consortia was itself a challenge of cooperation for both countries which added to the challenge of bilateral cooperation between them. Because much of the work entails the potential for developing commercial products, the development of understandings about intellectual property rights has consumed a fair amount of effort in the early stages of the Center. In many ways, the success of CERC will depend heavily on the ways in which the IPR issues are handled by the respective parties; there clearly is a need for CERC to help build and reinforce trust between the two sides so that meaningful collaboration can occur without apprehension about the disposition of new, commercially relevant IPR (CERC 2013).

5.2 Security Concerns

An enduring security concern and irritant in the relationship is the issue of US export controls. In the early years of the relationship, the export control question was a function of Cold War policies and multilateral controls exercised through CoCom. Over the course of the 1980s, however, China's status in the export control regime was changed and controls were liberalized substantially, and by the late 1980s, the two sides even were exploring the possibilities of transferring military technologies to China. The end of the Cold War and the abolition of CoCom might have led to an acceleration of liberalizing trends had these not coincided, roughly, with the June 4, 1989 events in Tiananmen and the imposition by the US of new sanctions against China. US-China cooperation in space, for example, was halted at the time and even today continues to be inactive as a result of Congressional budgetary mandates. Ironically, some have suggested that US refusal to cooperate with China's high priority space program has been one of the key factors behind China's rapid indigenous progress in space technology—with the help of Russia instead of the US.[6]

The gradual improvement of relations after the Tiananmen tragedy, especially with the rapid growth of FDI in the early 1990s, led to new thinking about export controls. On one hand, the deeper involvement of foreign corporations in the Chinese economy occasioned by the growth of FDI created pressures for further liberalization, especially as the share of high-technology industry in the foreign invested sector increased. On the other hand, forces resisting further liberalization within the US government meant that controls often still had teeth. As a result, the Chinese side has remained frustrated with US export control policies, as have some US companies who allegedly have lost business opportunities to European and Japanese firms who no longer are constrained by CoCom restrictions.

The export control question and the role of strategic technologies in the relationship more generally, acquired high political visibility following allegations in the mid-1990s that Loral Space and Communications Ltd. and Hughes Electronics Corp. had transferred sensitive technologies to China in connection with launch services provided by China for US satellites.[7] The issue was rapidly picked up by a Republican controlled House of Representatives which led to the establishment of a special committee, under the leadership of then Representative Christopher Cox, which rapidly expanded its mandate to the question of the transfer of strategic technologies to China more generally, including information relating to the miniaturization of nuclear warheads.

The Cox report, released publicly in redacted form in 1999, suggested in multiple ways that the PRC has been engaged in a sustained effort to gain access to a wide range of sensitive, controlled military and dual use technologies.[8] As a result of the Commission's work, to which the Chinese side took exception and offense, US export controls took on a new importance, and assumptions about the generally positive relationships between S&T cooperation and constructive

---

[6] From the US point of view, the effectiveness of export control policies is closely related to the availability of alternative suppliers; US unilateral controls, for instance, are recognized as being somewhat limited if advanced technologies are available from other countries. Less attention has been given to the question of whether the denial of technology through export controls has been a spur to successful indigenous technological development in China, as many Chinese observers allege.

[7] China was experiencing several commercial launch failures at the time having to do with the separation of the satellite from the launch vehicle. Allegedly, Loral and Hughes supplied critical information in attempt to solve the problem. The information was subject to export controls, but the companies failed to acquire the proper license.

[8] See US House of Representatives (1999). For a critical analysis of the work of the Cox Committee, see May et al. (1999)

political relations with China became reevaluated on both sides, and began to generate a serious trust deficit.

That deficit worsened after 2000. Although the 9/11 terrorist attack on the United States led to new forms of political cooperation with China, the changes in US immigration policy in response to 9/11 had significant impacts on Chinese citizens hoping to travel to the United States, including scientists and engineers. Security issues, especially US concerns about the leakage of strategic technologies to China through S&T cooperation as well as Chinese espionage, in short, were becoming more troubling for the relationship on both sides, and political opportunists in Congress and elsewhere have been only too happy to seize upon these, unfortunately often with some justification.

The work of the Cox Committee and the post-9/11 attention to immigration policy have led to an increasing focus on "human embodied" technology transfer resulting from professional visits and meetings and cooperative research activities. By the early years of the Bush administration, attention increasingly turned to what has come to be known as "deemed exports" involving the movement of people possessing technical knowledge across international borders. Export controls thus increasingly focused on the acquisition of technical knowledge and the travel patterns of people possessing knowledge thought to be sensitive, with the result that immigration policy, and the issuance of visas, came to play a far more central role in national security policy. Although progress has been made on resolving some of the visa issues pertaining to Chinese coming to the United States, concerns for "deemed exports" have made the process of getting a US visa often difficult for some travelers, have led to certain fields of research being off-limits to Chinese visitors, and in some cases have led the organizers of professional meetings to choose sites outside of the United States (especially Canada) for their gatherings.

5.3 Trade, Investment, and Competitiveness Issues

In addition to national security concerns, science and technology issues also have become far more politically prominent in economic relations between the two countries as well. China's redoubling of its efforts to foster scientific and technological development, especially with the launching of its Medium to Long-Term Scientific and Technological Development Plan (MLP) in 2006 has, in some ways, led to the further erosion of trust.[9] While the plan, and the increased R&D spending it entails, clearly offer new opportunities for cooperation, it also contains a variety of industrial policy tools that have often alienated China's key international partners (McGregor 2010; Atkinson 2012). More specifically, many Western observers have viewed China's emphasis on strengthening its indigenous innovation (*zizhu chuangxin*) capacity as a statist, neo-mercantilist type of policy which ignores the

[9] For discussions of the MLP, see Cao et al. (2006), Schwaag Serger and Breidne (2007).

principles of comparative advantage in favor of an attempt to establish "absolute advantage" (Atkinson 2012). Whether these views represent an embellishment of the efficacy of Chinese technology acquisition efforts or strategic technology policymaking has remained less important in gaining momentum in the media and business circles, especially in view of the highly-charged "anti-China" political environment emerging in Washington DC over the last 4–5 years.

While there is no doubt that Chinese officials are seeking to enhance the performance of their R&D system and to get more Chinese enterprises to focus resources on generating commercially viable innovations indigenously, there are many factors driving Chinese behavior. Some of these factors are historical in origin dating back to the rupture in Sino-Soviet relations in the late 1950s and others have to do with the fact that the current manufacturing oriented model that has driven Chinese economic development for the last three decades is no longer sustainable from a cost, environment, and energy perspective. Even more important, however, is a third factor, namely the recognition among China's top political and S&T leaders that innovation driven competition is becoming the paramount key factor for determining and sustaining economic success across the globe. Based on their reading of the economic tea leaves, Chinese officials have stated that if their country does not evolve into a knowledge economy driven by a high level of innovative performance, it will sit at the margins of the international economy for much of the twenty-first century (Suttmeier and Cao 1999). Chinese policies to realize these objectives, however, have nevertheless caused considerable conflict with its trading partners, and have had the effect of politicizing the S&T relationship with highly contentious trade policy concerns. In the process, new actors from the trade policy bureaucracies across both the OECD countries and China's Asian economic partners have become growing participants in the management of the S&T relationship.

Although the China of the first two decades of the twenty-first century is a far cry from the China of the 1980s in terms of scientific and technological capabilities, a substantial degree of asymmetry in those capabilities persists. PRC officials remain chagrined that despite the substantial increase in resources being made available for S&T upgrading, the results, especially on the innovation side, have been largely disappointing. Chinese public policy is committed to overcoming current defects in their national innovation ecosystem, and in doing so have introduced a series of new policies to foster the development and acquisition of advanced technologies. These include policies for the development of technical standards and patents, configuring the foreign investment regime to maximize technology transfer, and apparent support for illicit technology acquisition strategies through human and cyber espionage having both commercial and national security implications (McGregor 2010).

6 Science, Technology and Ethnic Identity

As noted at the outset, ethnic Chinese scientists and engineers, naturalized citizens and long time residents in the United States, all have played important roles in reestablishing relations in the 1970s, as did senior Chinese scientists who had been trained in the United States prior to 1949. The importance of these ties of ethnic identity has continued over the years, though with the passage of time, the nature of the Chinese diaspora itself changed.

Whereas leadership in promoting expanded cooperation between China and the United States in the 1970s and 1980s came from well-established ethnic Chinese scientists and engineers who came to the US largely before 1949, beginning in the late 1980s, Chinese scientists and engineers who came to the United States after 1978, and established careers in the US, became increasingly important actors in the development of the S&T relationship. These exemplars of the "brain drain" became tenured university professors, high-tech entrepreneurs, key members of the engineering staffs of major corporations, and in some cases, government officials, establishing families in the US and usually becoming US citizens. Since the 1990s, they have had an important part in building bridges to China for universities, corporations, and professional societies.

As a measure of the growth of scientific cooperation, there has been a steady increase in Sino-American co-authorship of professional papers since the mid-1990s, and a major share of these involves collaboration between ethnic Chinese investigators on both sides of the Pacific (Suttmeier 2008; Jin et al. 2007). While ethnic networks have played a very positive role in building constructive relationships between the technical communities of the two countries, questions about ethnic ties also have contributed to the elements of distrust, noted above. This is especially true with regard to high profile cases of espionage involving ethnic Chinese, and more generally, concerns about accelerated flows of science-based technology facilitated by ethnic Chinese scientists and engineers wanting to see a strong and prosperous China. It also has been exacerbated by Chinese government "talent recruitment" programs such as the One Thousand Talents Program which seeks to attract highly experienced Chinese scientists working in overseas universities, technology-based companies and think tanks to return back to China on either a full time or part time basis. Many ethnic Chinese faculty members working in the US and other countries have active laboratories in China with their own local Chinese graduate students supported with substantial funds from the PRC government. Issues have been raised about the time allocations of these faculty members as well as the likelihood that the handling of intellectual property within such "transpacific" networks can be rather relaxed, to say the least, and can contribute to the further erosion of trust.[10]

[10] See, for instance, Larson and Xin (2013) and Hannas et al. (2013).

7 Political Visibility and Political Commitment

In interesting ways, the 35 years of S&T cooperation between China and US has led to complex patterns of growing interdependence. The S&T activities of the two countries already have become steadily intertwined at multiple levels and across multiple domains. However as China increasingly approaches the status of a peer competitor, in which scientific and technological development is a critical component for achieving that status, the US appears to be reconsidering the wisdom of its 35 year engagement with China on science and technology. Increasingly, therefore, questions of trust and intent as the basis for sound political relations have come to shape the S&T relationship. There is an appreciable sense in which the prospects for building trust are not bright. While both sides approach the relationship with a sense of enhancing national interests, the extent to which those interests are served by positive, as opposed to zero sum interactions, seemingly differs between the two countries. Yet, the complexities of the relationship makes the drawing of accurate conclusions about this matter continuously difficult; there are many stakeholders on both sides, but their objectives and expectations differ considerably.

The ways in which national interests are served by the S&T relationship is not independent of the institutional configurations in the two countries and the ways in which they approach the management of the overall relationship. While China indeed is not without its problems of institutional fragmentation and lack of coordination, it is nevertheless the case that over time China has had significant institutional assets committed to a successful exploitation of the relationship. These include the dedicated funding streams, noted above, and a bureaucratic structure and continuity of personnel that have served it well. On the other hand, as we have seen, the US in general has been reluctant to make major investments in the relationships in terms of funding, institutional structures, and staffing. In general, staffing at the OSTP largely has been inadequate and staffing the executive secretariat in the State Department also has been a low priority.[11] This is especially problematic at a time when the multi-dimensional complexities of the Sino-US S&T relationship have been increasing.

For the US, however, it has been difficult to develop coordinated strategies for dealing with China in science and technology and for responding to the challenges that China's progress has created. For some, this is as it should be. Relations with China in science and technology should be driven by clear scientific opportunities and the interests of science are served best when the relationship does not acquire a high political profile. Within the current environment that exists in both countries at the moment, however, this seems almost impossible.

[11] In a comparative study of how six nations manage the challenges of reconciling science and technology policies with foreign affairs, Tim Flink and Ulrich Schreiterer identify a number of weaknesses in the US approach which accord with the more specific details of the US-China relationship (Flink and Schreiterer 2010).

To cite one specific constraint, the recent conflicts between Congressman Frank Wolf and OSTP illustrate the ambiguities associated with political visibility. Congressman Wolf, based on his human rights and national security concerns, has been a longtime critic of China. More recently, he has used his influence to forbid NASA from having contacts with China in the areas of space science and technology, and has worked to limit OSTP flexibility in dealing with China by proscribing OSTP expenditures involving China (See Mervis 2012). For increasing numbers of members of Congress, US engagement with China must result in a series of "wins" or policy victories to be recorded on a scorecard of Sino-American interactions; it is no longer politically acceptable for US policy initiatives to be seen as somehow "helping China." The increased centrality of China within both US domestic and foreign policy invites the higher political visibility associated with greater Congressional activism, and this, in turn, works against the building of a higher level cooperative relationship with China in S&T affairs. In this sense, Congressman Wolf's assault, intended to constrain OSTP activities with China, may actually work against the ability of the US to exploit the relationship more effectively for national interest.

8 Conclusion

The case of US-China relations in science and technology over the past 30 plus years points to a number of interesting, but complex, aspects of science, technology and international relations. In many ways, S&T relations normally have not been the stuff of "high politics" in relations between nations and in the foreign policy machinery of individual countries. On the other hand, S&T are not entirely isolated from high politics either. Science and technology issues, for instance, received high-level attention preceding and following diplomatic normalization with China, and were part of the formula for normalization, yet the convergence of political interests between China and the US vis-a-vis the former Soviet Union seemingly was a precondition for the rapid growth of relations in S&T.

As time has passed, the convergence of political interests that marked the early years of the relationship has given way to a far more complicated picture characterized by its share of divergence in many spheres of activity. In some ways, the strength of the S&T relationship provided a degree of continuity and cohesion when political relations became especially strained. Yet, it would seem that there are firm limits to the positive contributions of science and technology under conditions where political tensions rise (cf. Skolnikoff 2002). This is especially true where tensions over science and technology matters become the basis for the rise in political tensions and mistrust.

The US-China case, thus, points to a highly complex pattern of dynamic interactions. Political interests and common political understandings provide an indispensable framework for the development of S&T relationships. The strength of the latter, though, once developed, can contribute notably to the integrity of the framework, rather like a strong strand can help maintain the integrity of a web. The strength of the S&T relationship and the integrity of the framework however, are, as illustrated in the US China case, contingent on several factors.

The first of these is the evolution of science and technology itself. It has been argued that international cooperation in basic science can be furthered with minimal political constraints. As one moves toward areas of applied research, engineering, and technological development, the mix of cooperation and competition changes as the implications of scientific and technological development for national security and economic competitiveness loom larger (Suttmeier 1998). In a world where there is now often little space between basic science and new technology, and where the stakes of the latter are high for a nation's sense of well-being, it is not surprising that competitiveness and, possibly, conflict between nations arises. In this sense, science and technology issues again can rise to the level of high politics, as we have seen them doing increasingly over the past half-century in some issue areas.

Of course, the intensity of competition is related to the issues of asymmetry noted above. The notable increases in Chinese scientific and technological capabilities since 1978 have altered the asymmetries in significant ways and have contributed to changes in the political understandings constituting the core framework conditions.

Both the framework conditions and the strands of the relationship are also influenced, as we have seen, by a variety of transnational factors as well. It is remarkable that the original government to government agreement so quickly facilitated the growth of nongovernmental interactions between the two countries, involving corporations, universities, and NGOs. The clearly most powerful transnational force, however, has been that of Chinese ethnicity which has facilitated the growth of cooperative ties through governmental, corporate, and academic channels. But, as the political interests shaping the framework conditions have changed, as the asymmetries have been altered, and as continued scientific and technological development comes to be seen in zero sum terms vis-a-vis national security and economic competitiveness, ethnicity has become a far more complicating factor, contributing both to enhanced cooperation and possibilities for new forms of conflict and mistrust.

The state of US China scientific and technological cooperation in the second decade of the twenty-first century, thus, has evolved into a multifaceted and complex relationship. In many ways, the complexity has developed—and continues to develop—more rapidly than innovations in the mechanisms for governing the relationship, in spite of progressive efforts at institutional innovation; witness, for instance, the lack of meaningful engagement within what should have been an inspired US-China Innovation Dialogue. In a variety of ways, both sides exhibit notable institutional deficiencies for achieving enhanced cooperation and the furtherance of mutual interests. In addition, bilateral ties are only as good as the respective China watchers and America watchers inside and outside of government can effectively communicate accurate pictures of what is happening in both countries. Too often, bilateral communications have also suffered from excessive hyperbole in the media, especially with respect to S&T issues.

The lag in the creation of new mechanisms for governance, inaccurate mutual understandings, and media hyperbole are all evident in current discussions of cyber security. The steady intensification of serious bilateral tensions over cyber security issues in the relationship, especially in the first half of 2013, highlights the degree to which political trust issues—even among existing stakeholders on both sides—have begun to erode the once solid, but now increasingly fragile foundations for enhanced S&T cooperation between the two countries.

As the cyber security issues illustrate, there are interesting questions as to whether the search for institutional innovations in the governance of the S&T relationship in a more globalized world should be, or can be, conducted solely in a bilateral framework. These questions are likely to increase in salience throughout the current decade. At the same time, the broader implications of a failure to construct a new US-China bilateral foundation for responding to the key global issues of the twenty-first century will have a critical impact not only on these two countries but also on the economic, environment, and technological futures of many other nations in the coming years.

References

Atkinson, R. D. (2012). *Enough is enough: Confronting Chinese innovation mercantilism.* http://www.itif.org/publications/enough-enough-confronting-chinese-innovation-mercantilism. Accessed July 24, 2013.

Cao, C., Suttmeier, R. P., & Simon, D. F. (2006). China's 15-year science and technology plan. *Physics Today, 59*(12), 38–43.

CERC. (2013). *US-China clean energy research center*. http://www.us-china-cerc.org/. Accessed July 24, 2013.

Chang, I. (1995). *Thread of the silkworm*. New York, NY: Basic Books.

Department of State. (2012). *United States – China science and technology cooperation: Biennial report to the United States Congress*. http://www.state.gov/documents/organization/197119.pdf. Accessed July 24, 2013.

Flink, T., & Schreiterer, U. (2010). Science diplomacy at the intersection of S&T policies and foreign affairs: Toward a typology of national approaches. *Science and Public Policy, 37*(9), 665–677.

Hannas, W. C., Mulvenon, J., & Puglisi, A. B. (2013). *Chinese industrial espionage: Technology acquisition and military modernisation*. London: Routledge.

Jin, B., Rousseau, R., Suttmeier, R. P., & Cao, C. (2007). The role of ethnic ties in international collaboration: The overseas Chinese phenomenon. In D. Torres-Salinas & H. F. Moed (Eds.), *Proceedings of the ISSI 2007* (pp. 427–436). Madrid: CSIC.

Larson, C., & Xin, H. (2013). Divided loyalties land scientists in hot water. *Science, 340*(6136), 1029–1031.

May, M. M., Johnston, A., Panofsky, W. K. H., DiCapua, M., & Franklin, L. (1999). *The Cox Committee report: An assessment*. Palo Alto, CA: Center for International Security and Cooperation, Stanford University.

McGregor, J. (2010). *China's drive for 'indigenous innovation:' A web of industrial policies*. Washington, DC: US Chamber of Commerce. http://www.uschamber.com/reports/chinas-drive-indigenous-innovation-web-industrial-policies. Accessed July 24, 2013.

Mervis, J. (2012). Wolf, OSTP settle China spat. *Science INSIDER*. http://news.sciencemag.org/scienceinsider/2012/04/wolf-ostp-settle-china-spat.html. Accessed July 24, 2013.

National Science Board. (2012). *Science and engineering indicators, 2012*. Arlington, VA: National Science Foundation (NSB 12-01).

OECD. (2008). *OECD reviews of innovation policy: China*. Paris: Organisation for Economic Cooperation and Development.

Schwaag Serger, S., & Breidne, M. (2007). China's fifteen-year plan for science and technology: An assessment. *Asia Policy, 4*(1), 135–164.

Skolnikoff, E. B. (2002). Will science and technology undermine the international political system? *International Relations of the Asia-Pacific, 2*, 29–45.

Smith, K. (1998). The role of scientists in normalizing US China relations: 1965–1979. In A. L. C. De Cerreno & A. Keynan (Eds.), *Scientific cooperation, state conflict: The roles of scientists in mitigating international discord* (pp. 114–136). New York, NY: New York Academy of Sciences.

Springut, M., Schlaikjer, S., & Chen, D. (2011). *China's program for science and technology modernization: Implications for American competitiveness*. Report prepared for the US-China Economic and Security Review Commission. http://origin.www.uscc.gov/research_archive?page=3. Accessed July 24, 2013.

Suttmeier, R. P. (1998). Scientific cooperation and conflict management in US-China relations from 1978 to the present. In A. L. C. De Cerreno & A. Keynan (Eds.), *Scientific cooperation, state conflict: The roles of scientists in mitigating international discord* (pp. 137–164). New York, NY: New York Academy of Sciences.

Suttmeier, R. P. (2008). State, self organization, and identity in the building of Sino US cooperation in science and technology. *Asian Perspective, 32*(1), 5–31.

Suttmeier, R. P. (2010). From cold war science diplomacy to partnering in a networked world: 30 years of Sino-US relations in science and technology. *Journal of Science and Technology Policy in China, 1*(1), 18–28.

Suttmeier, R. P., & Cao, C. (1999). China faces the new industrial revolution: Research and innovation strategies for the 21st century. *Asian Perspective, 23*(3), 153–200.

U.S. House of Representatives. (1999). *Report of the select committee on US national security and military/commercial concerns with the People's Republic of China*. 105th Congress, 2nd Session, Report 105-851.

White House. (2012). *Fact sheet: U.S.-China science and technology cooperation highlights: 32 years of collaboration*. http://www.whitehouse.gov/sites/default/files/microsites/ostp/st-fact-sheet.pdf. Accessed July 24, 2013.

Wilsdon, J., & Keeley, J. (2007). *China: The next science superpower?* http://www.demos.co.uk/publications/atlaschina. Accessed July 24, 2013.

Part II
Co-Productionist Techno-Politics

Making and Remaking the Measurement of Science and Technology: The International Dimension

Benoît Godin and Joseph Lane

Abstract In most if not every discussion of policy, statistics play a leading role. Statistics serve as evidence, so it is claimed. Such is the case with statistics on science and technology. This chapter is concerned with documenting how, over history, the international dimension got into statistics on science and technology. More specifically, the chapter documents the policy issue that has influenced the development of official statistics after World War II—industrial competitiveness between countries or "gaps" between Europe and the United States—and the representation of science and technology involved in the statistics.

Using the OECD as an emblematic example, we suggest that the measurement of the international dimension of science and technology has not progressed much in the last 40 years. The same indicators that defined the phenomenon in the 1970s are those included in the more recent standardized manuals of globalization.

Keywords Science and technology • Internationalization • Statistics • History • OECD

1 Introduction

The OECD has been very influential on the conceptualization and articulation of national policies for science and technology among member countries (Salomon 2000). The interest of the organization in these matters goes back to the OEEC, the predecessor to the OECD. In 1958, the Council of Europe asked a working party to

B. Godin (✉)
INRS, Montreal, QC, Canada
e-mail: benoit.godin@ucs.inrs.ca

J. Lane
Center for Assistive Technology, School of Public Health and Health Professions, University at Buffalo, Buffalo, NY, USA
e-mail: joelane@buffalo.edu

M. Mayer et al. (eds.), *The Global Politics of Science and Technology - Vol. 2*, Global Power Shift, DOI 10.1007/978-3-642-55010-2_10,

examine the activities of the European Productivity Agency, where the main activities for science and technology were conducted at the OEEC. To the Council, there was a "scientific research crisis in Europe" (OEEC 1959a: 2–3):

> Between the highly developed, science-based industries of the United States and the explosive development of Russian technology, Europe sits uneasily. (...) True, Europe has the great advantage of the tradition and maturity of its scientific institutions, and particularly those for fundamental research. (...) But this is not enough. (...) Europe has, as a region, been slow to exploit in production the discoveries of its laboratories. It is no longer possible for each of its constituent countries to undertake the amount of research necessary for its security and prosperity ... Most of our governments have evolved little in the way of a coherent national science policy, while the concept of scientific research and development as an important and integral feature of company investment is foreign to the thought of most of European industry.

Following the working party report, Dana Wilgress was asked by the Secretary-General to visit member countries to discover their approaches to science and technology. He reported, "It is in Western Europe that most of the great scientific discoveries have taken place (...) but in the race for scientific advance, the countries on the Continent of Europe stood comparatively still for more than two decades while the Soviet Union and North America forged ahead" (OEEC 1959b: 14). The sources of the problem were many: the educational system was "better fitted for turning out people trained in the liberal arts than in science and technology"; there were prejudices against those who work with their hands, and few applications of the results of science; there were also a lack of resources for science, too great an emphasis on short-run profits and not enough on investment for the future, small-sized firms that were not so science-minded, and inadequacy of university facilities and technical training.

It was in this context that the newly created OECD, via a Directorate for Scientific Affairs, turned to the promotion of science and technology policies. From the very beginning international considerations moved the efforts. As an international organization, the OECD's purpose is international in scope. In fact, international cooperation was the object of the very first science and technology policy document produced by the OECD (OEEC at the time): *Co-operation in Scientific and Technical Research* (OEEC 1959b).

> While scientists have co-operated on a regular basis without regard to national boundaries, there are few co-operations between governments in science and technology ... Each European country has an interest in assuring that Western Europe as a whole does not fall behind in the race for scientific advance between North America on the one hand and Russia and China on the other ... The OEEC is the only international organization that is in the position to develop co-operation between the countries of Europe.

International co-operation was only one of the items on the early agenda of the Directorate. A second was industrial competitiveness between countries and the role of science and technology on international competitiveness. The study of competitiveness at the OECD was put entirely under the auspices of statistics. It led to the very first statistics on the internationalization of science and technology in the Western world.

This chapter is concerned with how the representation of science and technology since World War II was translated into economic terms and statistical metrics for domestic and international policy analysis. The OECD is studied as an emblematic example of this representation. As a matter of fact, OECD statistics rely on what the organization call "best practices" and experts are involved in their development as consultants. In turn, social researchers use the official statistics to study science, technology and innovation.

The first section of the chapter presents a brief chronology for the emergence of statistical comparisons of science and technology activity between countries, and the indicators generated for measuring international relations in science and technology over the past 50 years. The second section offers a critical look at research and development (R&D) expenditures as the main indicator used to compare nations, and explains how policy based on this metric affected the realization of socio-economic impacts from investments in science and technology.

2 The International Dimension of Statistics on Science and Technology

2.1 International Comparisons

Statistics on science and technology emerged in the second half of the nineteenth century (Godin 2007). At the time, the statistics came from scientists themselves, and were concerned with measuring the number of "men of science" (scientists): their demography and geography (Table 1). The socio-political context explains the kind of statistics produced. It was a time when the "progress" of civilization, later called culture, and the contribution of "great men" to it (including scientists), were central to many discussions. In fact, many thought that the lower social classes and "unfits" reproduced more than men of science did, representing a danger to the human race. This led to the idea of measuring the number of men of science a nation produces (Francis Galton; Alphonse de Candolle).

Soon, the idea came to some of using the numbers as an indicator of how a nation supports science. In fact, one of the first widespread uses of statistics (first half of the twentieth century) would be to contribute to the advancement of science and to improve the social conditions of scientists. Scientists complained that they received too little recognition for their work: low salaries, few prizes. The number of men of science came to serve as indicator of the social condition of these scientists and of the culture of science: a low number of men of science in a society was seen as indicative of insufficient public support to the scientific profession, and to a low culture of science (James McKeen Cattell).

From the mid-twentieth century onward, the statistics being collected changed completely. Thereafter, the main producers of statistics were governments and their statistical bureaus, followed by international like the OECD and UNESCO

Table 1 Historical development of statistics on science (nineteen to twentieth century)

Stages	Source	Main statistics
Emergence (1869–c.1930)	Scientists (Galton, de Candolle, Cattell)	Number of scientists
Institutionalization (1920–c.1970)	Governments and national statistical offices (pioneering role of the United States)	Monetary expenditures (and rates of return)
Internationalization (1960 and after)	International organizations (UNESCO, OECD, European Commission)	Technological innovation (indicators and international comparisons)

(Godin 2005). The most cherished statistics changed from the number of scientists to the money spent on R&D activities. The statistics were supported by a methodological manual, known as the Frascati manual (OECD 1962), adopted by OECD Member countries in 1963 (see Appendix 1). It allowed governments, so it is said, to manage their investments and assess the results (with what was called a "national budget on research") and compare their efforts to each other in a standardized way. Again, the socioeconomic context explains the situation. Science and technology were increasingly recognized as a factor, if not the main factor responsible for military power, economic progress and international competitiveness. Countries were classified and analyzed according to their performance, or differences in terms of (both production and use of) science and technology. In fact, the only way for a country to assess its performance is by comparing its efforts to those of other countries—or to those of the past. In the last 20 years or so, statistical scoreboards have been developed to this end (Appendix 1).

"Differences" in scientific and technical potential or "disparities" between countries, as measured early on at OECD (Freeman and Young 1965; OECD 1967), was an influential concept for decades to come. Yet, from a descriptive concept, disparities turned normative and came to be called "gaps" in a matter of a few years. In response to French discourses among others (like Pierre Cognard from the *Direction générale de la recherche scientifique et technique* and Jean-Jacques Servan-Schreiber, editor of the weekly *L'Express*), the OECD launched a study on *Gaps in Technology* which, supposedly, calculated the hegemony of the United States in science and technology and the dependence (gaps) of Europe on the United States (OECD 1968, 1970). From then on, the competitiveness between countries, and ways to maintain or improve their positions in world trade became a leitmotiv of discourses and statistics at both the national and international levels. Differences of industrial productivity between countries have been much studied in the following decades at the OECD, leading to the *Growth Project* in the 2000s on the role of information technology in explaining productivity. In the following pages we concentrate on international relations between countries (exchanges and flows of science and technology).

2.2 *Competitiveness Between Countries*

As mentioned above, the numbers coming from the very early statistical comparisons conducted at the OECD in the 1960s documented differences between countries on a number of dimensions of science and technology, including the international dimension, thus supporting early OEEC's fears. Certainly, the numbers were not un-contested, quite the contrary, but it gave rise to a series of indicators which are still central to the measurement of the international aspects of science and technology.

2.2.1 The Technological Balance of Payments

One of the oldest indicators of international trade in science and technology, and one of the first to undergo detailed scrutiny at the OECD, is the "technological balance of payment"—a term coined by the French in the 1960s. The technological balance of payment measures the technological flows between countries in terms of patents, licenses and know-how. It includes statistics on receipts and payments, the balance between the two, a coverage ratio and a figure on total transactions.

The indicator has always fared poorly in the mind of analysts, for methodological reasons. As a consequence, international harmonization and preparing a compendium of methods was suggested early on. A methodological manual would indeed be drafted in 1983, revised and distributed in 1990. That same year, the OECD launched its first technological balance of payments survey and began publishing regular statistical series. The results of the surveys were quite dissatisfying, however. A majority of countries did not adequately detail and document their statistics, and there were differences ranging from 60 % to 120 % between the amounts declared by the money recipients and by funders. A revision of the manual was therefore envisaged in 1994, but never conducted.

2.2.2 Trade in High-Technology

The main indicator of international relations in science and technology is trade in "high-technology". In direct response to the debate on gaps, an indicator on trade in high-technology between countries was developed: a ratio of investments in R&D to sales. Industries were analyzed according to three classes: high, medium and low-technology. High-technology is defined as those industries that have a ratio of R&D to sales over the average. It includes industries such as aerospace, electronics and drugs.

To be sure, the indicator had precursors back to the 1930s, such as "research intensity", "science-intensity" and "technology intensity". But it is the US Department of Commerce that has suggested the concept under this name in the early 1970s, in a study produced as a follow-up to the debates on gaps. From then on, high technology became the symbol of an 'advanced economy'.

The OECD adopted the concept a decade later, under a request of a Council of ministers for the study of high-technology industries. A methodological manual was envisaged for some time but never produced. Nevertheless, the indicator is now part of every measurement of the international dimension of science and technology. The first international statistics were published in 1986, then on a yearly basis in *Main Science and Technology Indicators*.

2.2.3 The Internalization of R&D

Following a reorientation of its whole statistical program on science and technology in the early 1990s (the *Technology and Economy Program*), the OECD has proposed a "framework" on economic globalization, until then a "fuzzy, subjective, badly-defined and a hotchpotch of ideas" (OECD 1992: 232). Regular indicators on globalization in science and technology started to be included in diverse statistical publications in 2001. To the OECD, globalization is defined as "the increasing internationalization of financial markets and of markets for goods and services ... whereby national resources become more and more internationally mobile while national economies become increasingly interdependent" (OECD 2005b: 11).

The framework on globalization gave impetus to changes in the main indicator on science and technology until then: R&D. Although the indicator on R&D is the oldest indicator on science and technology at the OECD, dating back to the 1960s, it had to wait the 1990s for its 'internationalization'. Until then, for example, government could not make a distinction between R&D conducted in national firms and in foreign affiliates. In 2005, the OECD published a *Handbook of Economic Globalization*—initially a manual.[1] The Handbook recommends that a firm's nationality takes precedence in statistics, rather than the country in which its research activities are carried out. Since 2005, numbers on the internationalization of R&D appear in a companion statistical report to the Handbook, published twice (2005 and 2010).

The three indicators discussed above (technological balance of payments, trade in high-technology, internationalization of R&D) constitute the core of indicators on the international relations, or globalization in science and technology at the OECD (see Appendix 2).[2] To be sure, other indicators like scientific collaboration

[1] The term "Manual" was changed to "Handbook" early on because the OECD judged the document not mature enough.

[2] The early draft versions of the Handbook covered five forms of globalization with regard to science and technology: R&D, patents, trade in disembodied technology (technology payments and receipts), international technology alliances between firms, and trade in high technology products. It finally centered around three: R&D, technological balance of payment, trade in high-technology products. The rationale for deleting the other two series of indicators was that the data are collected mainly by private sources, not governments. While the argument is true for technological alliances between firms, it is not entirely true for patents, which come from public patent offices in several countries.

have been added at the OECD in the last decade (co-signed publications), as well as indicators on the international mobility of human resources (highly skilled workers) and patents (triadic patents). Yet, these indicators are not discussed in the Handbook, but appear rather in the statistical publications of the Directorate for Science, Technology and Industry (DSTI), like *Measuring Globalisation: OECD Economic Globalisation Indicators*.[3] At the time, these indicators were not yet available in standardized form, as stated in the handbook. They do not have—not yet—the "status" that strictly economic indicators have.[4]

The three indicators discussed above are exactly the indicators used in *Gaps in Technology* (OECD 1968, 1970) to measure differences between the United States and European countries. A sign that indicators evolve very slowly. Many *new* dimensions of the internationalization of science and technology remain to be conceptualized and measured. Yet, *renewing* the oldest indicators is an option too, as the internationalization of R&D shows. Before going into this issue, let's mention a further dimension of the internationalization of statistics on science and technology.

2.3 Competition Between International Organizations

Not only do firms compete internationally. Countries do too, although doubts are often expressed on the concept of competitiveness as applied to countries. Equally, international organizations do compete between themselves. The history of statistics on science and technology is witness to such a competition.

In the 1960s and after, the OECD's Network of Experts on Science and Technology Indicators (NESTI) was not alone in the field of statistics on science and technology. The UNESCO's Division of Science Statistics attempted standardization of statistics too—and produced the first ever statistical studies of R&D at the international level (UNESCO 1969, 1972). Yet, UNESCO faced challenges foreign to the OECD, first of all the fact that it had a far more diverse membership that the OECD in terms of economic development (South economies) and economic systems (Eastern Europe). It required important adaptations to the OECD standards. To take just one example, an issue what that of classifying the R&D conducted in the non-profit sector, a sector in its own right according to the OECD, but

[3] The Handbook is a work of compromise from a diversity of OECD Directorates—Directorate for Financial and Enterprise Affairs, the Statistics Directorate, the Trade Directorate and the Directorate for Science, Technology and Industry—and international organizations like the International Monetary Fund (IMF). The latter could not imagine that the OECD instead of the IMF was responsible for such a handbook.

[4] The OECD Handbook limits the measurement of globalization of science and technology to indicators of *economic* globalization. Yet, two of these indicators—the technological balance of payments and trade in high-technology—are not un-contested, for methodological reasons, and they have no methodological manual.

distributed into the business, government or university sectors at UNESCO. The debate rested in the end on a difference in 'philosophy': an economic organization (OECD) competed here with a 'cultural' organization (UNESCO).

On every front, UNESCO lost the battle, and today the organization limits its measurement to R&D mainly. The reasons are many. First, the OECD offered resistance to every UNESCO's recommendations. It refused to depart from its practices. The OECD feared the "empty internationalism" (*sic*) of UNESCO which would, in fact, results in the OECD loss of control on statistical standards and methods (OECD 1977: 18). Second, UNESCO had a diversity of Member, a majority of which had few or no experience in statistics (in fact, many countries lacked interest in science and technology statistics). Third, UNESCO had limited financial resources to conduct an ambitious statistical program imagined in 1978 (UNESCO 1978)—due to the United States' withdrawal from the organization in the early 1980s.

3 A Critical Appraisal

3.1 Revisiting Standard Indicators Further

The above section described how post-World War II European discourses on science and technology focused on international comparisons in terms of gaps and lags. International organizations reported gaps between Europe and the Unites States on every established science and technology indicator. In the context of the R&D indicator specifically, the OEEC characterized the key issue as a lag in *development*, defined as insufficient activities "to exploit in production the discoveries of laboratories." The ensuing analysis by Dana Wilgress that "European research (basic science) is excellent but development lags," reinforced the point that excellence in research (science) is necessary but not sufficient to compete globally for socio-economic benefits (OEEC 1959b: 14). These noted international gaps were due to lags in the downstream application of development methods (i.e. engineering) which are designed to bring new conceptual discoveries to tangible forms. In other words, there were lags in the transition from discovery to technology then to commercial exploitation or industrial production.

Unfortunately, elected officials, government staff and their academic advisors didn't implement policies and measures consistent with these conclusions. That is, they did not establish separate data series to track three different activities: scientific research, engineering development, and industrial production (Stone and Lane 2012). Nor did they create unique metrics for the three distinct outputs generated by their underlying methodologies: conceptual discoveries from science, tangible prototypes from engineering, and commercial products/services from production. That oversight eliminated the option of collecting data in a manner permitting analysis of the differential effects of investments and expenditures separately or in aggregate. This, in turn, precluded the ability to actually determine the cause and effect relationships between these three activities, methodologies and outputs.

Instead, government and academic advisors chose to conceptually conflate scientific research and (engineering) development in one single expenditure category: Research & Development (R&D). Industrial production—equally critical for actually transforming knowledge into products and services—was simply omitted. Godin (2006) explains that it was convenient for statisticians to combine research and development because the available data was already co-mingled. In essence, the data compilation protocol drove the conceptual categories from which were derived metrics.

Over time, this integrated concept of R&D became the main indicator of international leadership in what was called technological innovation. One example shows the core problem with the "R&D" metric. Assume three countries have equal amounts of Gross Domestic Production and all three invest equal percentages of GDP in R&D activities. Therefore all three countries are deemed equal in technological innovation. However, in reality country A expends 100 % of R&D funding on scientific research (R) which results in high numbers for scholarly publications and citations; country B expends 100 % of R&D funding on engineering development (D) which results in high numbers for invention claims and patents filed; while country C expends 100 % of R&D funding on industrial production (P) which results in high numbers for units of goods manufactured and services rendered. Is there really any basis on which to claim that the three nations are equally innovative?

To be sure, public policy documents such as the Frascati Manual (OECD 1962) attempted to preserve conceptual distinctions between scientific research and engineering development. However, subsequent statistics continued to combine them in definitions of R&D as: "creative work undertaken on a systematic basis to increase the stock of technical knowledge . . . *and* [italics added] the use of this stock of knowledge to devise new applications." Therefore, the phrase research and development (R&D) was defined as collectively encompassing three categories of activity, tenuously defined by the intentions of the actors initiating the activity: basic research, applied research and experimental development.

These categories persist to the present time despite the fact that the actor's intentions do not actually determine the eventual application of the outputs. For example, projects defined by the investigators as basic research sometimes find their way into applications, while projects defined as applied may never actually be adopted by external stakeholders. The category of experimental development exists in a neither region, because government and academics actors still view this as yet another research activity. In contrast, professional engineers define experimental development as something quite different, where conceptual discoveries are reduced to a practical form to demonstrate a proof of the concept operating in reality.[5] Co-mingling research with development permanently drew the latter under

[5] On the controversial relations between development (engineering) and basic research in terms of public funding, see Belanger (1998).

the former, so that the D simply disappeared in statistics and future discussions of R&D (Godin and Lane 2012).[6]

Compounding the problems with the construct of R&D, policymakers and scholars adopted another contraction—Science and Technology. As explained above, the term science represents a specific type of knowledge and a specific method (or methods). In contrast, the term technology represents the output of a different methodology (engineering development).[7] So the phrase Science & Technology mixes a process with an output. Even within the framework of public policy, R&D and science and technology were supposed to reflect two different sets of data: R&D represented expenditures, while Science and Technology (originally Science, Engineering and Technology) represented the portion of professionals employed in those categorical professions (Thompson 1988).

Compounding the absence of conceptual distinctions was the lack of differentiation between the outputs of two methods: conceptual discoveries from scientific research versus tangible prototypes from engineering. Nor was there an effort to track the outputs from these activities as they moved into the production activity of firms. Industrial production as a category of investment, expenditure, process or output is largely missing from indicators labeled as science and technology. To be sure, the OECD has added indicators on innovation—defined as the commercialization of technological inventions—to econometric analyses and statistical series in the last 20 years. Nevertheless, as will become clear in the next section the analysis is missing the critical element of industrial production because links between R&D and commercialization are absent and commercialization is not measured at all.

3.2 Taking Concepts Seriously

Technological innovation has always been defined as the commercialization of discoveries and inventions in the form of new or improved products or services. To the OECD Oslo manual, "an innovation is the implementation of a new or significantly improved product (good or service), or process, a new marketing method, or a new organizational method (in business practices, workplace organization or external relations)" (OECD 2005a: 46). In the case of products (technological innovation) implementation means "introduced to the market" or commercialization (OECD 2005a: 47).

Given this official definition of technological innovation, one would expect, logically, to find some indicators of commercialization in the statistical series addressing Science and Technology. What one find instead is everything but commercialization. Take the OECD as example again. In the recently published *Measuring Innovation: A New Perspective*, there is one and only one indicator on

[6] One had to wait UNESCO (2010) for the very first statistical survey of engineering development.

[7] In contrast, in the nineteenth century technology meant "methodology" or *technique*.

commercialization (or rather one indicator *called* commercialization): patents. Yet, a patent is not commercialization at all. Instead, it is a national certification of ownership over a claimed invention. The OECD states that patent is included for the "likelihood [of a patent] of being commercialized" (OECD 2010a: 104–5). But probability is not certainty. The OECD ignores its own basic definition of technological innovation as the actual commercialization of technological invention. It offers no direct indicators on commercialization in the statistical scoreboard.

Since the 1990s, the OECD professed an intention to broaden the scope of this representation of innovation by articulating a systemic approach. "It is necessary", claims the OECD, "to analyse the role of public policies, economic incentives, organisations, market structure, foreign trade and investment, and other institutional factors, along with the complementarities and synergies (OECD 2010b: 29)." Right. However, in practice the organization is only adding indicators of activities peripheral to indicators on commercialization. The systemic approach shifts (unconsciously?) the study (and measurement) of technological innovation away from market commercialization. To be sure, commercialization is not everything. For example, as previously noted three related methodologies and associated outputs are necessary to accomplish technological innovation. But it is surprising that, given the decades-old definition of technological innovation as commercialization, there exist not one direct indicator of commercialization in the OECD statistical toolbox and that few efforts are devoted to improve the situation.

Some dimensions of commercialization are mentioned in the OECD Oslo manual as factors (objectives and effects of innovation activities), but not as indicators of commercialization. Above all, these factors rely on subjective reporting from managers, including yes/no answers or ordinal multiple choice scales (OECD 2005a: 107–8). Licenses and royalties exist as indicators. Yet, they are proxies of commercialization rather than direct indicators.

Direct indicators are:

- New products/processes (count rather than subjective answers)
- Production of these new products/processes (costs, machineries, tooling, instrumentation, machining, assemblies, design and testing, production in quantity)
- Distribution (diffusion)
- Sales
- Revenues
- Marketing
- Market value
- Market share
- Use (by customers)

Despite all the limitations of the early surveys of technological innovation like *Gaps in Technology* (counting the origin and diffusion of technological innovations from a selected list), the statisticians of the time were really measuring technological innovation as production and commercialization. The measurement of science and technology must perhaps return to this example for inspiration and for developing the statistical series further in a context of international relations.

4 Conclusion

The emergence of official statistics on science and technology in the 1940s was national in scope. As a consequence, international organizations like the OECD took on the task of cross-national standardization to permit comparisons between and among countries. The first such numbers revealed gaps between European nations and the United States. Indeed, as the recent battleground for two world wars, European countries had collectively suffered extraordinary costs and casualties. Nations accustomed to global influence experienced severe declines in their wealth, infrastructure and stature. To make a bad situation worse, instead of benefitting from an enlightened Pax Americana, Europe was partitioned as the demarcation line in the Cold War between the United States and the Soviet Union. This *political* context led to diverting resources to both reconstruction and defense, among others. The United States thus had every opportunity to maintain its international leadership in science and technology, which explains the then-proclaimed gap.

The measurement of globalization was not an issue on the agenda of the OECD program on science and technology at the time. It emerged in the 1990s, at the same that the organization reoriented its whole statistical program on science and technology toward economic issues, namely industrial productivity and economic growth. The pressure from policy-makers to measure the phenomenon led to a series of indicators not quite "original". Old indicators, namely those developed during the debates on gaps, were re-interpreted as indicators of globalization (technological balance of payments, trade in high technology) or revised (R&D). To be sure, new indicators were imagined, but they were contested among the diversity of statisticians involved in the OECD Handbook, who refused to include them. As a consequence, they were introduced within other statistical series (see OECD 2010a).

Throughout the entire timeframe some basic assumptions behind the indicators were never questioned. Two issues related to these persistent assumptions were addressed here. First, what interest are policymakers serving by focusing on science to the detriment of development and production? Second, if the creation of new net wealth (social and economic) is the core issue, why is there no indicator or statistics on engineering development and industrial production in the statistical series? If policy-makers want to close the gaps in societal welfare and economic performance, they need to focus their analyses on the methodologies and outputs that ultimately deliver on those goals.

The statistics evolving over time to measure international relations or globalization remain primarily national in scope. From a statistical point of view, this chapter prompts a subsequent question: Is there really an international dimension to science and technology for consideration by governmental policymaking bodies—as there is for climate change?

Appendix 1: OECD Methodological Documents (year = first edition)

Manuals

The Measurement of Scientific and Technical Activities: Proposed Standard Practice for Surveys of Research and Development (Frascati manual) (1962).
Proposed Standard Practice for the Collection and Interpretation of Data on the Technological Balance of Payments (1990).
Proposed Guidelines for Collecting and Interpreting Technological Innovation Data (Oslo manual) (1992).
Data on Patents and Their Utilization as Science and Technology Indicators (1994).
Manual on the Measurement of Human Resources in Science and Technology (Canberra manual) (1995).
Measuring Productivity – OECD Manual (2001).

Handbooks

Handbook on Economic Globalisation Indicators (2005).
Handbook on Deriving Capital Measures of Intellectual Property Products (2010).
Handbook on Measuring the Space Economy (2012).

Guides

Guide to Measuring the Information Society (2005).

Frameworks

A Framework for Biotechnology Statistics (2005).
Framework for Nanotechnology Indicators and Statistics (2008).
A Conceptual and Methodological Framework for Emerging Technologies Indicators (2011)

Others

Bibliometric Indicators and Analysis of Research Systems: Methods and Examples (1997).

Appendix 2: OECD Handbook of Economic Globalisation (2005)

Proposed Indicators of International Dissemination of Technology

Internationalization of R&D of Multinational Firms

Share of R&D expenditure and of the number of researchers of foreign-controlled affiliates
Share of industrial R&D expenditure financed from abroad
Share of value-added, turnover and employment attributable to foreign-controlled affiliates whose main domain is R&D
Share of parent companies in R&D expenditure and the number of researchers

Internationalization of the Diffusion of Technology

Technology payments and receipts as a percentage of GDP
Technology payments and receipts as a percentage of R&D expenditure

Internationalization of Trade in Technology-Intensive Products

Share of foreign-controlled affiliates' high-technology manufacturing exports in high-technology manufacturing output
Share of foreign-controlled affiliates' and parent companies' high-technology manufacturing exports and imports in total high-technology exports and imports

References

Belanger, D. O. (1998). *Enabling American innovation: Engineering and the national science foundation*. West Lafayette, IN: Purdue University Press.
Freeman, C., & Young, A. (1965). *The R&D effort in European Union, North America and the Soviet Union*. Paris: OECD.

Godin, B. (2005). *Measurement and statistics on science and technology: 1920 to the present*. London: Routledge.
Godin, B. (2006). Research and development: How the "D" got into R&D. *Science and Public Policy, 33*(1), 59–76.
Godin, B. (2007). From eugenics to scientometrics: Galton, Cattell, and men of science. *Social Studies of Science, 37*(5), 691–728.
Godin, B., & Lane, J. P. (2012). A century of talks on research: What happened to development and production? *International Journal Transitions and Innovation Systems, 2*(1), 5–13.
OECD. (1962). *The measurement of scientific and technical activities: Proposed standard practice for surveys of research and development*. Paris: OECD.
OECD. (1967). *The overall level and structure of R&D effort in OECD member countries*. Paris: OECD.
OECD. (1968). *Gaps in technology: General report*. Paris: OECD.
OECD. (1970). *Gaps in technology: Comparisons between countries in education, R&D, technological innovation, international economic exchanges*. Paris: OECD.
OECD. (1977). *Responses by the secretariat to the questions of the ad hoc group*. DSTI/SPR/77.52. Paris: OECD.
OECD. (1992). *Technology and the economy: The key relationships*. Paris: OECD.
OECD. (2005a). *Guidelines for collecting and interpreting innovation data*. Paris: OECD.
OECD. (2005b). *OECD handbook on economic globalization indicators*. Paris: OECD.
OECD. (2010a). *Measuring innovation: A new perspectivity*. Paris: OECD.
OECD. (2010b). *The OECD innovation strategy: Getting a head start on tomorrow*. Paris: OECD.
OEEC. (1959a). *A programme for European co-operation in science and technology*. C/WP26/W/4. Paris: OEEC.
OEEC. (1959b). *Co-operation in scientific and technical research*. C(59)165. Paris: OEEC.
Salomon, J.-J. (2000). L'OCDE et les politiques scientifiques. *Revue pour l'histoire du CNRS, 3*, 40–58.
Stone, V. I., & Lane, J. P. (2012). Modeling technology innovation: How science, engineering and industry methods can combine to generate beneficial socioeconomic impacts. *Implementation Science, 7*, 44.
Thompson, C. (1988). Some problems with R&D/SE&T-based definitions of high technology industry. *Area, 20*(3), 265–277.
UNESCO. (1969). *Bilateral institutional links in science and technology*. Science Policy Studies and Documents no. 13. Paris: UNESCO.
UNESCO. (1972). *R&D activities in international organizations*. UNESCO-ECE/COM-76/CONF.711/3.
UNESCO. (1978). *Recommendation concerning the international standardization of statistics on science and technology*. Paris: UNESCO.
UNESCO. (2010). *Engineering: Issues, challenges and opportunities for development*. Paris: UNESCO.

Wither Anarchy? Harvesting the 'Global' Bio-tech Body, Indian Markets and Biomedical Technologies

Anna M. Agathangelou

Abstract The co-emergence of life and value, bodies and the body politic is a major aspect of world politics today. This chapter, first, frames key debates in IR on anarchy, order and postcolonial understandings of the 'corporeal' and the 'international' with a focus on debates of biocapital and biovalue in STS. Second, I grapple with how biological sciences are simultaneously contesting and facilitating global biotechnology ventures, and how the 'international' and 'corporeality' co-emerge. I argue that what counts as corporeal and what counts as international must be critically examined in order to break away from the delirious and omnipresent re-inscriptions of imperialism and its dominant presumptions of anarchy and order that come with the imaginary, the thinking, and praxis of bio-value. In the attempt to craft a distinctive geopolitical niche, states and markets bio-innovate the making and (un)making of living beings and their distribution as symptomatic of practices, discourses, and strategies that define, zone, and make possible the appropriation and governing of life. The emergence of infrastructures of biotechnology and 'lively capital' debates in India and the play *Harvest* orient us at what is at the forefront of claiming and constituting 'global' power. Reading these debates in India, I want to argue, opens up the space for articulating analytics grounded in the empirical-as-'material-semiotic configurations' and 'orientations' that offer lessons and methods for IR and STS by challenging strategies of zonings (i.e., the 'international' and the 'corporeal,' theory and practice, bioeconomy and capital) upon which a geopolitically spatiotemporal order of modernity depends. I conclude with some insights into the ethical imperative to read ontologies and epistemologies that transgress and alter the hierarchies and disciplinary formations that come with anarchy and order.

Keywords Corporeality • Anarchy/Order • Tech-nobodies • Bio-tech body • Bio-value • Biomedical technologies • Life sciences • The international

A.M. Agathangelou (✉)
Department of Political Science, York University, Toronto, Canada
e-mail: agathang@yorku.ca

M. Mayer et al. (eds.), *The Global Politics of Science and Technology - Vol. 2*, Global Power Shift, DOI 10.1007/978-3-642-55010-2_11,

At first glance, it is difficult to see the connective threads between the debates about order and anarchy in IR,[1] new technologies (gas, oil, drones, etc.) and recent biomedical breakthroughs (reproductive technologies, regenerative medicine). IR debates with a focus on science and technology have focused on knowledge as power in the military and US security and the ways science can sustain in place the hegemonic role of the US and the primacy of its markets in the world order. (Paarlberg 2004; Fukuyama 2014). 'Critical tools' (Boero 2010; Hayden 2003, 2005; Helmreich 2007a, b) as well as insights from Science and Technology Studies (STS) have been appropriated to engage with these shifts and 'to question scientific facts reopen[ing] black boxes' (Roosth and Silbey 2008: 15). Yet, these insights have been displaced in the IR debates whose focus has been on order instead of asking the question: 'how shall [we] live together in the polis?' (Roosth and Silbey 2008).

Critiques of such IR approaches (Irwin 2008; Krige 2006; Jasanoff 2004, 2005; Guston 2001) are marginalized, thereby diffusing their impact on dominant narratives (Connolly 2002; Callon et al. 2009). Such IR critiques are pivotal in shaking dominant categories that limit querying "a plurality, a plurality of stakes that are anything but technical' (Callon et al. 2009: 14). What connects these sets of debates, I argue, is the meaning and contestation of corporeality. In this chapter, I am stretching these critiques to engage corporeality in the discourses, practices, and strategies that entangle the anarchy and order problematic. In this entanglement of anarchy and order corporeality does not register and if it does it remains descriptive. Jasanoff (2004) has pioneered a critique of these analyses arguing that they have missed that science and the social order are co-produced. In their rush "STS analyses of 'law, science and technology,' reproduce and sustain deep-seated cultural commitments to particular forms of evidence and reasoning, particular notions of liberty, and particular understandings of what constitutes a well-ordered society' (Jasanoff 2004; Murphy 2012) missing this co-production. Jasanoff's co-production is a useful conceptual tool for studying science, technology, society and the emergence of the global order. In this framework, co-production articulates that science 'is neither. . .constituted by interests alone nor. . .an unmediated reflection of nature.' Rather, "knowledge and its material embodiments are products of social work and, at the same time, constitutive of forms of social life" (Jasanoff 2004: 2). These studies highlight a dynamic approach to understanding the relationship between science and society where the social plays a pivotal role in shaping what comes to be regarded as knowledge of the natural world. Importantly, this approach allows us to nuance the relations between power and governance, power and the human, and being and value. This critical lens of co-production allows us to see how problematic and limiting it is to be thinking the global or order in terms of the dichotomy between domestic and international. A consideration of practices (Jasanoff 2004, 2005), that is, the starting of analysis from the event, not from the

[1] Kaplan (2005: 13 citing the Oxford English Dictionary) says anarchy refers to a 'state of lawlessness due to the absence or efficiency of the supreme power; political disorder.'

object renders corporeality a finite figure of determined practices, analogously the distinction between anarchy and order is a finite object of a determined practice. Importantly, what do such epistemologies abstracted from events and their methods and ethical-oriented direction do for us in thinking the relation of nature and interstate system when anarchy (in the desire to imagine order) becomes the “single most important characteristic underlying international relations” (Milner 1991: 67)? Does this dichotomy play into interests that desire terrains (i.e., places, humans, and animals) in a future ‘ordered’ interstate systems that require ‘domesticated’ states? Fanon and postcolonial STS scholars think so. Fanon, for instance, orients us to think about ‘social justice for all’ (Hobbes 1997; Fanon 1967) by engaging science and dereliction, state and national, capital and the corporeal. He challenges the underlying assumption that informs the zoning of the world and its bodies: anarchy and order connection and questions our notions of the corporeality that underlies this assumption of anarchy/order (1967; Agathangelou 2013b). What gives life to this imaginary of anarchy and order in world politics as well as some STS theorizations (Russell 2001; Ranger 1999) is the evasion of the ‘technological junction,’ that is, ‘the space, time, and place where one mobile workshop meets another’ (Mavhunga 2008: 87)[2] to co-produce the ‘order.’ This epistemological assumption, along with philosophies of violence, hierarchize spaces, places, and corporealities (i.e., ordered vs. anarchical; human vs. animals, culture vs. nature, institutions vs. state of nature etc.) thereby, presuming that ‘technology and nature have remained subaltern to humans, when in fact they are what makes humans people’ (Mavhunga 2008:87).

Different IR doctrines (human security or foreign policy) presume this ‘body’ differently: in one instance, it is a political subject that has reason and whose organic matter is not relevant for politics; in another, it refers to populations and is mere organic matter. Others borrow from biological studies ‘to support their claim that human beings are not individuated, conscious, and rational, but rather enmeshed in sensations and contagions, and shaped by affective and non-cognitive force fields’ (Rose 2013: 7–8). Such approaches presume a division between the humanities and the social sciences which has problematic effects on understanding the nuanced entanglements of social and human relations. STS has been more innovative, here, using ‘a form of argument that links to, but goes beyond, the important recognition that human capacities such as cognition and affect are “distributed”—not the individuated property of singular organisms, but constitutively dependent on the webs of interactions among multiple organic processes within and between organisms and other entities in a locale’ (Rose 2013: 7–8). Rose shows how biology and scientists are opening this ‘body’ for questioning, making it

[2] Mavhunga has coined the idea of a mobile workshop. For him, it means “the artifacts, skills and socio-technical relations that surround border-crossing people, nature, and technology as they move through time and space. These artifacts, skills and socio-technical relations are the very same ones scholars have used to define a workshop (be it an engineering plant or laboratory). Mobility renders the workshop portable and capable of operating on the move or being shifted from place to places” (Mavhunga 2008: 11).

possible for us to challenge prevailing ideas of anarchy and order. Conjuring anarchy 'as a haunting specter that must be subdued and controlled' (Kaplan 2005: 13) by any means possible, including scientific breakthroughs and technologies, misses out on crucial insights that are emerging in world politics today. Simply stated, IR critical theorists must not presume Hobbes' or even Locke's mythos of a state of nature (i.e., what have some called anarchy in IR) as a structural principle. While different IR theorists may disagree about the priority of the 'international' (i.e., as sociability), they nevertheless agree on the priority of the question of 'a state of nature' association as the basis for the subsequent theorization of politics proper. With the epistemology of the 'international' definitively shaped by the idea of 'primitive accumulation' of capital and the encounters of the Europeans with the 'others' (read 'wild Africans' and 'primitive') (Wright 1972 cited in Mavhunga 2008), any critical mode of analysis or line of inquiry can no longer address any social and institutional formation without recognizing that the basis of politics itself is imagined as anarchical and primitive thereby, making possible appropriation, thefts and a clash between capitalism and those economies which would challenge accumulation.

This chapter thinks through and alongside India's participation in the harvesting of organs, biotechnology innovations and life sciences. I probe both the dichotomy between and the connection of order (history) and anarchy (wild, untamable, 'mere nature').[3] How does a 'postcolonial economy' imagine and constitute itself as an innovative power in the emerging matrix of technology and scientific structure? How do Science and Technology facilitate understandings of anarchical regions, subjects and things as struggles of innovation, design and liberation that bring 'together different ways of knowing in creative, eclectic, unpredictable but also very violent ways' (Olwage 2013, citing Mavhunga 2013)? What are the stakes in taking seriously corporeality and its entanglements with Technology and Science as co-constituents of the historicity and materiality of the 'international?' What would an exploration of the basis of politics through the anarchical and the empirical as a set of insights and assumptions about a 'Modern Constitution' (Bruno 1993) and order force methodologically and conceptually for us?

Common to postcolonial approaches of STS is the idea that the scientific and the social/political are always already entangled with each other. One is not prioritized over the other. Drawing on the idea dominant in STS (Harding 2011; Hobson 2011; Carney 2011; Haraway 2008) that the world is made up of humans, non-humans, and objects (Bennett 2010; Murphy 2012) who act upon one another in mutually constitutive ways, I stretch this logic to approach and engage this 'international' order to explore the ways in which the anarchical can be imaginable and doable and constitutive of the international (Jasanoff 2011; Wynter 2001, 2003; Cooper 2008; Hecht 2012) and its multiple markets. In conversation with theorists who argue from the vantage point of the slave and the colonial (Fanon 1967), theory

[3] Hegel describes Africa as having 'no movement or development to exhibit' and belonging to 'the Unhistorical, Undeveloped Spirit still involved in the conditions of mere nature' (1992: 177).

of the South (Comaroff and Comaroff 2011; Okri 1993; Nyamnyoh 2012; Mbembe 2001) and the bush[4] to understand 'geometries of power' and 'cultural topographies' (Mavhunga 2013), I explore emerging biotechnology industries in India and debates in IR and STS to show how accepted canons presume slavery and colonization as analogies (Agathangelou 2013a, b) and the inside as ordered and civilized and the outside as anarchical and violent. Mavhunga's idea (2008: 16) of "operationalizing Jessica Dubow's framing of the map, which she calls '[those] place[s] found by way of the body,'" (Dubow 2000: 89) is an important tool in helping us problematize this dichotomization of the world and zoning of corporealities into anarchy and order. Dubow orients us to read the maps as she recognizes that bodies and sensations precede the assemblage of any map. Thus, reading for corporealities, those sets of "sensate bodies" (Dubow 2000: 89) in the fiction of anarchy and order allows us to question this territorial dichotomization that underlies the world bodies or structures: the 'interstate system' and the 'state.' Reading for the anarchical figure and recognizing it as a "corporeal being, a sensate body who alongside discursive accretions must also be seen to live in place and through perception" (Dubow 2000: 89) is a way of operationalizing, if you will, the global challenging of this dichotomization that underlies the majority of IR theories. Reading the segregation of the world that also comes in the form of the conceptual and the empirical could allow us to highlight that it is this zoning of the world that constitutes and is symptomatic of IR discourses, strategies, and practices that set in motion myths and fantasies that argue for a transcendence from our multiple worlds (Agathangelou and Ling 2009; Latour 2005). Thus, the International is in question.

Mavhunga's work orients us in addition to reading the "maps human bodies draw" and "the body-in-motion across places as a body-at-work, mapping or drawing lines of traversal between departure and destination" (Mavhunga 2008: 16) to think of the relationships among people, nature, and technology simultaneously. For him, these three are not separated. "Traces technology leaves as the body moves through place" (Mavhunga 2008: 16) are equally important to the traces that people leave. In his words,

> All these [three: people, nature and technology] are bodies or systems, be they respiratory, digestive, neurological, cardiovascular, social systems, ecosystems, and sociotechnical systems. They are also entities in space, a space in motion. Never mind where the force is coming from, technology, nature and people are not immune to the temporal and spatial motions of their time. Each is at work, the action self-driven or caused by other entities. I call such a space or mode of power and agency where nature, technology, and bodies come together and form alliances, a mobile workshop. Bodies (of people, animals, inanimate things, and places) become members of a community, and as such, interact, break through, and affect members of another. Mobile workshops are means through which humans extend (or try to extend) their agency (through technology, or technological work, or by making alliances with animals, insects, other humans, and so on). In that space, intentionalities get configured and reconfigured. The mobile workshop is a way of referencing the mobilities of bodies across (against) conventional boundaries (Mavhunga 2008: 21).

[4] Special thanks to Siba N'Zatioula Grovogui for conversations about his mother and the bush.

If these bodies are spaces in motion and in *mobile workshops* then their contribution to change is ongoing, making it possible for us to put pressure on underlying ideas of STS that presume science, technology, and peoples and their laboratories so clearly defined and delineated in western and non-western spaces. Activities of what is science, what is technology and what are spaces are in question. What are places and bodies are also in question.

Reading the segregation of the world that also comes in the form of the conceptual and the empirical along with the traces that bodies and technologies leave behind could allow us to highlight that it is this zoning of the world that constitutes and is symptomatic of an IR (as a discipline, practice, and set of institutions) that argues for a transcendence from our multiple worlds (Agathangelou and Ling 2009; Latour 2005) setting in motion myths and fantasies about the agency of who, when, what and how in world politics.

The chapter tracks a range of politics, texts, policies, spaces, and theories that articulate and engage an emerging understanding of a *global bio-tech-body*. Its reconfiguration comes in the form of capacity and humanity, as well as the capability of taking risks, thereby turning itself into an enterprise that can generate its value transnationally by participating in venturist operations. While laying out the structural aspects of biotech growth in India, I engage with the fundamental contention of the new political sociology of science that existing institutional power relations and resource asymmetries make a significant difference to how scientific developments play out in varied sites. In short, from the vantage point of the anarchical corporeal and its contingent and multiple forms and its problematization that comes from synthetic biology, engineering, and digital technologies, I rethink modern technologies and processes, especially the articulation of their value and meaning in 'science cultures' of modernity.

1 IR and STS: Life Sciences and Biotechnologies

In a brilliant introduction to *Becoming Bio-subjects: Bodies, Systems and Technologies* Gerlach et al. say social sciences or 'the discursive frames that are used to describe the social or institutional effects of an imagined but not quite yet realized technology' (2011: 4) are fictions that frame and locate 'unfamiliar, underdeveloped, or as yet unknown genetic technologies'. For them, 'it is important to analyse not only the technologies themselves, but the social science fictions in and through which we understand and talk about them' (Gerlach et al. 2011: 4). Others who focus on biomedicine engage with the production of biovalue (Cooper 2008; Dummit 2012; Sunder Rajan 2012). Bio-value here 'refers to the plethora of ways in which vitality itself has become a potential source of value: biovalue as the value to be extracted from the vital properties of living processes' (Rose 2007: 32; Wadlby 2002). Sunder Rajan (2007) notes the implications of an ever-tighter alliance between new technologies in life sciences and capital accumulation; the ontological question of what constitutes life is systematically being oriented toward a predictive teleology which, in turn, informs the production of scientific facts.

Biotechnology emerges from a triangular relationship among the life sciences (labs, test tubes), ICTs (computer hardware, database structures, algorithms, fiber networks) and capital flows (market forces, speculative capital, venture activities) (Sunder Rajan 2007). In today's markets, the life sciences consolidate proprietary claims over abstracted or extracted entities beyond the physical and sovereign bodies containing them. The rapid development in genomics opens up the space for capital to constitute new forms of value and to redirect resources to further develop technoscience.

Sunder Rajan points to historical, geographical and institutional trajectories through which 'this crisis manifests itself' (2012: 343), including how pharmaceutical industries are turning themselves into sites of investment and speculation (2012: 326). In Canada, Merck committed $35 million to a Biosciences Fund to fuel future pharmaceutical innovation, support early stage biotechnology companies and attract more investors. Today's global biotech industry is estimated as worth $305.7 billion, an increase of 41.3 % since 2008, with around 60 % based in the US, but China, India, Japan and Brazil rapidly gaining ground (Rose 2013: 3).

Sunder Rajan notes the uneven distribution of bio-value production within emerging economies. For him, the ensuing 'surplus health' is analogous to the surplus labor discussed by Karl Marx: 'It is the locus of materialization of value in biocapital. Value is always future oriented' (Sunder Rajan 2012: 326). He stretches debates in life sciences to account for how epistemic and practical structures of biomedicine and value logics of capital are geopolitically reconfigured to secure value as the 'ontology of capital': 'One could (and should) contribute to the generation of value, to the generation of Health-as-value' (Sunder Rajan 2012: 326).

This moral expectation for capital is shifting our understanding of healthiness and, of course, value and its relation to corporeality (Agathangelou 2013a, b). Many more states (India, China, South Africa, Nigeria) are investing in life sciences as a way of developing the infrastructure for research and development, including clinical trials and genetically modified crops, to renew capitalism (Cooper 2008). Business models in the life sciences now operate on financial speculation. Thus, capital health becomes the contestation site of generating assets and value, pushing to 'undo' anything in the way: bodily intolerance to drugs, lack of consumers, etc.

Through a focus on drug company executives, marketers, researchers, doctors and patients, Joseph Dumit assesses the industry's strategies for expanding its markets; he asks whether the huge growth in medication ties us to a radically new conception of ourselves as intrinsically ill and requiring treatment (Dumit 2012). By tracing the development of drugs through pharmaceutical companies, he shows how they have introduced the idea of *risk* as something to be treated without being eliminated, thus redefining the subject and his/her 'body'. In an interesting use of Karl Marx, he says:

> I have attempted to channel Marx, twenty-first century Marx, using twenty-first century technology. Karl Biomarx is the author of a future automatic translation of Capital into Biomedicine. I use biomedicine as the general term because I think this analysis has

> relevance to the broad set of health industries that are science, statistical information-based, and for-profit. (Dumit 2012: 51)[5]

In turn, biomedicine 'calls to life the powers of science (through cooperation and social force) in order to create wellness, but …wants to use Treatments as the measuring rod for those giant health forces, and confine them within the limits required to maintain the already created Health as Health' (Dumit 2012: 53).

Of course, in reading bio-Marx and bio-medical understandings of illness as value in the world today, Dumit and Rajan want to explain "the entrenchment of market structures within contemporary life science research programmes and institutions" (Sunder Rajan and Leonelli 2013: 8). Sunder Rajan's biocapital (2007), Landecker's capitalization of biology, the commodification of bodies in Adriana Petryna's take on health and clinical-trial trafficking (2005), Lawrence Cohen's take on organs (2005), along with Sarah Franklin and Margaret Lock's capitalization of life (2003) contribute toward a proposed rewriting of *Capital* (Dumit 2012: footnote 6, 47) an approach that pierces the dominant understandings of the circuits of biomedicine. All these authors want to problematize the ways biomedical regimes generate profits based on the 'catastrophe' of corporeality. Clearly, for these authors, genomics, brain sciences, biological sciences enhance, and manage the body rather than addressing health problems and their "language of market-driven biomedicine and the terms of an established canon of political-economic thought" (Meyer 2013: 707). I agree; pharmaceutical firms and their investment in clinical trials have to be engaged critically. Unbinding "biotechnology" from the adhesive quality of "newness" is pivotal. These biomedical practices can be articulated as "emergence" which "is not simply a theoretical gesture, but a reordering of the conditions of possibility in time, as these possibilities are continually contested and reformed over time" (Meyer 2013: 707 citing Sunder Rajan 2012). A focus on ethnographies and the role of promise that comes with the emergence of different institutional formations is of utmost importance (Fortun 2012: 334) especially for those of us who care to rupture the violences that come with such projects.

In sum, Dumit and Sunder Rajan focus on biovalue and the concomitant corporate culture generating biovalue through risk reduction. While tracing this convergence between corporate culture and societal formations reveals the affective economies of health in the production of bio-value, especially in emerging powers such as India, this intellectual trajectory still sustains in place a "sorting of biocapital of species…which hold[s] stable the item against which different species of biocapital exist at all–capital itself" (Helmreich 2008: 474) and the way value is made possible. Such a focus on the capitalization of biology as well as the primacy of productivity of the "essence of their species being" (Helmreich 2008: 474) misses other possibilities that come with translations that challenge the daily corporeal tending of imperial-centrism.

[5] Translation is covered in the next section.

2 Translations

Nikolas Rose explores the 'politics of life,' reworking the concept of 'biovalue' by saying that 'more generally, we can use the term to refer to the plethora of ways in which vitality itself has become a potential source of value: biovalue as the value to be extracted from the vital properties of living processes' (Rose 2007: 32). If vitality refers to 'the ability to sustain life' (Oxford Encyclopedia) or the 'capacity to live, grow and develop' (yourdictionary.com), how do we understand life that is slowly deteriorating? Rose asks us to look at conditions within biology such as genomics and neuroscience, as well as synthetic biology which makes use of animal models, so that we may explore human capacities and pathologies that 'open up the working of our bodies and our minds to new kinds of knowledge and intervention' (Rose 2013: 2). He asks us to consider the capitalization of biological knowledge which shapes and creates biological truths through 'promises and predictions of the biovalue to be harvested: enhanced crop yields, bioenergy, bioremediation, and, of course, advanced medical and health technologies based on biology' (Rose 2013: 3). The ensuing technologization of vitality and life sciences makes life seem amenable to intervention and open to projects of control, such as cloning (Rose 2007; Haraway 2008).

Such thought is entangled with a capitalized and corporealized state and a global political economy of biological knowledge and practice (Rose 2013: 3). Companies, states and regions that are working to generate assets through leveraging what STS calls bio-value are venturing into experimentation and the creation of knowledge about the emergence of a *global-bio-tech body*. This 'global political economy' underpins the revolution of the pharmaceutical industry through systematically informed state interventions, such as international agreements on patents, or private and public sector agreements on funding for genomics and neuroscience. It also regenerates our positionalities as knowledge producers; as Rose claims, there is 'a lot of value in the work we do to maintain our selves' (2012).

While Rose sees biology as a crucial actor in the formation of our social life by remaking what is human and what is 'nature,' he contends that sociology is entangled with the nature of human beings. Drawing on Karen Barad, he shows the shift from capacities with socially constructed features to a focus on non-human entities and their active role in human history. This shift highlights that social sciences should be deeply concerned with biology. Of course, his focus is sociological, while I consider the ways biologies in multiple sites are constituted as international without disassociating order from the notion of anarchy (i.e., those zones that 'primitive accumulation' is extracted with force and theft). Such a disassociation fetishizes those 'wild' ones whose 'life' cannot even be at the "explicit center of political calculation" (Foucault 1978, 1990]). This fetishization makes possible the harnessing of the 'wild' flesh and environments towards the preemption of the biological finitude of those bio-empowered to participate in the configuration of corporeality as a biotechnological asset. Should a discipline like IR, whose major site of analysis is the international and capital formation, that is, the state and the market system and its contingent institutions, be concerned with biology? If so, how? In fact, various scholars already envisage the corporeal and the architecture of the international on a radical new level (Ong 2013; Mavhunga 2013; Jasanoff 2011; Helmreich 2008). And the answer is yes: let's talk.

I urge attentiveness to 'elsewhere' environments. Ong sees a 'growing network of contingent associations among a constellation of scientific objects and values [thereby] transmitting feelings of corporeal vulnerability and expectations for a new biopolitics of security' (2013: 69). Thus, 'elsewhere environments' become the new sites of 'order' and surplus generation. In the making of biotechnological networks (MIT research labs, Biopolis in Singapore) degeneracy seems to be prescribed on some bodies in the Global South (spaces in Africa, Asia etc.) and animated by the necropolitical conditions that render such life disabled, thereby articulating it as anarchical (where brutality unfolds all in the name of humanity). These corporealities are used as raw material or in the words of biologists bio-matter to generate and order 'projects' for the regeneration of what comes to be a 'new' global-bio-tech body.

Briefly stated, the postcolonial is being refigured in the world of technoscience in the life sciences. Since the 1990s, Haraway has argued that life sciences in the southeast and south Asia are playing a major role in reconstituting biocapital powers. She is arguing that the 'new' moves of biocapital is leading to "the enclosure of our own bodily commons" (Haraway and Harvey 1995: 510). Mavhunga (2013) pushes this argument to engage with technology and Africa. As states, corporations, other organizations, and people turn 'the itinerant into a knowledge resource and vessel for the legal and ethical movement of technologies and ideas across the globe' (Mavhunga 2013: 3) a different approach will be required; we cannot pretend that history does not matter in the 54-country diversity of Africa (Mavhunga 2013: 1).

Grappling with infrastructural changes resulting from neoliberal shifts that demand new approaches to deal with debt economies, technological and scientific undertakings, and nuclear disasters, Mbembe, Mavhunga, Helmreich, Dumit, Sunder Rajan, Haraway, Ong, Rose, and Agathangelou problematize the infrastructural changes of the life sciences whose central focus is now the 'health' outcome rather than the destruction of physical bodies, the environments of these bodies and their entanglements (Agathangelou 2013a, b). Emerging economies and the info-tech revolution intensify the effects of acceleration on health and on the body (Dumit 2012) for several reasons, including the multiple crises of capital and its contingent institutions, cutting-edge designs and innovations in biology and technology, and the emergence of the life sciences. The resulting 'epistemic shift in our relationship to the human' provides us with insights into a 'new figure of living' (Rose 2013: 17), opening up a space for capital and biomedical institutions to focus on *symptoms* as value generating mechanisms. The common-sensual, counterfactual reality that the symptom is the possible site of benefit or value generation is made present, real, and innovative, albeit with a necessary *translation* in different parts of the world as 'technology isn't the same everywhere for everyone and at all moments' (Mavhunga 2013: 14).

In a final analysis, this reading of symptoms as sites of fungibility and value generation orient us toward an approach that displaces (Agathangelou 2013a, b) entanglements, networks, and 'deep anxieties about the precariousness of civilization' (Mavhunga 2013: 11). Like Mavhunga who problematizes the way of thinking and engaging with life sciences and Africa, I think the debates about the symptom and its possibility for generating value allow us to replace counterfactual claims with the precariousness of modern ethical life:

> What happens to these properties when the incoming thing comes into contact with Africans?.... What if the 'laboratory' is no longer the Western building where science is practiced, but the crop field, the forest, and other 'open' places where knowledge is made and turned into tangible practice outcomes? (Mavhunga 2013: 13)

Mavhunga reminds us that science and technology are not spatially neutral. He challenges the idea that the European colonization overdetermined existing ways of thinking, doing and being. For him, the relations of animals, inanimate objects and bodies and places rupture such an idea as well as other racist ideas about Africans being wild or backward.

> Bodies (of people, animals, inanimate things, and places) become members of a community, and as such, interact, break through, and affect members of another. Mobile workshops are means through which humans extend (or try to extend) their agency (through technology, or technological work, or by making alliances with animals, insects, other humans, and so on) (Mavhunga 2008: 21).

This idea of the mobile workshop ruptures the idea that 'Africa' is a wild and nonmodern space thereby, showing that Africa and the nonmodern are "constituted through-rather than prior to–translocal encounters and entanglements with modernity, science, and biomedicine" (Zhan 2013: 243). The mobile workshop itself as a 'material-semiotic configuration' and 'orientation' (ibid) problematize and accord us simultaneously critical insights about biocapital and bio-value studies. It allows us to orient ourselves toward an anti-imperial-anti-African centric material-semiotic configuration and orientation that does not presume that 'Africa' is the space and site of anarchy (*αναρχή*) (i.e, a site that appropriations and thefts can happen with impunity).

Agathangelou and Ling (2009) argue that the international is constituted through multiple world-making projects. They coin the concept of worldism to show that making, unmaking and even generating anew relations are based on the contestation of histories, routes of power, and ideas that may 'in due course, take on new and sometimes unexpected meanings and forms' (Zhan 2009: 172). As Agathangelou and Ling see it:

> [Worldism is] a site of multiple worlds. These refer to the various and contending ways of being, knowing, and relating that have been passed onto us from previous generations... Worldism registers not only the 'difference' that comes from multiple worlds, but also their entwinements...Worldism as everyday life enacts self-other reverberations and syncretic engagements, especially by communities at the margins. (2009: 85–86)

In making this argument, we push theorists who articulate the global as a set of 'universal markers...articulated locally' (Zhan 2009: 171). We challenge studies of world systems (Harvey 2005; Wallerstein 1974; Dirlik 1994) that draw on 'totality' and presume that global capitalism is shaped as a 'compressed global timespace, organized by the new global finance and division of labor, and propelled by the regime of flexible production and accumulation' (Agathangelou and Ling 2009: 171). While productive, these approaches leave much to be desired in understanding how the making, unmaking and generating anew (Agathangelou 2013a, b) of histories of corporeal relations unfold.

By way of contrast, analytically and methodologically, worldism makes possible a close examination of the processes through which various worlds and corporealities are constituted and experienced in everyday knowledge and practices, including those whose design and innovation is marginalized in the name of being uncivilized, backward, anarchical and without reason (Agathangelou 2013b).

Along with others who challenge the assumption that 'the global is a finished product, a monolithic, expansive economic force, or the sum of geographic locales' (Zhan 2009: 171), I turn biotechnology's corporealized sanctioned relationship between global and local, body and body politic, order and anarchy inside out. In focusing on the 'symptom' it is important to unpack its 'creativities,' its effect on different bodies including those who become sites of experimentation of how to deal with the symptom. Perhaps tech-neo-imperialism includes and exceeds critical frameworks like Marx's and postcolonial analyses whose major focus is introducing the 'other' as an equivalent subject (Said 1979). Perhaps this imperialism has in its arsenal the 'innovation' of life sciences as high-speed, laboratory designed and operated symptoms and things, producing new knowledge about life and the constitution of twenty-first century bodies.

3 *Translational Brokers* or *Rhetorical Crutches*?

In their 'Introduction: Biomedical Transactions, Postgenomics, and Knowledge/Value' Sunder Rajan and Leonelli (2013) engage with the crucial component of life sciences, namely, translational research, touched on above. They say the US National Institutes of Health introduced this idea in the 1990s with a focus on biomedical research of cancer. Their chief concern is not how research becomes translated, however, but the 'sites and contexts of the materialization of translational research in its myriad meanings in contemporary landscapes of biomedicine':

> In other words, what does the importance of something(s) called 'translational research' tell us about biomedicine today? And what do the configurations of biomedicine today tell us about knowledge, value, and their relationships? What kinds of science are actually being imagined, created and performed within translational research regimes? (Sunder Rajan and Leonelli 2013: 464).

In engaging with technoscientific emergence, Sunder Rajan and Leonelli set the ground for IR's simultaneous consideration of biomedicine and the international. They ask, 'How are knowledge and value related when the meanings of both terms are at stake?' (2013: 465). This question brings to the fore that knowledge and value are entangled with each other. If that is the case, then their question about what kind of science is being imagined, created and performed within transnational research regimes allows us to consider what kind of corporeal-international is being imagined, created and performed through these transnational research regimes. Do these imaginaries, creations and performances open or foreclose other possibilities of corporeal 'projects'?

Their incisive analysis highlights a different orientation of thinking and orients us towards a crucial site of analysis of monetizable/fungible/tradable kinds of value in a central commercial infrastructure. It 'forc[es] us to ask what contemporary biomedical "knowledge" might mean and to whom, given the variety of domains of circumstances under which it is produced, circulated and used' (ibid).

The idea of 'mobility' (Sunder Rajan and Leonelli 2013; Meyer 2010; Keramaris et al. 2008) and the theorization of translation in STS evoke actor-network theory. Sunder Rajan and Leonelli say:

> We are interested in how something like translation emerges as an actor's category in biomedicine, most recently in the guise of translational research. In this regard, it is worth asking whether translational research, in fact, is concerned with translation at all. ...We locate our concern with translational research, then, not in an analytic that is confined to translation but in one that is more broadly concerned with a biomedicine 'in the trans' – what might be called, following Sarah Franklin (2006), 'transbiology.' We wish to situate this concern in a conjuncture of globalizing, postgenomic biocapital. (Sunder Rajan and Leonelli 2013: 467).

Sunder Rajan and Leonelli problematize translation as a process that includes the use of animal models to gain insight into how to treat symptoms, biobanks with DNA data for private use, and race-based diagnostics which reanimate ideas about human biological variation whereby diseases can be directly correlated with race. Importantly, they employ the problematic of 'trans'[6] to trace the circuits of this mobility and its effects on different peoples.

These endeavors are the focus of intense marketing and capital. Biological sciences are facilitating understandings of biological matter including the human body as a resource to be capitalized on, mapped, invaded, miniaturized, disaggregated and reworked as product or disaggregated as multiple symptoms. Biological materials and knowledge are understood as commercial resources normativized in biomedical structural formations and orienting us to the 'trans' (cited above) as a problem of time (Agathangelou and Killian 2014).

India's National Biotech Development strategy explicitly connects innovation with translational research (NBD 2009: 12). It calls for the establishment of an

[6] Following Sarah Franklin's idea of transbiology 'trans' refers to "the movement across, through, over to, from one place, person, thing or one state to another." This mobility compels nuanced changes in our understandings "of species and speciation and also challenges understandings of purification" and segregation practices. Importantly, trans implies that "corporeal experiences and events are blended at multiple levels, resulting in chimeric entities and liminal states of existence" (Kelley and Hayward 2009).

interdisciplinary center for translational research into 'technology policy for public health, molecular pathogenesis of disease, technology development, scale up, and product evaluation and technology diffusion' (ibid: 33). The logics that centralize the imaginary of a translational research that entails moving research products to the market (Keramaris et al. 2008) foreclose the space for asking questions about new visions of corporeality of a multi-world, a democratic world of sciences.

While the NBD report claims to be 'the result of intensive consultations' (Sahai 2009: 1) it indexes only one possibility for the future in the present world formation (s). Its proposed center of 'translational research,' sustains the idea that the capitalist trajectory requires mediation of all transactions. In an interesting way, this logic reifies transactions and registers biomedical trajectories whose goals become concessions to the private sector. This national development strategy prioritizes the private sector's interests without considering, for example, rural and working class livelihoods (Sahai 2009: 3). Rajan Sunder and Leonelli attempt to problematize this logic:

> The question of capital is only one side of the structural coin. The other side is the globalization of biomedicine. We believe that there is not yet a body of work that is speaking in a consistent manner to question the globalization of the live sciences or biomedicine. (2013: 469–470)

They feel much has been missed about the registers of the 'global' and the contestations of the social, and I stretch their logic to argue that the ways this "globalization of biomedicine" emerges evades the connections of financialization and biopolitical technologies and imaginaries about the reinventions of the corporealities of the 'global' (Fanon 1967).

Seeing though capital and biomedicine as two sides of the same structural coin does not allow us to question the items within which biomedicine and value exist: capital and global raciality regimes. The social concept of raciality, for instance, is transformed into a genetic category mobilizing and assembling normative understandings of racial difference, allowing biomedicine to use this entanglement for profits. For instance, how do states take up genetic technologies to understand racial differences? How do they use these racial differences to generate profits? Is biomedicine using biological nature based on racial differences as an asset thereby, generating profits without turning it into some form of commodity product?[7]

IR and STS focus on how innovations and designs in clinical and laboratory practices, including reproductive technologies, DNA testing, cognitive testing, markets, and industrial networks, are entangled, leveraging different bodies (material, physical bodies and their abstraction) in different places as forms of knowledge

---

[7] Birch and Tyfield (2013: 313, 316) argue, "Instead, the bioeconomy, as it currently exists, is centered on the changing patterns of appropriation of value constructed elsewhere in the economy (e.g., Chinese factories). What the STS perspectives... miss is the fact that (biological) nature is an important asset, even if it is not a source of latent value from commodity production."

and potential (potential value, potential capacity). Such entanglements include connections with legal regimes, patenting practices, international negotiations of bodies and biology, professional standards and protocols for bioproducts, and collaborations and competitions in industrial networks.

Sunder Rajan and Leonelli focus on postgenomics, 'to signpost the most recent period in the history of the life sciences, where several key developments.... have coincided with the completion of the first genome sequencing projects in the late 1990s and 2000s.' (2013: 471–472). They comment on the contemporary integration of disciplines due to 'the intrinsic relationality of biological objects and processes,' adding that 'the crucial integration in the postgenomic era is ... understanding the complex temporality of organisms' (ibid), and noting 'a renewed attention to systems and to processes of synthesis aimed to create life... evident [in] the current prioritization of funding devoted to systems and synthetic biology across the globe (ibid).

Put otherwise, postgenomics pushes for the integration or assembling of sites in ways we have not previously imagined. This integration centralizes imaginaries and regimes that can unseat the 'natural' property of the organism, but it does not easily unseat the effect whereby economies of commodification and commercialization through translation result in a globalized and racialized 'personification' of the corporeality of experimentation/the experimenter/buyer/commercial profit maker. Rather, capital, biomedicine, and race are entangled economically and regionally (Africa vs. Asia vs. North America vs. Europe; urban vs. rural; areas of disasters vs. areas of innovation) and naturalized as innovation. From this point of view, capital and biomedicine and their racist imaginaries are already immanent in the fiction of the form of the economies of property and commodity, even the biomedical one as place-based.

The reinvention of biomedical sites like MIT or Silicon Valley in the US is a capitalist move to invent new circuits of capital in the form of the biopharmaceutical sectors and their adjacent clusters. But these shifting sites and their 'worlding' (an in-depth discussion follows) are rife with *tensions* even when they are embedded and articulated through various world-making projects as 'integrations.' While biomedicine may conjure up dreams of capital and profit-making, its integration in capital's and global raciality's frames requires mediations and multiple brokerings (Meyer 2010; Mavhunga 2013). More crucially, it depends on an understanding of a function of value as a limit in order to continue its brokering and at different sites.

4 The State and Biotechnologies in India

The Indian state is undergoing a dramatic shift in biotechnologies. Given its struggling economy in the 1980s, the government of Prime Minister Rajiv Gandhi saw biotechnology (DBT Research and Development, 2009; Nyamnjoh 2012; Synopsis of Public IP in India, 2008) as a launching pad from which to refurbish its capital power, allowing an agricultural economy to become an information economy.

Historically, government, academia and industry in India have had structurally demarcated and functionally distinct imperatives. The government has concentrated on basic research and some socially oriented research without an explicit applications' focus, academic institutions have considered pedagogy, and industry has focused on low-risk commercial research, catering to the endemic needs of the Indian population. But in 1986, the government gave the Department of Biotechnology (DBT) independent status within the Ministry of Science and Technology and began to increase funding in the science and technology sectors, with support to the life sciences sector rising by 16 times (Natesh and Bhan 2009). Large-scale investments in education and research in the biological sciences became priorities, as the state moved 'to build an edifice of innovation and enterprise, modeled along the lines of American capitalism, upon a firm foundation of academic life sciences nurtured within publicly funded institutions' (ibid: 124).

In the pharmaceutical industry, by becoming a signatory to the WTO-imposed General Agreement on Trade and Tariffs (GATT), India was expected to transition over a 10-year period from a process patenting system, involving cheap bulk manufacture of off-patent generic drugs for a domestic market, to a WTO-compliant product patent regime, involving original research on drug discovery and development (Ghooi and Ravindra 2014; Shanker 2002; Rao et al. 2004). This required investment in R&D facilities; it also meant abandoning a production-based business model in favor of a potentially lucrative but risky acquisition-based model, in which Indian companies would be evaluated not just by the number of products they sold but by their potential value, all the while in direct competition with powerful North American and European companies. Many Indian companies retooled their operations; the stakes are not just profits as such but the expansion of speculative economic practices of Indian firms in Africa, Southeast Asia, and Latin America.

The growth of health biotechnology companies gained momentum in the post-1990s with the emergence of various kinds of biotech concerns: small start-ups, large pharmaceutical firms adopting biotechnology, spin-offs of established pharmaceutical companies like Dabur, Ranbaxy, Reddy, and Cadila, and dedicated biotechnology players like Biocon and Shantha Biotechnics. The list includes a spin-off company from the Indian Institute of Science (IISc), Strand Genomics (Arora 2005). At the same time, MNCs like Eli Lilly, Glaxo Smith-Kline, and Pfizer gained entry into the commercial biotech sector. Renewed MNC participation stemmed from India's signing of the WTO treaty, and the country's comparative advantages in cost and skilled manpower in the healthcare domain.

In India, as in many developing countries, the number of financial institutions investing in a new technology is limited, and even those who do are risk-averse and bureaucratic in their approval processes. To remedy this, the Indian government created the Biotech Consortium of India Limited (BCIL) in 1990. This public body was to fulfill the same functions as venture capital companies in the US, albeit in relation with educational and other corporate bodies of the US, that is, promote the creation of firms by providing seed money and any complementary competencies required by scientists. Its role was to guide start-ups, support their efforts to locate funds, and arrange technology transfers. DBT set it up jointly with government sponsored financial institutions, including the Industrial Development Bank of India (IDBI), the Industrial Credit and Investment Corporation of India (ICICI), and about 30 industries, mainly in the private sector. To date, it has been involved in fund syndication, a technology scale-up, packaging technology, and transfer of technology from laboratories (Ramani and Visalakshi 2000).

Because of the primacy of state involvement in health research funding, health biotechnology firms in India have developed in clusters, with a concentration of companies in Andhra Pradesh, Maharashtra, Karnataka and Delhi. Various state governments are giving a special boost to biotechnology cluster development with favorable policies, including tax and excise concessions, and venture capital support. This has resulted in an upsurge of new biotechnology firms, especially in Andhra Pradesh and Karnataka. Firms in the healthcare sector are mostly medium-sized and newly dedicated biotechnology initiatives with flexible organizational structures. Their activity profile, though diverse, is concentrated on recombinant therapeutics, vaccines, diagnostics and antibodies (Arora 2005).The growth of biopharmaceuticals has been so important that some foresee that it will not only equal or increase revenues generated by the conventional Indian pharmaceutical generics industry, but will cause a paradigm shift from the development of chemistry-driven medicines to biopharmaceuticals.

India is party to several international treaties that directly impact biotechnology management (international trade law, international environmental law, intellectual property law, and international human rights law). The government is articulating its viability as a major player by emphasizing its emerging infrastructure of law and norms, including the 2008 Indian National Biotechnology Regulatory Bill which defines and endorses modern biotechnology. By so doing, it legitimates world-making projects whose goals are capturing profits from different markets, constituting Indian biomedical innovation as merely economic.

Importantly, however, commercialization or the capitalization of bio-matter has not flattened meanings.[8] Nor does it readily translate differences into neoliberal market logic and thereby subsume or obliterate them. On the contrary, as the play *Harvest* makes clear, familiar terms of differences are renegotiated and deployed through the commercialized infrastructure of India and other developing countries all the while pushing us to consider those practices and imaginaries that challenge and undo the global raciality of capital as the only "sign under which all of today's encounters of the economic with the biological must travel" (Helmreich 2008: 474).

---

[8] Michelle Murphy (2013) has brilliantly articulated this 'commercialization' within what she calls "'economization of life' (distinct from the commodification of life) which I define as the ways in which living-being is governed and altered for the sake of improving the macro-economy of nations, as is typically measured through GDP" (2013: 153).

5 Rupturing Through *Harvest* the Biomedical State

While India is accelerating its biomedical work, making projects as forms of innovation, many sites are simultaneously those of capture. 'Innovation' as articulated through biovalue generation is not free; acceleration is challenged by instabilities and catastrophes of corporealities and land formations (how are lands polluted and with what effects on health) at these sites. At the same time, biomedical radicalizations are intensifying the effects of acceleration on the corporeal, the corporeality of the state and the corporeality of the world politic. *Harvest*, by Manjula Padmanabhan shows these effects at the level of the physical body, pushing the boundaries of knowledge, valuation, and translation, challenging the idea of knowledge production and value as co-produced without place-based connections and creativities.

The Indian state seeks to become a market player in the global arena by leveraging assets from Indian genetic material, including harvesting human organs, as shown in *Harvest* where Indian characters are structurally positioned to sell, trade, and reproduce organs due to global economic disparity coupled with the lack of rights within nation-states. Written for the Onassis competition, which sought essays addressing problems in the twenty-first century, *Harvest* won the competition, was translated into Greek, and premiered at the theatre of Texhnis in Athens, directed by Mimis Kougioumtzis. It has been broadcast by BBC and made into the feature film *Body* by Govind Nihalini.

A satire on 'transplant tourism,' *Harvest* is set in a future where such global transactions are both legal and cannibalistic. The play dramatizes the changes in a lower-middle class family living in a low-income tenement building in Bombay. When Om Prakash, the eldest son and head of the family, cannot find a job, he signs a contract with the multinational corporation Interplanta Services to sell his body parts and organs to an unknown receiver in the US, a transaction made possible by the multiple international agreements between the US and India. In exchange, Om receives an exorbitant amount of money and material benefits, which all family members enjoy. Subsequent events lead the unknown American buyer not only to claim the rights to Om's body, but to invade the family's home and run their lives and bodies through a centrally placed panoptical device called the contact module (Chandran 1999). Jeetu, Om's younger brother, is a male prostitute whose eyes and body are taken away to prolong the lives of American recipients. The Guards and Agents, the executors of the recipient's will; called 'monster[s],' 'beast[s]' and 'machine[s]' (Padmanabhan 1997: 14), they conspire with those of high-purchase power persons in the Global North who desire an extension of their vitality, and capture poor Indian subjects involved in the illegal organ trade.

The play offers a plausible imaginary into the future. An emerging US and Indian enterprising subject in collaboration with the 'new' US institutional infrastructures of medical technology alter each other, the state, the market and the international. This trans 'project' produces alternative visions, understandings and experiences of order, state, infrastructures, species and bodies (i.e., racialized, sexual, and class). The American purchaser of Om's body, Virgil, describes the agenda of multinational corporations like InterPlanta Services:

> Virgil: We look for young men's bodies to live in and young women's bodies in which to sow their children –
> Jaya: Why! Don't you have your own?
> Virgil: We ... lost the art of having children.
> Jaya: How can that be?
> Virgil: We began to live longer and longer. And healthier each generation. We secured paradise – at the cost of birds and flowers, bees and snakes! We are determined to make our amends. So we designed this programme. In exchange for the life support we offer poorer sections of the world, we gain fresh bodies for ourselves. (Padmanabhan 1997: 96)

The play contrasts US aging bodies with poor Indians' as biomatter (organic energy) to be resourced and harvested for the design of fresh bodies for those who have the resources to increase their vitality. While the characters are introduced as types, with 'Indian' and 'American' worlds depicted through a discourse of binary division, 'perpetuating an artificial sense of difference between "self" and "other"' (Detsi-Diamanti 2002: 112), these dichotomizations are ruptured as the play progresses. Bodies born in one part of the world can move to other nations. Aging and dying bodies' vitality can be extended (and even become fresh) by the purchases of bio-matter from those that require life support in disadvantaged communities. Padmanabhan says: 'For the sake of coherence, this play is set in Bombay; the Donors are Indians and the Recipients, North American. Ideally, however, the Donors and Recipients should take on the racial identities, name, costumes and accents most suited to the location of the production' (Padmanabhan 2003: 6).

The play foregrounds changes in corporeal and body politic 'reproduction,' with bifurcation strategies that articulate such matter as components that presumably function anarchically until the primacy of the global-tech-body is established. If, for example, poor Indians or Africans are thought to be 'wild' or 'flesh,' it is mainly because such diverse components of the corporeal order (i.e., the corporealities of a certain class, nationality, and region 'tech-nobodies') (Silko 1991) will give up their places toward the emergence of global projects and zones for the emergence of tech-bodies of the international order proper. Such foregrounding implies naturalized simple dichotomies between mind and body, first-world oppressive-receivers and third-world victim-donors, rich North and poor South. In this dystopian future, advances in medical science globalization and organ transplant technologies reconfigure the colonial relations of metropoles and peripheries based on new technoscientific capability thereby rendering the violence that is pivotal in the emergence of fresh vitality as 'innovation.'

6 Back to the Prophecy

In the globalizing processes of innovation such as transplant tourism, national boundaries are ruptured; alliances are forged across nation spaces, and multiple actors who capture global markets of biology and life sciences along corporeal lines. Pravinchandra challenges '[t]he very idea of choice' (2006: 7) in most third world environments, and I agree.

The economization of body parts may stem from global inequalities of metropoles and peripheries or First and Third World colonial relations and environments, but these processes cannot be assumed or simply articulated as analytic categories (Latour 1987, cited in Sunder Rajan and Leonelli 2013: 467) or as structural questions of capital. Sunder Rajan and Leonelli suggest capital and globalization are two sides of a coin. While I agree, I think they must be thought together along with the imaginaries and practices of a global order whose assemblage depends on the disassociation between anarchical and ordered corporealized zones. Thinking this relation of globalization and capital from the the vantage point of the anarchical and ordered corporeality allows us to stage a limit to the particular nexus of knowledge and value that characterizes the 'modern' world. Extending though biocapital's boundaries by drawing the connection between biomedicine and globalization with the theory of the 'trans' the determinations of bio-value achieved by a shift of sovereign bodies to 'symptoms' still foreground *the fiction of capital productivity, thereby making it possible to associate* biological species with their productivity (i.e., those of value and those without). Hence, this conceptual shift makes possible thinking the emergence of biocapital by 'clinicalising' what becomes assembled as anarchical corporeality as a question of localisability, visibility, and audibility thereby, evading how some species, objects, technologies, environments are bracketed both empirically and conceptually even at the moment of becoming the 'raw material' in the process of generating value and knowledge. In trans-economies, the forces of authority and subalternity are involved at multiple registers, including emergent relationships of biomedicine in marginal spaces around the world.

In addition to the development of new biomedical technology in the life sciences, including medicines and medical equipment which facilitate organ trafficking in India, moves to generate the conditions and constitutionality for biomedical research to compete economically have made some bodies to be bioavailable, or 'available for the selective dis-aggregation of one's cells or tissues and their reincorporation into another body (or machine)' (Cohen 2005: 83). *Harvest* suggests the emergence of biomedical projects as commodifying and degrading women, including disassembling them to make organs and tissues available either for experimentation or for another's vitality. Jaya's wife is articulated as 'Justice,' because of her active resistance to colonization by US recipients (Padmanabhan 1994: 91); although Virgil claims otherwise, he really wants Jaya's womb.

While India's infrastructure remains dominated by a socialist model of state-sponsored R&D, the state is leveraging science and technology into biotechnology-worlding-projects to make its wealth and its raciality state power anew. These aggressive international/worlding market players include public-sector labs that work closely with academia and industry:

> Many government labs in India are quite literally involved in market ventures of their own, involving the creation, protection and exchange of intellectual property. ... that of high-tech

capitalism, emblematized by technologically intensive fields of the life sciences such as biotechnology. (Valdiya 2010: 122–123)

In the promise of biotech futures in India, a particular 'species of biocapital' articulates two central transformations in 'biotic substance' and 'economic speculation and sentiment' (Helmreich 2008). The desire to play a leading role in this market calls forth a discourse and set of imaginaries that embrace global raciality and techno-capitalism. Active participation in such international projects inspires a 'mania' for newer infrastructures whose links with other parts of the world (MIT) are entangled and unlike the US (tapping into investors and consumers) tap into the state and a re-configured public (Sunder Rajan 2007) whose 'corporeality' becomes a contestation site in the struggle to make claims and gain power in markets of 'innovation' and speculation.

Put otherwise, the bio-colonial-techno-capitalist matrix is constituent of both India and the US. The latter already inhabits the dominant inside in ways that problematize it, even if it does not turn it inside out. As for the former, although bio-technocapital India may be in a bio-colonial-techno-capitalist terrain not of its own making, the country leverages these projects with other worlding projects (ayuvedic technologies as 'traditional') challenging states and actors that presume it to be less powerful. Its tactics and investments in bio-colonial-techno capitalism allow it to re-script dominant imaginaries in unforeseen ways, facilitated by the emerging possibilities of technoscientific advancements, including the generica industry under the umbrella of WTO exceptions and compulsory licenses.

Sunder Rajan says this re-orientation in biomedicine and its multiple-economies 'is contained within this particular property of value itself—as simultaneously material and abstract, attribute and subject' (2012: 242). Because the 'logics of capital themselves come to define what value means, in all of its material, abstract, symbolic, and agential manifestations', we can no longer 'think of transcending the crisis simply through institutional responses' (2012: 342). Although Sunder Rajan isolates a crucial point about the crisis of value, we need to look deeper. 'Value' and its properties are always a fantasy whether surplus or lack, material or abstract, symptom or illness. What if we invert the subject and object of analysis, no longer focusing on capital's innovation (i.e., the relation of globalization and capital alone) and instead think this emergence from those whose lives and bodies make it possible and make other-worlds otherwise including the suturing value itself?

Leveraging bodies as knowledge and potential (potential value, potential capacity) is not a historical happenstance. International negotiations and legal infrastructures with bodies and biology as their central site of power and surplus, professional standards and protocols for bioproducts, and collaborations/competitions in industrial networks is the result of the naturalization of the violence on corporealities behind the 'law of value.' This erasure of conditions under which value is generated (family, inheritance, lineage) sets an imaginary of anarchy in motion whereby a hierarchy of 'value rests on an unlimited and always a soliciting groundlessness, a groundlessness marked, however, surreptitiously, by forms of violence' (Barrett 1999: 31).

Engaging fungibility from the vantage point of this anarchical flesh and the zones that is relegated reveals the terror of the ordering of world politics and exposes the limits of existing concepts and emergences of institutional formations including global political economy and STS. Terror enables value. To understand this, it is crucial to understand the basis of what becomes constituted as global capital power as an anatomization and an archive of not just marginalization and colonization, but also a configuration and orientation of anarchy, opening up the space to think the generative process and its contingent methods of constituting global-bio-tech bodies and institutions from the position of those whose lives and bodies are non-value within this emerging global raciality matrix. Fanon says: 'Decolonisation which sets out to change the order of the world, is, obviously, a program of complete disorder . . . it is the meeting of two forces, opposed to each other by their very nature' (1967: 41). He lays the groundwork for a theory of antagonism 'over and above a theory of conflict' (Wilderson 2010: 75) between masters and servants, as Hobbes would put it, thereby making it possible for us to stretch the debates of value (Sunder Rajan 2012; Murphy 2013) and presumptions and conceptualizations of fungibility. In contrast to dominant theories of economics and capital which define fungibility to be a standardization process that "prescribes that consumption decisions are based exclusively on the consumer's total wealth [its composition is irrelevant]" (Abeler and Marklein 2008: 2), I want to argue that the abstractness and immateriality and distribution of value zoning over corporeality, transforms bodies, human and non-human, into a force field of intensities which would undoubtedly undo the question of value. More importantly, turning symptoms and 'difference' into value (either use, exchange or surplus) through violence and reproduction/synthetic generation does not go unnoticed. In other words, what if, fungibility is not only the process that changes corporeality into a symptom for biocapital's expropriation but something that identifies and stages a limit to bio-value?

Projects of domination, such as turning corporealities into bio-matter fodder, and projects of justice in the form of innovations (Mavhunga 2013) inevitably become entangled with dominant social imaginaries whose goal is generating profits. While the trope of health research in the US focuses on individual empowerment and self-improvement within a dominant capitalist ethos, in India, it draws on narratives of social amelioration nurtured within a socialist ethos. Thus, one way to answer the question of uneven translation might be to recognize that while the components of an assemblage can be translated, it is not so easy for the hierarchized dynamics, uneven distributions or their structural conjunctures.

7 As a Way of Conclusion

In crafting distinctive geopolitical niches through biomedical innovations, different sites are fashioning a world expressed in the construction of an international biomedical leadership. As part of its world-making effort, the US and Indian governments are exporting cheap clinical trials and bodies that are an assemblage of a biomedical industry (innovation model) and biomedicine (mixture of socialist and capitalist), sustaining in place imaginaries of violence that 'capture' the vitality and corporealities (i.e., primitive accumulation) of those at the margins. As Fanon says, 'the replacing of a certain 'species' of men by another 'species' of men [in] a total, complete, and absolute substitution' (1967: 35) must happen without analogies and abstraction; only a 'radical decision' can abolish this biomedical innovation by turning its innovation against itself; only without analogies can earth and anarchy emerge, abolishing once and for all the chaos and the slaughtering that comes with the presumption of a bifurcation of the world into anarchical and ordered corporealities.

References

Agathangelou, A. M. (2013a). Neoliberal geopolitical order and value: Queerness as a speculative economy and anti-blackness as terror, for a special issue titled 'Murderous Inclusions'. *International Journal of Feminist Politics, 15*(4), 453–476.

Agathangelou, A. M. (2013b). The avenging of the Americas: Cholera matters, empire, and economies of blackness. *The Greek Review of Social Research, 140–141*((B–C)), 185–201.

Agathangelou, A. M., & Killian, K. D. (2014). *Time, temporality and violence in international relations: (De) fatalizing the present, forging radical alternatives*. London and New York: Routledge.

Agathangelou, A. M., & Ling, L. H. M. (2009). *Transforming world politics: From empire to multiple worlds*. London and New York: Routledge.

Arora, P. (2005). Healthcare biotechnology firms in India: Evolution, structure and growth. *Current Science, 89*(3), 458–464.

Barrett, L. (1999). *Blackness and value*. Cambridge: Cambridge University Press.

Bennett, J. (2010). *Vibrant matter: A political ecology of things*. Durham: Duke UP.

Birch, K., & Tyfield, D. (2013). Theorizing the bioeconomy: Biovalue, biocapital, bioeconomics or . . . what? *Science, Technology, & Human Values, 38*(3), 299–327.

Boero, F. (2010). Study of species in the era of biodiversity: A tale of stupidity. *Diversity, 2*(1), 115–126.

Bruno, L. (1993). *We have never been modern* (Catherine Porter, trans). Cambridge: Harvard University Press.

Callon, M., Lascoumes, P., & Barthe, Y. (2009). *Acting in an uncertain world: An essay on technical democracy* (G. Burchell, Trans.). Cambridge, MA: MIT Press.

Carney, J. (2011). Out of Africa: Colonial rice history in the Black Atlantic. In S. Harding (Ed.), *Postcolonial science and technology and studies reader* (pp. 140–158). Durham: Duke Press.

Chandran, M. (1999). The panoptic living room in Manjula Padmanabhan's harvest. *Journal of Literature and Aesthetics, 7*(1), 86–89.

Cohen, L. (2005). Operability, bioavailability, and exception. In A. Ong & S. J. Collier (Eds.), *Global assemblages: Technology, politics and ethics as anthropological problems*. Oxford: Blackwell.

Comaroff, J., & Comaroff, J. (2011). *Theory from the South: Or, how Euro-America is evolving toward Africa*. Boulder, Colorado: Paradigm.

Connolly, W. E. (2002). *Neuropolitics: Thinking, culture, speed*. Minneapolis, MN: University of Minnesota Press.

Cooper, M. (2008). *Life as surplus: Biotechnology and capitalism in the neoliberal era*. Seattle, Washington: University of Washington Press.

DBT Affiliated Autonomous Institutions. (2010). *Ministry of Science and Technology, India*. http://dbtindia.nic.in/Draft%20NBR%20Act_%2028may2008.pdf. Accessed February 2014.

DBT Research and Development. (2009). *Ministry of Science and Technology, India*. http://www.dbtindia.nic.in/uniquepage.asp?id_pk=21. Accessed February 2014.

Department of Biotechnology Ministry of Science and Technology Government of India (2009). National Development Biotechnology Strategy. http://dbtindia.nic.in/biotechstrategy/Biotech_strategy

Detsi-Diamanti, Z. (2002). Bio-slavery, or the cannibalistic quest for longevity: Harvesting for human organs in Manjula Padmanabhan's drama. In D. Pastourmazi (Ed.), *Biotechnological and medical themes in science fiction* (pp. 111–130). Thessaloniki: University Studio Press.

Dirlik, A. (1994). The postcolonial aura: Third world criticism in the age of global capitalism. *Critical Inquiry, 20*(2), 328–356.

Dubow, J. (2000). From a view on the world to a point of view in it': Rethinking sight, space and the colonial sublet. *Interventions, 2*(1), 89–90, 92.

Dumit, J. (2012). *Drugs for life: How pharmaceutical companies define our health futures.* Durham: Duke University Press.

Dummit, J. (2012). Prescription maximization and the accumulation of surplus health in the pharmaceutical industry: The_BioMarx_Experiment. In K. Sunder Rajan (Ed.), *Lively capital: Biotechnologies, ethics, and governance in the global market.* Durham: Duke University Press.

Fanon, F. (1967). *The wretched of the earth.* New York, NY: Grove Press.

Fortun, K. (2012). Biopolitics and the informating of environmentalism. In K. Sunder Rajan (Ed.), *Lively capital* (pp. 306–328). Durham: Duke University Press.

Foucault, M. (1978, 1990). *The birth of biopolitics: Lectures at the Collège de France, 1978–1979.* Picador.

Franklin, S. (2006). The cyborg embryo: Our path to transbiology. *Theory Culture and Society, 23*(7–8), 167–187.

Franklin, S., & Lock, M. (eds.) (2003). *Remaking life and death: Toward an anthropology of the biosciences.* Santa Fe, New Mexico: SAR Press.

Fukuyama, F. (2014). *Robert Wright interviews Francis Fukuyama on limits of science.* http://meaningoflife.tv/video.php?speaker=fukuyama&topic=limitsci. Accessed February 2014.

Georg, H. (1992). *Lectures on the philosophy of world history; introduction, reason in history.* Cambridge: Cambridge University Press.

Gerlach, N., Hamilton, S. N., Sullivan, R., & Walton, P. L. (Eds.). (2011). *Becoming biosubjects: Bodies, systems, technologies.* Toronto: University of Toronto Press.

Ghooi, R. B., & Ravindra, B. (2014). Institutional review boards: Challenges and opportunities. *Ethics, 5*(2), 60–65.

Guston, D. H. (2001). Boundary organizations in environmental policy and science: An introduction. *Science Technology & Human Values, 26*(4), 399–408.

Haraway, D. (2008). *When species meet.* Minneapolis, Minnesota: University of Minnesota Press.

Haraway, D., & Harvey, D. (1995). Nature, politics, and possibilities. *Society and Space, 16*, 507–527.

Harding, S. (2011). *The postcolonial science and technology studies reader.* Durham: Duke University Press.

Harvey, D. (2005). *Brief history of neoliberalism.* Oxford: Oxford UP.

Hayden, C. (2003). From market to market: Bioprospecting's idioms of inclusion. *American Ethnologist, 30*(3), 1–13.

Hayden, C. (2005). Bioprospecting's representational dilemma [Special issue on postcolonial technoscience, Maureen McNeil (ed.)]. *Science as Culture, 14*(2), 185–200.

Hecht, G. (2012). *Being nuclear: Africans and the global uranium trade.* Cambridge, MA: MIT Press.

Helmreich, S. (2007a). Species of biocapital. *Science as Culture, 17*(4), 463–478.

Helmreich, S. (2007b). An anthropologist underwater: Immersive underwater: Immersive soundscapes, submarine cyborgs, and transductive ethnography. *American Ethnologist, 24*(4), 622–641.

Helmreich, S. (2008). Species of biocapital. *Science and Culture, 17*(4), 463–478.

Hobbes, T. (1997). Of the laws of nature. In *Leviathan.* New York, NY: Touchstone.

Hobson, J. M. (2011). Discovering the oriental west. In S. Harding (Ed.), *Postcolonial science and technology reader.* Durham: Duke Press.

Irwin, A. (2008). STS perspectives on scientific governance. In *Handbook of science and technology studies* (pp. 39–60). Cambridge, MA: MIT Press.
Jasanoff, S. (Ed.). (2004). *States of knowledge: Co-production of science and social order*. London: Routledge.
Jasanoff, S. (2005). *Designs on nature: Science and democracy in Europe and the United States*. Princeton: Princeton UP.
Jasanoff, S. (Ed.). (2011). *Reframing rights: Bioconstitutionalism in the genetic age*. Cambridge, MA: MIT Press.
Kaplan, A. (2005). *Anarchy of empire in the making of U.S. culture*. Harvard UP.
Kelley, L., & Hayward, E. (2009). *Tranimals: Theorizing the trans-in zoontology*. November 5–8, 2009, from http://www.litsciarts.org/slsa09/archive/slsa09-1132.pdf
Keramaris, N. C., Kanakaris, N. K., Tzioupis, C., Kontakis, G., & Giannoudis, P. V. (2008). Translational research: From benchside to bedside. *Injury International Journal, 39*, 643–650.
Krige, J. (2006). *American hegemony and the postwar reconstruction of science in Europe*. Cambridge, MA: MIT Press.
Latour, B. (1987). *Science in action*. Cambridge, MA: Harvard University Press.
Latour, B. (2005). *Reassembling the social: An introduction to actor-network theory*. Oxford: Oxford UP.
Mavhunga, C. C. (2008). *The mobile workshop: Mobility, technology, and human-animal interaction in Gonarezhou (National Park), 1850-present*. Dissertation, University of Michigan.
Mavhunga, C. C. (2013). *What is Africa in technology? What is technology in Africa?* Keynote at MIT-Africa Interest Group, Massachusetts Institute of Technology, October 1.
Mbembe, A. (2001). *On the postcolony*. Oakland, California: Cornell University Press.
Meyer, N. (2010). *Modeling proteins, making scientists: Rendering molecular life in the contemporary biosciences*. Durham: Duke UP.
Meyers, T. (2013). Lively capital: Biotechnologies, ethics, and governance in global markets ed by Kaushik Sunder Rajan (review). *Bulletin of the History of Medicine, 87*(4), 706–707.
Milner, H. (1991). The assumption of anarchy in international relations theory: A critique. *Review of International Studies, 17*(1), 67–85.
Murphy, M. (2012). *Seizing the means of reproduction: Technoscience, feminist health, and biopolitics in the contradictions of American empire*. Durham: Duke University Press.
Murphy, M. (2013). Economization of life: Calculative infrastructures of population and economy. In P. Rawes (Ed.), *Relational ecologies: Subjectivity, sex, nature and architecture*. London: Routledge.
Natesh, S., & Bhan, K. (2009). Biotechnology sector in India: Strengths, limitations, remedies and outlook. *Current Science, 97*(2), 157–169.
Nyamnjoh, F. B. (2012). "Potted plants in greenhouses": A critical reflection on the resilience of colonial education in Africa. *Journal of Asian and African Studies, 47*(2), 129–154.
Nyamnyoh, F. (2012). Blinded by sight: Divining the future of anthropology in Africa. *Africa Spectrum, 47*(2–3), 63–92.
Okri, B. (1993). *Songs of enchantment*. London: Jonathan Cape.
Olwage, E. (2013). *'Growing together': The politics of knowing and conserving (bio) diversity in a small conservancy in Cape Town*. Unpublished master's dissertation, University of Cape Town, Cape Town.
Ong, A. (2013). A milieu of mutations: The pluripotency and fungibility of life in Asia. *East Asian Science, Technology and Society, 7*, 69–85.
Paarlberg, R. L. (2004). Knowledge as power: Science, military dominance, and U.S. Security. *International Security, 29*(1), 122–151.
Padmanabhan, M. (1994). *Harvest*. London: Aurora Metro Publications Limited.
Padmanabhan, M. (1997). *Harvest*. New Delhi: Kali for Women.
Padmanabhan, M. (2003). Harvest. In H. Gilbert (Ed.), *Postcolonial plays: An anthology* (pp. 214–249). London: Routledge.
Petryna, A. (2005). Ethical variability: Drug development and globalizing clinical trials. *American Ethnologist, 32*(2), 183–197.

Pravinchandra, S. (2006). The third world body commodified: Manjula Padmanabhan's harvest. *Un/Worldly Bodies, 8*, 1–17.

Ramani, S. V., & Visalakshi, S. (2000). The chicken or the egg problem revisited: The role of resources and incentives in the integration of biotechnology techniques. *International Journal of Biotechnology, 2*(4), 297–312.

Ranger, T. (1999). *Voices from the rocks: Nature, culture, and history in the Matopos Hills of Zimbabwe*. Bloomington: Indiana University Press.

Rao, R., Ratna, J. V., & Rao, T. M. (2004). Comparative bioavailability of four marketed Sparfloxacin formulation in healthy human volunteers. *E- Journal of Chemistry, 1*, 43–50.

Roosth, S., & Silbey, S. (2008). Science and technology studies: From controversies to post-humanist social theory. In B. S. Turner (Ed.), *Blackwell companion to social theory* (pp. 1–22). Oxford: Blackwell.

Rose, N. (2007). *The politics of life itself: Biomedicine, power, and subjectivity in the twenty-first century*. Princeton: Princeton University Press.

Rose, N. (2013). The human sciences in a biological age. *Theory, Culture & Society, 30*(1), 3–34.

Russell, E. (2001). *War and nature: Fighting humans and insects with chemicals from World War I to silent spring*. Cambridge: Cambridge University Press.

Sahai, S. (2009). India's national biotechnology development strategy: A policy mired with controversies. *Annals of Neurosciences, 16*(1), 2–3.

Said, E. W. (1979). *Orientalism*. New York, NY: Vintage.

Shanker, D. (2002). India, the pharmaceutical industry and the validity of TRIPS. *Journal of World Intellectual Property, 5*(3), 315–371.

Silko, L. M. (1991). *Almanac of the dead*. New York, NY: Penguin Books.

Sunder Rajan, K. (2007). *Biocapital: The constitution of postgenomic life*. Durham: Duke UP.

Sunder Rajan, K. (2012). Pharmaceutical crises and questions of value: Terrains and logics of global therapeutic politics. *South Atlantic Quarterly, 111*(2), 321–346.

Sunder, R. K., & Leonelli, S. (2013). Introduction: Biomedical transactions, postgenomics, and knowledge/value. *Public Culture, 25*(3), 463–475.

Synopsis of Public IP in India. (2008). *Central Ministry of Science and Technology*. Government of India.

Valdiya, S. (2010). *Neoliberal reform and biomedical research in India: A story of globalization, industrial change and science*. Unpublished dissertation, Rensselaer Polytechnic Institute, New York.

Wadlby, C. (2002). Stem cells, tissue cultures and the production of biovalue. *Health, 6*(3), 305–323.

Wallerstein, I. (1974). *The modern world-system*. New York, NY: Academic Press.

Wilderson, F. (2010). *Red, white & black: Cinema and the structure of U.S. antagonisms*. Durham: Duke University Press.

Wright, A. (1972). *Valley of the ironwoods: A personal record of ten years served as district commissioner in Rhodesia's largest administrative area, Nuannetsi, in the South-Eastern Lowveld*. Cape Town: T.V. Bulpin.

Wynter, S. (2001). Towards the sociogenic principle: Fanon, identity, the puzzle of conscious experience, and what it is like to be "black". In M. F. Duran-Cogan & A. Gomez-Moriana (Eds.), *National identities and sociopolitical changes in Latin America* (pp. 30–66). New York, NY: Routledge.

Wynter, S. (2003). Unsettling the coloniality of being/power/truth/freedom: Towards the human, after man, its over-representation–an argument. *New Centennial Review, 3*(3), 257–337.

Zhan, M. (2009). *Other-worldly: Making Chinese medicine through transnational frames*. Durham: Duke University Press.

Zhan, M. (2013). The empirical as conceptual transdisciplinary engagements with an experiential medicine. *Science, Technology, & Human Values, 39*(2), 236–263.

International Relations, Cybersecurity, and Content Analysis: A Constructivist Approach

Johan Eriksson and Giampiero Giacomello

Abstract The development and global diffusion of access to the Internet has—as is the case with most new and game-changing technologies—been accompanied with fears and threat perceptions. This chapter argues, on the one hand, that Constructivist IR theory is particularly suited for addressing and interpreting the threat discourses and identity issues which come in focus in cybersecurity. The ease with which cyber-culprits can hide their identity and location, operating through networks of hijacked computers across the world, makes fear-mongering threat and identity discourses a key issue in cybersecurity. On the other hand, this chapter presents content analysis—a set of quantitative methods focusing on key word searches—as a pertinent or even ubiquitous method for both the study and practice of cybersecurity. Through simple and globally accessible interfaces, the entire Web can be scrutinized using content analysis. Equipped with Constructivist theory and content analysis methods, the IR scholar stands prepared to uncover and better understand the massive discursive world of the Internet.

Keywords Cybersecurity • Constructivism • Content analysis • Search string

1 Introduction

What is the meaning of security in the digital age, and how can International Relations (IR) scholars address this topic? While there are as many answers to these questions as there are approaches in the empirical cyber-security literature, and in the IR literature (for an overview, see Eriksson and Giacomello 2006), there

J. Eriksson (✉)
Swedish Institute of International Affairs, Södertörn University, Stockholm, Sweden
e-mail: johan.eriksson@sh.se

G. Giacomello
University of Bologna, Bologna, Italy
e-mail: giampiero.giacomello@unibo.it

M. Mayer et al. (eds.), *The Global Politics of Science and Technology - Vol. 2*, Global Power Shift, DOI 10.1007/978-3-642-55010-2_12,

is an emerging consensus that cyber-security has become increasingly politicized, and even militarized (Dunn Cavelty 2012; Nye 2011). Digital-age security has moved out of the low-political and technical-managerial realm of cyber-geeks, to the high-political domain of foreign and security policy. The nature of the Internet implies endemic difficulties of uncovering and understanding cyber-attacks—the identity of culprits, their goals and capacities, and indeed whether an "incident" really is an attack or merely a systems failure. Moreover, the ever-growing Internet is made up of massive amounts of globally accessible text and images. Against this background, this chapter suggests that Constructivism as a theoretical paradigm and content analysis as a particular set of methods particularly pertinent for investigating security in the digital age. While Constructivism highlights issues of identity and the power of discourse and framing, content analysis, as a set of quantitative methods, can uncover patterns of meaning in massive amounts of text.

The "Constructivist turn" in International Relations (IR) is undeniable. Surveys show that Constructivism has surpassed both Liberalism and Realism as the most popular paradigm among IR scholars, and with Constructivist Alexander Wendt being seen as the individual with the strongest influence in the field of IR (Avey et al. 2012). Notably, Constructivism has percolated into the previously "undisputed turf" of Realist theory, i.e. security studies, thanks to path breaking work by early Constructivist scholars (Katzenstein 1996; Kier 1997; cf. Guzzini and Leander 2006). In particular, studies of threat construction, or "securitization", have become a major constructivist contribution to IR in general and security studies in particular (Balzacq 2010; Buzan et al. 1998; Eriksson 2001). Likewise, in cybersecurity studies, recent contributions demonstrate not only how important cyberwarfare has become for national security (Libicki 2007; Clarke and Knake 2010) but also how central the perception and construction of threats is in cyber-political discourse (Eriksson and Giacomello 2007; Dunn Cavelty 2008; Deibert 2013).

These considerations also apply to cyberterrorism, the bulk of which is made of propaganda (or perception management as it is now called), "hacktivism", recruiting, and intelligence-gathering. Cyberterrorism is more about instilling fear and winning the "hearts and minds" of people than about death, physical destruction and crippling of infrastructures (Conway 2002; Giacomello 2004; Weimann 2005). Perceptions of what a country is and what it stands for are crucial in international politics and the decision-making of groups and leaders quickly deteriorates under conditions of threat perceptions and stress (Huddy et al. 2007). Hence, propaganda and perception management (or psychological operations), widely used by cyberterrorists, are ideal tools to increase that sense of threat and stress. Constructivists emphasize the unavoidability and significance of interpretation, of how perceptions of reality are always "filtered" and shaped by particular values, identities and interests, especially with respect to the understanding of social and political activities such as decision-making. Hence Constructivism seems particularly suited for threat analysis and for grasping the centrality of the war of ideas in cybersecurity.

Research has shown that there are three types of stakeholders in security in cyberspace, namely (a) sovereign states, (b) private sector agents and (c) advocacy NGOs and users (Giacomello 2005; Libicki 2007; Dunn Cavelty and Suter 2009). All three types of actors should be fully considered to grasp the fundamental dynamics of cybersecurity. Undoubtedly, the state has had to "compromise" more in the realm of digital security than in other areas. But it is still one of the three crucial stakeholders and is far from being "demised" (Eriksson and Giacomello 2006, 2009). While cyberwarfare is generally considered as a capability of states, the rise of private authority also in this field cannot be denied. Indeed, a major part of critical infrastructures, such as ISPs (Internet Service Providers), are owned and operated by private actors whose business logic does not necessarily coincide with the security demands of the state (Giacomello 2005; Dunn Cavelty and Suter 2009).

One feature of cyberspace that is peculiar, for IR and security theories as well as for "real" life, is that it truly empowers individuals[1] or groups of individuals. For instance, in no other realm, the military is so concerned with what individuals may do as in cyberspace. This feature should, of course, not to be overestimated, as there is no way a person or a non-state group, no matter how highly skilled, could stand on par with the military of a modern state. But the level of "interference" with the national security of a country that that person or group could generate can range from noticeable to considerable, depending on the target-country and the skills of the attackers. This state of affairs is basically unheard of in all other realms that attain to national security. The modern language of international politics, early influenced by Waltz (1959) simply rejected the "individual" as playing any relevant role in international politics. Cyberspace, however, has demonstrated that it is time to reconsider such rejection (see JP Singh 2014).

While Constructivism has developed into a set of applicable IR theories, and Constructivist empirical research has grown considerably, there is still need for more systematic empirical inquiry, including research on cybersecurity. It is high time to "get our hands dirty" and increase the quantity and quality of applied research in the field of security and, more specifically, cybersecurity. The highly innovative volume by Klotz and Lynch (2007) provides valuable guidelines on how to move from "theory" to "practice". These "strategies" seem quite suited for IR and security in the digital age. In this chapter, we thus present examples of how to apply the Klotz and Lynch roadmap to this topic. One of the methods suggested is *content analysis* (Klotz and Lynch 2007: 52–55). It is only one of the many techniques explored by Klotz and Lynch, who in their volume address the whole community of Constructivist scholars, but we argue that this method, or rather this set of methods, is particularly suited for investigating International Relations and Security in cyberspace.[2]

[1] Such actors could be colloquially defined as the "super-individuals".

[2] For a wide discussion of "cyberspace" see Benedikt (1991) and Gibson (1984) for the original definition ("an infinite artificial world where humans navigate in information-based space").

In this chapter, we first provide a brief review of the “state of the art” of the literature on IR and security in the digital age. Then we offer an overview of content analysis techniques and show how these are crucial to examine Web texts (and images). Indeed, we argue that (a) a considerable amount of information exists on the Internet, and even more so on the Web is *text* and (b) such text can be effectively investigated using content analysis methods (e.g. Simmons 2011; Eriksson and Giacomello 2013). Content analysis methods are indeed among the principal methods of inquiry recommended, and preferred by constructivist scholars (Klotz and Lynch 2007). Hence, Constructivist research should actually be the preferred, or at least a principal, approach in investigating cyberspace and particularly the Web. Given these conditions, we conclude the chapter by arguing for the choice of a constructivist approach for those interested in pursuing cyberspace studies.

2 IR and Cybersecurity: State of the Art

What, then, can the IR literature say about digital-age security? This question was addressed in a review a few years ago, which discussed how a variety of bodies of literature dealt with the nature of security in the digital age (Eriksson and Giacomello 2006, 2007). This review looked first at three fields which could be expected to have something explicitly to say about digital-age security: the information society literature, the cybersecurity literature, and the security studies literature. The review also discussed what the “big three” IR paradigms (Realism, Liberalism and Constructivism), could say about digital-age security.

It was observed that the general literature on information society did not deal explicitly with cybersecurity issues, although the analysis of the perforation of national sovereignty and the empowering of alternative identities and communities or of relevance (Castells 1996, 2000; Mowlana 1997). Further, the literature on cybersecurity—which has grown considerably—was and still is largely atheoretical, as well as being dominated by technological and policy-oriented analyses. Many of these analyses, while providing useful information on particular technologies and empirical developments, have a tendency to be deterministic, as well as fear-mongering. Security studies, on the other hand, is a growing and diversifying subfield of IR, with strong ambitions to apply and develop theory, as well as to provide critical analyses of threat construction and policy responses. While globalization and its discontents are clearly in focus within contemporary security studies, it remains surprisingly silent on cybersecurity, with a few exceptions that prove the rule (Carr 2011; Dunn Cavelty 2008, 2013; Eriksson and Giacomello 2007).

Turning to the question of what the “big three” IR paradigms are able to say about cybersecurity, the following observations were made. First, Realism treats cybersecurity mainly as a force-multiplier in a traditional state-centric perspective. The rise of nonstate authority in cybersecurity, and the perforation of sovereign boundaries in cyberattacks, is given little or no attention in Realist analyses.

Second, Liberalism tends to avoid security issues, but when some of its inherent idealism and fear of treading on Realist ground (security studies) is removed, Liberalism helps explain the complexity of power and security in a digital age. Specifically, Liberal theory sheds light on both the rise of private authority in security and crisis management (including cybersecurity), and on the merging of the civil and military spheres, which is particularly evident regarding cybersecurity. Finally, Constructivism, the newest of the "big three", and today a very influential paradigm in IR, is particularly useful for shedding light on the "virtuality" of security in the digital age, i.e. significance of text, images and symbols for security in the digital age. Web defacing, for example, is an explicit attack on symbols—the online equivalent of flag burning. Moreover, constructivist analysis can explain the role of language in digital-age security, including the meaning and function of rhetoric such as "cyberwar" and "cyberterrorism". In addition, if war and terror in cyberspace is mainly about a battle of hearts and minds, then Constructivism has a lot to contribute, with its focus on ideas and identity in world politics.

While nearly a decade has passed since we published our review (Eriksson and Giacomello 2006, 2007), the literature which explicitly links IR and digital-age security remains small. Nevertheless, noteworthy contributions have been made, and some research trends can be observed. As noted above, studies of threat construction/securitization, arguably one of the major Constructivist contributions to IR has emerged in cybersecurity studies (Dunn Cavelty 2008, 2013: 106; Eriksson 2001; Hansen and Nissenbaum 2009). This particular field of research contains case studies of how particular states (and other actors) have put cybersecurity on policy agendas, and in so doing used various types of threat frames, mainly by invoking fear-mongering analogies from the "offline" world—from "viruses" and "worms" to "war" and "terror". These are problems that imply loss of bits, bytes and money, but rarely death and destruction, at least not in any direct sense.

Certain cyber-events have gained an incredible amount of attention, and spurred debates on "cyber war". Two of the most discussed events are the 2007 cyber-attacks on Estonia in 2007, and the 2010 Stuxnet worm, which attacked the Iranian nuclear industry. The Estonian cyber-attacks, which were a reaction to the Estonian decision to move a Soviet soldier statue from a park in central Tallinn, were a great number of so-called distributed denial of service attacks (DDoS), which made it impossible to access the online systems of many branches of the Estonian government, as well as banks and news media organizations. The Stuxnet worm was an advanced malware specifically designed to cripple certain industrial operating systems developed by Siemens, which were used by Iranian nuclear power plants. Both cases illustrate the near-impossibility of identifying attackers, although rumours quickly spread that Russian and US actors, respectively, were behind the attacks in Estonia and Iran. The few IR scholars who have studied these and other cyber-attacks have largely focused on threat discourses and political implications, arguing that the actual physical implications were quite limited, not warranting the over-dramatized rhetoric and political actions that followed (cf. Dunn Cavelty 2012; Hansen and Nissenbaum 2009). Some observers, however, have adopted a more alarmist approach, calling for strengthened international collaboration against cyber war (Herzog 2011).

A major new constructivist contribution to IR and digital-age security studies is Madeleine Carr's PhD dissertation from the Australian National University (Carr 2011). By applying a social constructivist perspective, Carr convincingly shows how both material and social forms of power are essential in conceptualizing power and security in the information age. This perspective allows for an understanding of how politicians give voice to a variety of often conflicting conceptualizations of the Internet, and, perhaps most importantly, how politicians' conceptualizations shape technology. Thus, technological (and historical) determinism is avoided; ideas and actors are considered to have "agency" shaping Internet politics. This does not only yield better insight than other perspectives into why the Internet has developed as it has, but also implies a politically promising perspective, emphasizing the ability and thus also the responsibility of actors. The future might largely be shaped by technological developments, but people—including elected politicians—play a much greater role in shaping technology, and thus the future, than commonly perceived. This observation, which is very convincingly and clearly presented in Carr's dissertation, deserves to get a deep and widespread impact within academia as well as among "practitioners" of Internet politics and governance.

Another trend in IR and digital security studies is the "Liberal discovery" of the rise of private authority. While there is still a clash of more traditional state-centric views and the Liberal focus on nonstate actors and the fragmentation of power (see Eriksson and Giacomello 2009; Mueller and Schmidt 2013), the tendency seems to be to go beyond the stalemated debate about the obstinacy or obsoleteness of the state. Focus is now more on interactions between public and private actors, and on the increasingly blurred boundaries between these domains, domestically and internationally. Significant contributions have been made on public-private partnerships in cybersecurity and in critical infrastructure more generally, although observations are mixed: some observe the growing power of industry in terms of development, ownership and increasingly policy and control (Dunn Cavelty and Suter 2009), while others observe that there is much talk about public-private partnership, but not as much private action as especially many politicians would like to see (Carr 2011). In Carr's case study of US cybersecurity policy, she shows how US politicians have embraced the view that the government has neither the capacity nor the responsibility to provide cybersecurity. US politicians believe that private actors can and must take on this role. Carr shows that this image of a lack of (governmental) power is not only rejected by private actors, but has also become a self-fulfilling prophecy: perceptions of insecurity in cyberspace prevail, and the much-wanted public-private partnerships have not materialized (Carr 2011).

Apart from these trends and observation, and despite noteworthy exceptions, the gap remains between on the one hand the large, hardnosed empirical, atheoretical and technologically oriented cybersecurity literature, and on the other the much smaller, theoretically oriented IR scholarship. The attention to cybersecurity in IR studies is likely to grow, however, given the growing militarization and noteworthy move from the low-political realm of technical issues to the high-politics of foreign policy (Dunn Cavelty 2012). Major new research programs, and perhaps a new book series on IR and digital-age-security, are warranted.

3 The Success of Content Analysis

In her 2011 "Presidential Address" at the annual convention of the International Studies Association, Professor Beth A. Simmons made a passionate and, at the same time, realistic, evaluation of what the global information age means for International Studies (Simmons 2011, 2013). Her main argument was that scholars will have to adapt and learn how to take advantage of the huge amount of information available online to support their research findings. Her conclusions were quite shareable, of course. But what really caught the authors' attention was her choice of examples for the "future" types of research, which, in our view, are all based on content analysis. Exploiting the opportunity to access and download large bodies of text that can be later scrutinized using keywords and concepts with appropriate software. Content analysis is or could be one of the principal methods of investigation for Constructivist scholars more broadly, and for studies of cyber-security in particular. Hence, allow us to briefly discuss some of the details and technical aspects of content analysis, which we think are warranted to be able to appreciate the value and functions of this set of methods. In the concluding section, we will bring the discussion back to the empirical topic of digital-age security.

Content analysis includes a broad collection of methods, ranging from the examination of text to images. It is mostly a *quantitative* methodology (Weber 1990; Berry and Kogan 2010; Leetaru 2011a, b; Statsoft 2013), but its flexibility make it suitable to qualitative applications as narrative or discourse analysis (Phillips and Hardy 2002; Renkema 2009). It is then hardly surprising that its application is so widespread in the social sciences (Phillips and Hardy 2002; Klotz and Lynch 2007). Furthermore, software developed specifically for content analysis has undoubtedly made research less time-consuming, especially regarding data-gathering. As an established research methodology, content analysis is not limited to the relatively simple examination of the "frames" in which actors operate, but it can be apply to causal investigation such as, for example, why cybercrime is underestimated by policy-makers (Bueno et al. 2011).[3] In another instance, scholars have applied content analysis also to study epistemic communities (Klotz and Lynch 2007: 52–53), while hackers definitively see themselves as an "epistemic community". Last but not least, the study of "culture", so central for Constructivist scholars as explanatory factor (e.g. Katzenstein 1996), has been

[3] For example, a quick analysis of policy-makers' public speeches on crime and security world-wide would reveal how relatively few times that term appears. A skeptical view of this fact can be found, for instance, in Florencio and Herley (2012). Thus, different hypotheses could be tested to explain which view is most correct.

transformed, by merging content analysis and computing, into "Culturnomics 2.0",[4] which uses large amount of online news to forecast "future economic events" (Leetaru 2011b).[5]

The main critique against these methodological tools is about lack of attention to the general context in which verbal interaction occurs. So, it all depends on *how* content analysis is conducted. Content analysis is a widely used method for studying political communication and culture (Holsti 1969; Weber 1990; Druckman 2005; Leetaru 2011a; Sjøvaag and Stavelin 2012). A text, whether of a lecture delivered or of a document drafted, is the *material* manifestation of speech. A "step further" is *discourse* analysis, where the discourse is no longer considered just a simple reflection of reality but seen as its essential constituent part. The way a political discourse is constructed carries a great weight, and here the analysis of language provides a better understanding of social reality at large (Phillips and Hardy 2002; Leetaru 2011b).

Generally speaking, content analysis research can be structured on three levels of analysis: (a) frequency of categories, (b) key-words in context and (c) word frequency list. The first level (frequency of categories) is about how many times a pre-defined category appears in a text. Categories, such as "censorship", "cyberwar" or "privacy", are also elements of wider frames. Frames are general codes through which the actors interpret complex issues. These interpretive schemes simplify external reality through a selective process in which the actors emphasize certain aspects rather than others (Snow and Benford 1992). The elements on which the actors put their focus are the most salient in the communication (Entman 1993). Different frames represent alternative ways to address a theme (for instance, "hacktivism") and, consequently, the policy-making process can be viewed as a struggle among different frames (Jordan and Taylor 2004).

The second level of analysis, that is the KWIC, or key-words in context, illustrates the extracted piece of text where the term is inserted for a length of three lines, allowing a better understanding of its meaning. Indeed, a word isolated from its context may cause some mistakes. For instance, identical words may take up completely different meanings ("controlling hate speech online" versus "controlling freedom of speech online") and some terms could be simply negations ("this is not a cyberwar"). Thus, interpretation and selection are essential undertakings for guarantying effectiveness of content analysis tools. Finally, the third level of analysis is the wordlist: the frequency of words used in the text. It provides a list of the terms beyond those included in the vocabulary and can provide a large spectrum of information and descriptive statistics that could be used, for example, in forensic analysis of cybercrime (Bueno et al. 2011).

The validity of the classification is obviously linked to the accuracy of correspondence between categories, words and concepts. The software has algorithms

---

[4] The original "Culturnomics" was limited only to the analysis of books in digital archives (Leetaru 2011b).

[5] More specifically, Leetaru (2011b) explores the "news tone and geographic location".

and mechanisms (such as the identification of additional terms) that help to change the vocabulary without altering the final outcome. The construction of the "coding scheme" moves from two elements: the research question and the texts on which the analysis will be performed. The use of dictionaries of course helps to provide consistency across the categories through synonyms and terms related to the shared theme. Finally, collection and selection of documents are the preliminary tasks before applying the software to the text according to the conceptual categories that have been created through a vocabulary of relevant terms.

All in all, as Weber (1990: 13) points out, "there is no simple right way to do content analysis". The methods must be defined according to the aims and problems of research. The content analysis can be used for many purposes: to identify the intention of those who communicate a message, to reflect the cultural patterns of groups and institutions, companies or to reveal the focus of certain actors. According to Holsti, the content analysis is defined as: "technique for making inferences by systematically and objectively identifying characteristics of specified messages" (1969: 25). Clearly, the amount of texts, whether official documents, speeches, declarations or personal letters, which could be analyzed and "coded" by a single scholar or even a team of researchers is relatively limited. Boredom of scrutinizing stacks and stacks of texts may feed sloppiness to a point that the overall value of an entire research might be compromised. The watershed in content analysis, as in many other fields, was the invention of the (personal) computer.

Computers, with their capacity to speed up, and avoid human errors in dull routine tasks have been applied with amazing efficiency to content analysis. Thanks to the Moore's Law,[6] the faster their processors become, the speedier such analytical processes will be. The merging of content analysis and computers has allowed scholars to exploit "advances in computational techniques to perform innovative, but computationally complex, forms of analysis on large warehouses of textual data" (Leetaru 2011a: 6). The power and capacity of computers and computer networks have evolved considerably, but their primary mission remains the same today. Computers were built to perform routine activities more efficiently than humans—who are not as efficient or reliable when it comes to repetitive and menial jobs. Early on, computers were linked together in networks, multiplying by several degrees of magnitude their resourcefulness and making them even more indispensable for our societies to function. Furthermore, these networked computers have allowed to remotely controlling processes over long distances, making them core elements of globalization.[7]

Upon the networked infrastructures so vital for our economies and societies, a World Wide Web of "contents", full of images but also of *words*, has been

[6] According to this "law", the number of transistors on integrated circuits doubles approximately every 2 years and more processors mean more computing power. The period was further "reduced" to 18 months.

[7] For example the SCADA (system control and data acquisition) controllers that allow operators to remotely check the status and functioning of various production or distribution processes.

superimposed. The importance of this relationship, between content and the Web, has long been recognized by scholars (Weare and Lin 2000; Srivastava and Sahami 2009; Liu 2011; Eriksson and Giacomello 2013; Halvorson and Rach 2012). The instances of users launching a "search" to find some kind of "content" on the Internet (and the Web) are simply so numerous that most people (we included) do not pay any notice. After turning a computer on, one of the first things a user has to do, most of the times especially if it connects to a network, is to type in a password, so that the relevant software application runs a search to find if the characters and symbols entered correspond (for position and format) or not to those stored in the password file. Google's search algorithms too rely on keywords to provide the user with guesses what he or she may look for. All search engines follow this pattern.

Today, most people would hardly stop to think (or know) that "the Web" was invented by Tim Berners-Lee as an environment dedicated to hypertext. This is a tool that allows the user to move from one concept to another that is logically "linked" to the former. Unsurprisingly, the World-Wide Web developer termed it, from the start, the "Information Universe" (Berners-Lee et al. 1992). The Web, email, blogs and other social media are now integral and indispensable elements of many people's lives as well as their main source of "content". This online content can be thoroughly analyzed to understand social phenomena (Liu 2011; Ajith et al. 2010; Weare and Lin 2000) or even for forecasting (Leetaru 2011b), which is seldom undertaken in the social sciences. Google is perhaps the most innovative enterprise when it comes to employing big data in order to predict social behavior or, for instance, outbreaks of flu (Cukier and Mayer-Schoenberger 2013).

Software programs for content analysis, such as *Hamlet II* or *Words in Context* (as many others)[8] allow the researcher to quickly assess the frequency of terms and their location within the text. Programs such as those mentioned and other search text files for words or categories in a given vocabulary list, count their joint frequencies within any specified context unit, within sentences, or as collocations within a given span of words and then provide the relevant figures.[9] Words are classified in the same categories because they have similar meaning—a "reduction" and classification process called "tagging" (Weber 1990). To operationalize abstract concepts such as "national interests" or "risk aversion" requires building a category, which is methodologically complex because it entails capturing the essence of the social phenomenon (Druckman 2005). In that sense, it is crucial to reach the right trade-off between "reliability" and "validity" of the categories (Weber 1990; Trochim 1999; Druckman 2005). In the end, the level of internal consistency and accuracy depends on the ambiguity in the meaning of words as well as by the choices made by the researcher in defining categories and coding rules.

[8] There are even good open source applications for content analysis such as KH Coder. More can be easily obtained from http://www.sourceforge.net; a Web site dedicated exclusively to Open Source software Including applications for forensic analysis of firewalls and the like.

[9] See the Web site of Hamlet II for example, available at http://apb.newmdsx.com/hamlet2.html.

The use of software can certainly avoid many common methodological problems allowing automatic application of coding rules, and ensuring accuracy for comparison and reproducibility (Weber 1990). Computers do not generate cognitive differences, as they would occur with "human coders". But the increasing simplicity, immediacy and flexibility of the available content-analysis software today can only facilitate to the goals of researching cyberspace and, more specifically, cybersecurity (Lee et al. 2003; Pera and Ng 2009; Srivastava and Sahami 2009; Liu 2011; Eriksson and Giacomello 2013; Purkait 2012).[10] For example, technical security reports rely on text analysis and keywords (Bueno et al. 2011), "brute force" attacks on trying all of the words of entire vocabularies (in different languages) to guess passwords and there are plenty of other examples of how "text" still dominates the Web, despite all references to the visual experience of the Web. Indeed, whether it is applied to implement the filtering of Web information (Lee et al. 2003), to counter phishing (Purkait 2012), to stop Web spamming (Pera and Ng 2009) or to sort through online news (Sjøvaag and Stavelin 2012), computerized content-analysis is the tool of choice of scholars and researchers working on cyberspace and cybersecurity.

Text mining will, most likely, be the future for content analysis (Berry and Castellanos 2008; Leetaru 2011a; Halvorson and Rach 2012; StatSoft 2013). Text mining is defined as the process of transforming "unstructured (textual) information" into "meaningful numeric indices" (StatSoft 2013).[11] The "processed" text (as indexes) could then be more easily manipulated with other quantitative techniques. The vast text warehouses already available on the Web will further grow in size and number and most of them will be available to researchers (Berry and Castellanos 2008; Liu 2011; Halvorson and Rach 2012; Leetaru 2011a, b). More specifically, the indexes obtained via text mining could already be examined or fed into data mining programs for further analysis. In fact, text mining algorithms can process open-ended survey responses, emails, insurance claims, diagnostic interviews that earlier content analysis software found too varied and contrasting to compute simply because the authors of those texts wrote instinctively or without much structure. Mastering the skills of computerized content analysis is clearly become the hallmark of those researchers and scholars who want to bring social science research on cyberspace to a higher level of accuracy and reliability. It is fairly easy to extrapolate what this methodological innovation could entail not only for content analysis but also for Constructivism, as research findings will gain increased robustness. Fittingly, the limitations of Constructivism in applied research were one of the criticisms raised by Klotz and Lynch (2007) as in need of correction.

---

[10] The Web site (http://www.kalevleetaru.com) of Kalev H. Leetaru, a recognized expert in computerized content analysis, offers several examples of that type of application.

[11] A shorter definition could be "turning text into numbers". For more technical details see the "text mining" section in the Statsoft Web site at http://www.statsoft.com/textbook/.

4 Concluding Remarks on Cybersecurity and Content Analysis

Professor James Forsyth (2013: 109) aptly observed that "the extraordinary puzzle of cyberspace can be made less daunting by conceptualizing it in the ordinary language of international politics". We agree, but *cum grano salis*. In fact the "traditional" international politics language is wholly unequipped to deal with the "individual" level of analysis, which is (along with the private sector, as cybersecurity stake-holder) the great novelty of cyberspace. To better incorporate this innovation in the IR and security language we turn to Constructivism that is, among the various IR theories, the best equipped to this task; and to content analysis as principal methodology to conduct research in cyberspace and cybersecurity. In this chapter we argued that Constructivist theory is particularly useful for uncovering the text-based meanings, identities, symbols and ideas, which constitute the "virtual" world of cybersecurity.

Let us briefly comment on the (cybersecurity) functions of content analysis, which are many. What the different tools of content analysis have in common is that they are all about selectively and systematically processing text. Beyond this technical level, the functions of content analysis are plentiful. It is noteworthy that content analysis is not simply a scientific set of methods for textual analysis, but should be understood as a much broader approach to quantitatively structuring and studying text, far beyond scientific analysis. Whether explicitly or implicitly, content analysis is applied in, for example, cyber-attacks (e.g. spamming, identity thefts, computer break-ins); intelligence and surveillance; crime investigation; market research; academic research; journalistic investigation; and for personal purposes (hobbies, travel planning, etc.).

Cyber-attacks such as denial-of service attacks and break-ins into computer networks are seldom defined as content analysis, but they eventually are forms of content analysis (Amoroso 2011; Eriksson and Giacomello 2013). Spamming—unsolicited e-mails messages sent out in huge numbers—is a main type of "cyberplague". By letting the search engine look for e-mail addresses including the @ sign, spammers are conducting content analysis. Hackers run "password-guessing" software that typically rely on dictionaries. This is why people are recommended to use complicated passwords, which are not found in any dictionary. Then there are also "hybrid" attacks, which can insert numbers (0–9) and special characters (such as # or ∞). Password guessing is a type content analysis, which, if successful, can give access to e-mail, bank accounts, medical records, military plans, and systems of operation for example airports and power plants. Moreover, intelligence, crime investigation and surveillance increasingly rely on (online) content analysis. These systems use satellites as well as online tools for retrieving communication. Despite the complexity and advanced computing power of such systems, they rely on a basic form of content analysis: keywords.

One may argue that there should be some ethical considerations to make, as proficiency in content analysis may be exploited for the "wrong" reason, that is, to exploit weakness in computer security. Yes, the risk is there and other research methodologies (such as case-studies) do not present the same problems. But this is an even more compelling reason for scholars working on cyberspace and cybersecurity to be well-equipped to grasp this methodology and be able to master it to augment knowledge. Last but not least, while we have mainly advocated the strengths and uses of Constructivism and content analysis, there are obvious limitations as well. In particular, both Constructivism and content analysis tend to focus on the most common and significant patterns, while neglecting the less common, oddities, and arguably the less significant. While Constructivist analyses often focus on dominant discourses—master frames and paradigms—content analysis has a similar in effect in that it is all about frequency (word-count). This is indeed a limitation in terms of scientific scope: How are we going to understand the less frequent, what may seem odd and obscure, but what might potentially have great significance? This applies not least to Western cybersecurity discourse, which tend to focus on the usual suspects (China and Russia): Who knows what the next cyber-plague will be, from where it will come, and who will have instigated it?

References

Ajith, A., Hassanien, A.-E., & Snášel, V. (Eds.). (2010). *Computational social network analysis: Trends, tools and research advances*. London: Springer.

Amoroso, E. G. (2011). *Cyber attacks: Protecting national infrastructure*. Burlington: Elsevier.

Avey, P. C., Desh, M. C., Long, D. D., Maliniak, D., Peterson, S., & Tierney, M. J. (2012, January/February). The Beltway vs. the Ivory Tower. *Foreign Policy*. http://www.foreignpolicy.com/articles/2012/01/03/the_beltway_vs_the_ivory_tower?wp_login_redirect=0. Accessed January 28, 2014.

Balzacq, T. (Ed.). (2010). *Securitization theory: How security problems emerge and dissolve*. London: Routledge.

Benedikt, M. (Ed.). (1991). *Cyberspace*. Cambridge, MA: The MIT Press.

Berners-Lee, T., Cailliau, R., Groff, J.-F., & Pollermann, B. (1992). World-wide web: The information universe. *Internet Research, 2*(1), 52–58.

Berry, M. W., & Castellanos, M. (Eds.). (2008). *Survey of text mining II: Clustering, classification, and retrieval*. London: Springer.

Berry, M. W., & Kogan, J. (2010). *Text mining: Applications and theory*. Chichester, UK: Wiley.

Bueno, P., Dirro, T., Greve, P., Kashyap, R., Marcus, D., Paget, F., et al. (2011). *McAfee threats report: First quarter 2011*. Santa Clara, CA: McAfee Lab.

Buzan, B., Wæver, O., & de Wilde, J. (1998). *Security: A new framework for analysis*. London: Lynne Rienner.

Carr, M. (2011). *The irony of the information age: US power and the internet in international relations*. PhD dissertation, Australian National University, Canberra.

Castells, M. (1996). *The rise of the network society* (The information age: Economy, society and culture, Vol. 1). Malden, MA: Blackwell.

Castells, M. (2000). *Information age: Rise of the network society* (2nd ed., Vol. 1). Malden, MA: Blackwell.

Clarke, R. A., & Knake, R. K. (2010). *Cyberwar: The next threat to national security and what to do about it*. New York, NY: HarperCollins.

Conway, M. (2002). Reality bytes: Cyberterrorism and terrorist 'use' of the internet. *First Monday*, 7(11). http://doras.dcu.ie/498/1/first_mon_7_11_2002.pdf. Accessed January 28, 2014.

Cukier, K., & Mayer-Schoenberger, V. (2013). The rise of big data: How it's changing the way we think about the world. *Foreign Affairs, May/June, 92*(3), 28–40.

Deibert, R. (2013). *Black code: Inside the battle for cyberspace*. Toronto: Signal.

Druckman, D. (2005). *Doing research: Methods of inquiry for conflict analysis*. London: Sage.

Dunn Cavelty, M. (2008). Cyber-terror—looming threat or phantom menace? The framing of the US cyber-threat debate. *Journal of Information Technology and Politics, 4*(1), 19–36.

Dunn Cavelty, M. (2012). The militarization of cyber security as a source of global tension. *Social Science Research Network* (SSRN). http://papers.ssrn.com/sol3/papers.cfm?abstract_id=2007043. Accessed January 28, 2014.

Dunn Cavelty, M. (2013). From cyber-bombs to political fallout: Threat representation with an impact in the cyber-security discourse. *International Studies Review, 15*(1), 105–122.

Dunn Cavelty, M., & Suter, M. (2009). Public–private partnerships are no silver bullet: An expanded governance model for critical infrastructure protection. *International Journal of Critical Infrastructure Protection, 2*(4), 179–187.

Entman, R. M. (1993). Framing: Toward clarification of a fractured paradigm. *Journal of Communication, 43*(4), 51–58.

Eriksson, J. (Ed.). (2001). *Threat politics: New perspectives on security, risk and crisis management*. Aldershot: Ashgate.

Eriksson, J., & Giacomello, G. (2006). The information revolution, security and international relations: (IR)relevant theory? *International Political Science Review, 27*(3), 221–244.

Eriksson, J., & Giacomello, G. (Eds.). (2007). *International relations and security in the digital age*. London: Routledge.

Eriksson, J., & Giacomello, G. (2009). Who controls the internet? Beyond the obstinacy and obsolescence of the state. *International Studies Review, 11*(1), 205–230.

Eriksson, J., & Giacomello, G. (2013). Content analysis in the digital age: Tools, functions and implications for security. In J. Krueger, B. Nickolay, & S. Gaycken (Eds.), *The secure information society: Ethical, legal and political challenges* (pp. 137–148). Dordrecht: Springer.

Florencio, D., & Herley, C. (2012, April 14). The cybercrime wave that wasn't. *The New York Times*, SR5.

Forsyth, J. W., Jr. (2013). What great powers make it: International order and the logic of cooperation in cyberspace. *Strategic Studies Quarterly, 7*(1), 93–113.

Giacomello, G. (2004). Bangs for the buck: A cost-benefit analysis of cyberterrorism. *Studies in Conflict and Terrorism, 27*(5), 387–408.

Giacomello, G. (2005). *National governments and control of the internet: A digital challenge*. London: Routledge.

Gibson, W. (1984). *Neuromancer*. New York, NY: Ace Books.

Guzzini, S., & Leander, A. (Eds.). (2006). *Constructivism and international relations: Alexander Wendt and his critics*. London: Routledge.

Halvorson, K., & Rach, M. (2012). *Content strategy for the web* (2nd ed.). Berkeley, CA: New Riders.

Hansen, L., & Nissenbaum, H. (2009). Digital disaster, cyber-security, and the Copenhagen school. *International Studies Quarterly, 53*(4), 1155–1175.

Herzog, S. (2011). Revisiting the Estonian cyber attacks: Digital threats and multinational responses. *Journal of Strategic Security, 4*(2), 49–60.

Holsti, O. R. (1969). *Content analysis for the social sciences and humanities*. Reading, MA: Addison-Wesley.

Huddy, L., Feldman, S., & Weber, C. (2007). The political consequences of perceived threat and felt insecurity. *The Annals of the American Academy of Political and Social Science, 614*, 131–153.

Jordan, T., & Taylor, P. A. (2004). *Hacktivism and cyberwars: Rebels with a cause?* London: Routledge.

Katzenstein, P. J. (Ed.). (1996). *The culture of national security*. New York, NY: Columbia University Press.

Kier, E. (1997). *Imagining war: French and British military doctrines between the wars.* Princeton, NJ: Princeton University Press.

Klotz, A., & Lynch, C. (2007). *Strategies for research in constructivist international relations.* New York, NY: M. E. Sharp.

Lee, P. Y., Hui, S. C., & Fong, A. C. M. (2003). A structural and content-based analysis for web filtering. *Internet Research, 13*(1), 27–37.

Leetaru, K. (2011a). *Data mining methods for the content analyst: An introduction to the computational analysis of content.* London: Routledge.

Leetaru, K. (2011b). Culturomics 2.0: Forecasting large-scale human behavior using global news media tone in time and space. *First Monday, 16*(9). http://firstmonday.org/ojs/index.php/fm/article/view/3663/3040. Accessed January 29, 2014.

Libicki, M. (2007). *Conquest in cyberspace: National security and information warfare.* Cambridge: Cambridge University Press.

Liu, B. (2011). *Web data mining: Exploring hyperlinks, contents and usage data* (2nd ed.). Berlin: Springer.

Mowlana, H. (1997). *Global information and world communication: New frontiers in international relations.* London: Sage.

Mueller, M. A., & Schmidt, A. (2013). Internet security and networked governance in international relations. *International Studies Review, 15*(1), 89–104.

Nye, J. S. (2011, February 27). Cyberspace wars. *International Herald Tribune.* http://belfercenter.hks.harvard.edu/publication/20794/cyberspace_wars.html. Accessed January 28, 2014.

Pera, M. S., & Ng, Y.-K. (2009). A structural, content-similarity measure for detecting spam documents on the web. *International Journal of Web Information Systems, 5*(4), 431–464.

Phillips, N., & Hardy, C. (2002). *Discourse analysis.* Thousand Oaks, CA: Sage.

Purkait, S. (2012). Phishing counter measures and their effectiveness – literature review. *Information Management and Computer Security, 20*(5), 382–420.

Renkema, J. (Ed.). (2009). *Discourse, of course: An overview of research in discourse studies.* Amsterdam: John Benjamins.

Simmons, B. A. (2011). International studies in the global information age. *International Studies Quarterly, 55*(3), 589–599.

Simmons, B. A. (2013). International relations in the information age. *International Studies Review, 15*(1), 1–4.

Singh, J. P. (2014). The power of large-scale interactions through information technologies and changes in cultural identity politics. In M. Mayer, M. Carpes, & R. Knoblich (Eds.), *The global politics of science and technology* (Vol. 2). Heidelberg: Springer.

Sjøvaag, H., & Stavelin, E. (2012). Web media and the quantitative content analysis: Methodological challenges in measuring online news content. *Convergence: The International Journal of Research into New Media Technologies, 18*(2), 215–229.

Snow, A. D., & Benford, R. D. (1992). Master frames and cycles of protest. In A. D. Morris & C. McClurg Mueller (Eds.), *Frontiers in social movement theory* (pp. 133–155). New Haven, CT: Yale University Press.

Srivastava, A. N., & Sahami, M. (Eds.). (2009). *Text mining: Classification, clustering, and applications.* London: CRC Press.

StatSoft, Inc. (2013). *Electronic statistics textbook.* Tulsa, OK: StatSoft. http://www.statsoft.com/textbook. Accessed January 28, 2014.

Trochim, W. M. (1999). *The research methods knowledge base* (2nd ed.). Ithaca, NY: Cornell University Custom Publishing.

Waltz, K. (1959). *Man, the state and war.* New York, NY: Columbia University Press.

Weare, C., & Lin, W.-Y. (2000). Content analysis of the world wide web: Opportunities and challenges. *Social Science Computer Review, 18*(3), 272–292.

Weber, R. P. (1990). *Basic content analysis.* London: Sage.

Weimann, G. (2005). Cyberterrorism: The sum of all fears? *Studies in Conflict and Terrorism, 28*(2), 129–149.

Information Technology Challenges for Energy and Environmental Policy Research

Sean S. Costigan and Erica M. Dingman

Abstract Advancements in the fields of computer science and the culture of transparency run contrary to the prevailing standards of weak data sharing and information silos. Nowhere is this more evident than in the field of climate change policy as makers and researchers need better access to data. While complexity in climate science will continue to challenge policy makers, journalists and researchers, basic steps should be undertaken by all organizations operating in these fields to improve access to their raw data, to share that same data or the methods that they used to collect it, and to extend services to the international community. Recognition of the problem is a good start, but pressure is mounting for change in the present policy analysis and creation system and time is at a premium, in particular for dynamically changing environments like the Arctic. The chapter closes with recommendations for using information technology and big data relevant for global energy and environmental policy and international relations at large.

Keywords Arctic • Big data • Arctic Council • Transparency sharing • Climate change • Information technology • Information literacy

1 Introduction

We have a practical problem: Data is necessary for policy decision-making, but data alone is never sufficient. In the field of environmental policy, successful outcomes require both excellent data and detailed contextualization for a systems-thinking approach to adequately project and understand risk. Furthermore,

S.S. Costigan (✉)
The New School, New York, United States
e-mail: sean_costigan@post.harvard.edu

E.M. Dingman
World Policy Institute, New York, United States
e-mail: emdingman@gmail.com

M. Mayer et al. (eds.), *The Global Politics of Science and Technology - Vol. 2*, Global Power Shift, DOI 10.1007/978-3-642-55010-2_13,

good policy depends on harnessing today's prodigious intellectual and technological capacity through identifying, collecting, extrapolating and sharing data from disparate sources. Yet even with present computing and people power, enormous challenges are the norm: Empirical data and crucial background information remain thinly spread over a wide range of knowledge sources or are to be found in deep and inaccessible pools.

Though widely criticized for the estimate, in 2010 Google's Eric Schmidt stated that humans create every 2 days as much data as was created up to 2003 (Siegler 2010). In 2011, Mashable created an annual estimate and accompanying infographic that depicted the human creation of 1.8 ZB of data that year (Catone 2011). In the past decade, computer and cognitive scientists have made great strides and have also created emergent fields of inquiry and production, such as big data where emphasis is now shifting from human-made information to machine-generated information. Despite such focused effort, a core question remains how to take all this "data dust" and turn it into gold dust of the sort that would make systems thinking possible. As we will show, the filter failure challenge has specific relevance in the realm of international relations, security and the environment, in which poorly conceived or executed responses may lead to potentially severe cascading consequences locally and globally.

In addition to filter failure and instances of "too much information" there is another side of the coin: a paucity of access to valuable information since so much remains siloed in inaccessible or costly databases. Data producers have a mixed record of data sharing, fearing negative economic effects, political disagreements and even, at the most basic level, the affordability of sharing. To give one a sense of how much data exists in databases that are not easily accessed from the so-called "surface web", the 2010 Independent Oracle Users Group (IOUG) Database Growth Survey, conducted in July-August 2010 put the estimate at 1 ZB of data existing in databases, and that 16 % of organizations surveyed reported a data growth rate in excess of 50 % annually (Hamby 2011; McKendrick 2010).

On the technological end, there are concrete ways to close the gap between data creators and consumers through the novel use of existing technology and well-defined user design. To that end, systems should be envisioned that would help increase knowledge exchange and creation through contextualization and visualization of complex source data for multilingual access and presentation. The requisite raw information pool would be substantially deepened by expanding data exchange partnerships, alongside which an international project would aim to create or refine a standard for how such data should made sustainably accessible. In this space public-private partnerships are often one good way to proceed, being Google Flu Trends the most well-known example (Carmody 2013). By extension, a portal that provides guided, "smart" usage together with difficult to access data would constitute an enormously valuable contribution to good decision-making and research. It would not only become a trustworthy source of data on the Internet, but also create knowledge and raise awareness of issues of data relevance, validity and reliability.

On the educational side of the equation, training in information literacy remains woefully inadequate for proper use of data about international relations, security and environmental concerns. In North America, Masters degree programs generally require methodology courses that aim to improve scholarly techniques, but rarely do these courses go into detail on information literacy, which leads to an increasing gap between those who can use data effectively and those who can barely use information technologies. To make matters more pressing, libraries (where such training most often takes place) are under increasing budgetary pressure, leaving today's students on their own (Kelley 2012). Yet given the interconnectedness of environmental, economic and security concerns, the proper use of data and statistics is vital.

2 Databases, Policy and Climate Change Research

Changes to the natural environment have revealed infrastructure frailties and potential supply chain routes once thought impassable. These changes may have exigent implications for energy security and interstate relations, possibly altering our present notions of winners and losers. At present, political elites appear locked in a discourse based on traditional understanding of geography and security. But to what degree are scientists, academics and decision-makers able to share their work in one field with another person for the purpose of comparison or proposed action? Is data made available in a coherent and contextualized manner or are silos the norm? Are academic reports, scientific and government data easily accessible to journalists, other media outlets and the public at large?

We would propose that the current system of relatively weak sharing owes to several concerns, including inherent complexity in scientific research, "filter failure" (Trapani 2008), crisis fatigue (Nettler 2013), professional incentive structures, and a reliance on a "business as usual" approach. However, these are not exclusively academic concerns, as climate change is a field of study where the "known unknowns" likely have considerable implications for human security writ large and so require more concerted and novel efforts. Such efforts should have already started, as the developed world has perhaps only just begun to experience the effects of climate change, while Small Island Nations and indigenous communities in locations such as the Arctic are already at a heightened level of concern as most of their socioeconomic activity takes place along coastlines.

Concurrently, knowledge gaps may be the result of long term strategic planning subverted by near term decision-making, potentially because of inabilities to plot futures based on shared data. Denial may well be an evolutionary adaptation (Varki 2009). In any event, people often make good short-term decisions that have unknown long-term consequences. To put this in perspective using an example from another field of study, the apparent upsurge of cyber attacks has created considerable and costly challenges for governments and financial institutions, as well as to business and to the general public, all of which was unwittingly unleashed by the advantageous invention of interlinked and ubiquitous computing. Likewise, the impacts and costs of climate change to infrastructures and systems that society has come to depend upon may elude even the best of strategic planning. At a minimum, then, actions related to the environment must be based on accessible, timely and relevant data.

3 Government Initiatives and Big Data

With Web 2.0 now over a decade old, a generation of web users has grown up with the idea of accessing databases and mashing up content in new ways, on their own. Companies have developed business models that are based, at least in part, on allowing people to extend their data into new and novel creations through APIs and other modes of access. Some successful examples of innovation are ESPN and Cisco, with the former seeing 50 companies using their data in new ways and the latter realizing monetary gains through improvements in customer communications made possible by opening up their APIs (Kamenetska 2013). At present, there are some 13,000 open APIs available, creating an "API economy" and an expectation of greater utility (McKendrick 2013).

In the United States, the creation of Data.gov in May 2009 was heralded as a major improvement in transparency and data access. The purpose of Data.gov is to "increase public access to high value, machine readable datasets generated by the Executive Branch of the Federal Government" (Data.gov). Some critics were concerned that there was no coherent framework to the data, making it a "data dumping ground," but the crucial point that data.gov underscored is that space must be created to share data and access should be granted to data to improve outcomes. A similar project was undertaken in the United Kingdom, under the guidance of Tim Berners-Lee (2010). In both cases, still better systems could be built with tools to compare, analyze and grow new ideas from the outset. In terms of the UK efforts, Sean Maguire (2011) argues that despite the UK government's stated aim for global leadership in data transparency, data "remains frustratingly opaque."

In 2012, the Obama administration announced the creation of the Big Data Research and Development Initiative, which with $200 million USD in funding was to "to extract knowledge and insights from large and complex collections of digital data" through notable programs (The White House 2012) that address big data throughout the federal government. It is noteworthy that well prior to the inception of data.gov and the Big Data Research and Development Initiative, excellent tools were crafted to take advantage of government created and administered data with end-user researchers in mind.

The combination of continual reductions in costs of storage, computing power increases and the vast collection of data in most domains has enabled an emerging science of Big Data. If used effectively, big data can significantly contribute to organizational productivity and delivery of services in both public and private sectors (Yiu 2012; Bollier and Firestone 2010). Scientific discovery, once reliant on a small sampling of data now benefits by the technological power of big data collection (Bollier and Firestone 2010; Marks 2012). Government productivity and delivery of services have already improved as a result of big data (Yiu 2012; Maguire 2011). Visualizations created as a result of big data have greatly improved the interaction between the government and private sectors with that of the consumer (Bollier and Firestone 2010; Yiu 2012).

However, numerous hurdles remain. Analysis shows that government and society are yet to resolve legal and moral questions, particularly regarding questions of personal privacy (Cate 2008; Cukier and Mayer-Schoenberger 2013). Fred Cate (2008) suggests that because of the "absence of a legal regime governing data mining . . . fuel[ing] privacy concerns;" and government officials lack the criteria by which to judge "acceptable conduct" (Cate 2008: 437). Another primary concern is that data quantity does not necessarily equate to the quality of the final product. Data without context or understanding creates its own problems and vulnerabilities, potentially leading to erroneous conclusions or the spread of misinformation. Additionally, data may just be descriptive. Kenneth Cukier and Victor Mayer-Schoenberger (2013) reason that big data may provide information on 'what' happened but that information will not necessarily tell us 'why' something occurred (Cukier and Mayer-Schoenberger 2013: 40).

The techniques being created today are set to fundamentally reshape many fields of inquiry, not just in the sciences, and we are at the very early stages of understanding the transformative power of big data. Rufus Pollock, co-founder and director of Britain's Open Knowledge Foundation states: "We are still at the beginning. We are getting the data. Now it's about building the tools and communities to use it" (Maguire 2011: 525; citing Pollock).

4 Information and the Case of the Arctic

While the Arctic was largely ignored until recently, the region is now an epicenter of activity marked by dramatic climate change, natural resource development, increasing commercial and leisure marine traffic, which is both affected by and affects governments and local indigenous communities, among others. Research into the present day challenges and changes in the Arctic has been particularly affected by data gaps and information silos. Mounting regional and international interest in the Arctic has resulted in considerable academic and scientific research seeking to better explain a region undergoing extraordinary environmental and socio-economic change. In tandem, political and commercial interests from places as distant as India and China have intensified the demand for cohesive data management. But because such information emanates from disparate sources—where it often remains cloistered—the processes related to analysis and decision-making may be predisposed to inaccurate assessments of present and future challenges. Integrated data that is further contextualized is more likely to reflect the full spectrum of issues and perspectives, as an holistic approach to Arctic challenges has not only regional but ultimately global value.

In far-off locations beyond the Arctic, development and environmental change in the Arctic is compelling non-Arctic states to consider the broad spectrum of security implications to their own national interests. Melting sea ice will continue to affect sea level rise potentially eroding economic, environmental and societal security. It will also affect industrial development which is dependent upon reliable and timely cargo shipments; and it will affect the routes through which ships pass and the related economies of ports of call. In the final analysis it affects people and national political systems worldwide.

These concerns call into question the selective application of knowledge and the gaps between data creators and those who may need to act on data. In the Arctic specifically, well-funded economic interests such as commercial shipping and the cruise trade, or hydrocarbon exploitation and mining traditionally outweigh the interests of local indigenous communities and a digital divide is evident (Subramony 2010). Indeed, the challenge of data collection and sharing has played a significant role in the thinking of numerous Arctic researchers. For example, at the 2012 International Polar Year Conference (IPY), data and information accessibility was deemed consequential enough that it warranted a programmed topic area and entailed 20 papers authored by 80 researchers in total, all of whom directly treated the legacy of accessing, sharing and preserving data (IPY 2012).

For example, the title of an IPY 2012 oral presentation given by Yasuhiro Murayama, "The ICSU World Data System Tackles IPY Data Challenges," depicts the extent to which coordination of Arctic data is now considered as a global challenge. With initial funding of 5 years, the ICSU World Data System has "undertaken to address issues raised by the data sets collected during IPY" (Murayama 2012). Established in 2008, the International Council for Science (ICSU), which has a global constituency, advocates for long-term open access to data and information amongst scientists and social scientists. And precisely because polar science is considered to have a global impact, an ICSU statement "Universality of Science in the Polar Regions" states (ICSU 2010):

It is now clear that environmental changes in the polar regions are not only indicators of the effects of human activities elsewhere on Earth, but also that these changes will have profound effects on human society around the world if they continue. Observation, understanding, and sustainable management of the polar regions are therefore in the common interest of all humanity.

While Arctic commercial activity is still relatively minimal, the existing climate trends strongly suggest that it will increase exponentially in the foreseeable future. Insofar as the eight Arctic nations—Canada, Denmark including Greenland and the Faroe Islands, Norway, Iceland, Sweden, Finland, the Russian Federation and United States—strive to gain knowledge pertaining to internal security writ large, long term forecasting is potentially attenuated by scattered data. Yet at present access to such data is extremely limited even though scientists, academics, policy-makers and observers are best served by timely data placed in a framework that is both simple and comprehensive for vastly diverse audiences. Importantly, advancing challenges spawned by climate change in the Arctic can only be understood with complete information associated with many data points, for example: burgeoning shipping routes and cruise line activity, hydrocarbons and mineral exploitation and its implications for northern indigenous peoples as well as governments.

In Towards a Common Strategy for the Arctic: The Nordic Countries Can Lead the Way, Per Augustsson of the Center for Transatlantic Relations states:

In the end, it is not really possible to separate discussions on the Arctic environment from resource extraction, shipping, research, or the situation for indigenous peoples. Energy resources, shipping and security are equally interlinked. Today, therefore, we need to focus not only on a range of different complex Arctic issues, but also on the inter-linkages and 'the big picture.' If we do not, we risk fragmented Arctic policies (Augustsson 2011).

The interconnection between these seemingly disparate categories of study cannot be underestimated. Yet information sources have not yet met the needs of analysis. The often-used phrase of 'an ecosystem approach' to Arctic governance could well be employed to development of a systematized information highway for Arctic analysis. What is desperately needed is a system where information is aggregated into a coherent system, mapping concepts and allowing for mash-ups that make new insights possible and potential actions steps easier for all to see.

5 Data Collection for the Arctic Council

As the primary regional decision-making body, the Arctic Council (AC) was formed in 1996 as an intergovernmental forum that provides a means for promoting cooperation, coordination and interaction amongst Arctic states and the involvement of the Arctic Indigenous Communities on common Arctic issues. To meet this mandate the Council created six working groups each of which is tasked with executing projects directed by Arctic Council Members. Success of a working group project is highly dependent upon the collection of data from a diverse set of sources and locations, although the outcome is likely complicated by methodological inconsistencies among data sets specific to the originating nation or organization. Despite this complexity the Arctic Council has collected a vast amount of data which it shares on the Council's website ArcticData.is.

ArcticData.is was developed by two Council working groups "as a means of allowing data developed through their activities to be accessible in a format and manner to allow them to be easily accessible and useable" (ArcticData: a). As others have found, the working groups noted that data is often hidden in documents, graphs, and maps, making it virtually inaccessible to most people. AC members deemed it essential to streamline the data into one location for future analysis by AC members and outside researchers, which would ultimately benefit from the addition of more knowledge.

Precipitated by extensive datasets such as those created for the Arctic Marine Shipping Assessment and the Oil & Gas Assessment, the ACs initial plan was to collect "different kinds of information [which] could be used more effectively, [and] allow for the development of new data products, and for a more cohesive observation of trends and patterns" (CAFF and PAME). However, despite this desirable accomplishment, input of new data input is slow in coming.

In 2004 rapid and unexpected changes to Arctic sea ice conditions resulted in increased economic activity, prompting Arctic Council Members to initiate a comprehensive study of Arctic marine shipping. The designated working group, Protection of the Arctic Marine Environment (PAME), conducted circumpolar analysis, which was compiled in the 2009 Arctic Marine Shipping Assessment (AMSA). AMSA is an extensive compilation of raw data, maps, informative photos and information associated with historical, legal, environmental, infrastructure and other perspectives.

As expected, information needs associated with marine use were significant. To that end, in 2006 a questionnaire was distributed to each Arctic state requesting information for the year 2004, except when data was insufficient and data from a later year was provided. Responses varied—some countries provided detailed information while others submitted only basic or incomplete information. PAME found inconsistencies. For instance, variations in how states defined each vessel type such as some states defining oil carriers as tankers. Inconsistencies were handled as best possible in the final report (Arctic Council 2009). For the reporting year 2004, AMSA reported that approximately 6,000 vessels (covering all types of vessels) travelled through Arctic waters, of which slightly less than 50 % were fishing vessels. Of the remaining vessel types, most were for community re-supply, tourism and bulk cargo vessels carrying oil, gas and iron ore. However, many vessels made multiple voyages and "probable underreporting bias and obvious data gaps in many areas and vessel types" most likely resulted in an underestimate of the number of Arctic voyages (Arctic Council 2009: 72).

The resulting report made 17 recommendations to promote safety and environmental awareness of current and future Arctic shipping activity. The executive summary stated (Arctic Council 2009):

The result of the AMSA data survey effort produced a comprehensive estimate of how many ships (less naval vessels) operated in the Arctic for a given year. This survey represents an historic capture of information from the Arctic states that can be used as a long-term data-base against which to measure future Arctic marine traffic levels.

As per Council plans, raw data from the final 2009 AMSA report was uploaded to ArcticData.is, which provided a significant baseline. However, since the initial data input no additional data or information reflecting subsequent years of shipping activity has been added (Arctic Council 2009).

6 Arctic Marine Data: Other Sources

As is the case with AMSA research, a primary reason for research into marine activity is the associated threat that increased exploitation will have on the already vulnerable Arctic environment, its people and infrastructure. While the Arctic has already undergone considerable change brought on by industrial emissions produced elsewhere, the World Wildlife Fund (WWF) states: "As the region loses its sea ice cover, it is open not only to new development opportunities but to an increased scale of existing threats, including industrial developments from the energy sector and transportation" (World Wildlife Fund 2012; 3). From the time of its first workshop on Arctic tourism in 1998, the WWF has done extensive research on the effects of increased Arctic activity from tourism to fisheries and marine transport. Accordingly, the 2012 WWF report Important Marine Areas in the Arctic, which is focused on conservation, includes data on oil and gas development and shipping. It represents a sampling of how research has moved forward since the 2009 AMSA report. As a Permanent Observer at the Arctic Council, WWF data would have an influential role in the decision-making process.

The report is primarily concerned with identifying ecologically or biologically significant marine areas, visually depicted through the use of maps, and a few maps show shipping and energy development overlapping with areas of conservation. In the report, the map titled "Overlap in Ecologically and Biologically Significant Areas/Priority Areas/Marine protected Areas as defined by various groups, and AIS shipping data" is derived from current data and depicts density of shipping traffic in 2011 inclusive of all vessel types and sizes. However, it is noteworthy that another map "Commercial Shipping Activity from A Global Map of Human Impact on Marine Ecosystems," also showing shipping activity is based on data retrieved from October 2004 to September 2005. In both cases the report provides no further detail as to the data associated with each map. For instance, in both maps red represents high-density traffic whereas blue represents low-density but a numerical component associated with the color-coding is not included.

The report states (World Wildlife Fund 2012: 5):

This document is intended to further facilitate discussions and work within the international community, Arctic Council and among Arctic stakeholders, provide for a full scientific description of important arctic marine areas, and help inform the CBD [Convention on Biological Diversity] process for the description of EBSAs [Ecologically or Biologically Significance].

While the purpose of the report was to inform Arctic decision-makers who were presumed to have a significant background in Arctic issues, other less-informed observers would be hard pressed to find the data that informed the report. Although a mass media article that appeared in IPolitics, a subscription-based Canadian online news publication, provided some additional clarity into the WWF data, the article stated that WWF reported that 46 passages occurred through the Northern Sea Route (NSR) in 2012 compared to 34 in 2011 (Munson 2013). However, no further detail was included on ship density. A map included with the article was derived from information processed by Det Norske Veritus and WWF, adding clarity that the WWF maps lacked. The article also reported on a recent study by John Higginbotham et al. (2012), the lead author of "Canada-US Arctic Marine Corridors and Resource Development," that argued for a joint Canada-US marine strategy in order to facilitate development of the Northwest Passage's commercial potential. The report stated that in contrast to Canada and the US, Russia has a well-developed fleet of Arctic-worthy vessels that includes 36 icebreakers with 5 additional vessels under construction and 8 in the planning conveys a stark advantage over Canada and the US. In comparison Canada owns six icebreakers and the US has five, many of which are in need of repair; both have plans for one additional vessel each.

There is no doubt that because of advancements in today's technologies most now have access to vast quantities of data. However, as Eric Guimont, a government employee who is the director of strategic research at Aboriginal Affairs Canada said, "The problem is no longer the availability of data, but its accessibility. It's out there, but it's scattered across hundreds of organizations" (Lenihan 2013). As the cases above demonstrate, data is often inaccessible to those who depend on accurate and timely information and is often not directly shared by the data's creators. A wide range of sectors including commercial, investment, diplomatic, legal, scientific and academic are expected to hold expertise, but each sector's knowledge may understood only by those most closely affiliated with originating information. Indeed increasing interest in the Arctic region will demand vast improvements not only in data but in the communications of data for a wide range of specialists engaged in vastly different types of work. While colorful and informative maps are useful for raising awareness, an ideal system would provide both ready-made information and infographics for immediate use and, critically, access to the data that produced those graphics.

7 Communicating Results

Much to the consternation of scientists, a common refrain is the difficulty of communicating data to a broader audience. This point was made particularly clear at a conference convened by the Pace Law School, in Spring 2012. Founder of 350.org, Bill McKibben, also known for his mass appeal, opened the event with a short video to an audience consisting mostly of academics, government officials including a representative of the Environmental Protection Agency and others, all with a degree of expertise in the effects of climate change. Yet, the overriding concern of the conference was how best to communicate data and scientific findings to politicians and industry decision-makers who are positioned to implement change, a daunting task when communicated information may lead not to desired results but to further politicization, creating more public confusion and drift. From shifting weather patterns that affect farming practices, to coastal flooding that may have a crippling effect on city infrastructures, expert speakers captured the extent of climate impacts on the issues. While the content was substantive, particularly remarkable were the number of speakers that strove to communicate with clarity, passion and humor without compromising their integrity and data.

In one notable presentation, Dr. Lewis Ziska, research plant pathologist for the U.S. Department of Agriculture (USDA), lamented the common inconsistency between scientific research and dissemination. For instance, in 2012 the USDA released a new plant hardiness zone map updating information from 1990. Although many zones have shifted one-half zone warmer since 1990 the USDA website states that, "changes in zones are not reliable evidence of whether there has been global warming" (United States Department of Agriculture). Rather, the USDA claims it was a result of more sophisticated mapping. However, in an email exchange on 11 March 2013, Ziska said that in his personal opinion "we need to do more to address global uncertainty in the food supply posed by climate disruptions, and that this uncertainty extends not only to production, but to nutrition, and distribution." Financial support for Ziska's lab in Beltsville, Md., may help explain some of the problem, too: Ziska had to reduce his research staff from a high of ten in 1999 down to three in 2008 as a result of budget cuts (Christopher 2008).

Although the participants at the Pace University conference were refreshingly clear communicators, most often information is conveyed through lengthy reports that for most to understand would require a massive investment of time and concentration. In an interview with Christoph Seidler, science editor for Spiegel Online, who covered the 2013 Arctic Frontiers conference said: "A lot of overview reports produced with several hundred pages each and no one including policy makers have the time to read that" (Arctic Frontiers 2013). In some cases scientists have dispelled the notion that they are poor communicators with well-written articles for mass media (Radford 2011). At other times mass media journalists report on the latest scientific findings or decipher those lengthy reports generated by government or think tanks for public consumption and decision makers alike. Yet with the increasing volume of information, concise and factual translation is critical.

8 Recommendations for Government and Academe

While many see too much data as a fundamental problem, inundating researchers, when it comes to the Arctic, filter failure and data overload are secondary concerns: the real issue is fragmentation of data and a lack of neutral space to collect, share and communicate. To overcome the challenges we have detailed, there is pressing need to move ahead on several fronts simultaneously. The following measures render information technologies related with big data effectively supporting global environmental policy in particular, and increasingly informationalized international relations at large:

- The extension of government services like data.gov for environmental science and policy research
- Public-private partnerships to increase innovation and data sharing
- Creation of a global initiative to inculcate and improve information literacy in university students
- In academe, development of a "Global Systems Science" (Helbing 2013) to uncover and bridge gaps in knowledge.
- Data transparency initiatives in commercial and governmental concerns to ensure equal access to economic and environmental information.
- For foundations and grant-giving bodies, funding should be tied to transparency and access.
- Make the linkage between information technologies, big data, and public policies a central concern for the analysis and theoretical understanding of future international relations

References

Arctic Council. (2009). *Arctic marine shipping assessment 2009 Report* (second printing). Accessed 30 January, 2014, from http://www.pame.is/images/stories/AMSA_2009_Report/AMSA_2009_Report_2nd_print.pdf

ArcticData (a). *What is ArcticData?* Arctic Council. http://www.arcticdata.is/index.php?option=com_content&view=category&layout=blog&id=15&Itemid=151. Accessed January 30, 2014.

Augustsson, P. (2011). Towards a common strategy for the Arctic: The Nordic countries can lead the way. In R. Nurick & M. Nordenman (Eds.), *Nordic-Baltic security in the 21st century: The regional agenda and the global role*. Washington, DC: Atlantic Council.

BBC News. (2010, January 21). Tim Berners-Lee unveils government data project. *BBC News*. http://news.bbc.co.uk/2/hi/8470797.stm. Accessed January 30, 2014.

Bollier, D., & Firestone, C. M. (2010). *The promise and peril of big data*. Washington, DC: The Aspen Institute.

CAFF & PAME. ArcticData (www.arcticdata.is) accessibility to Arctic Council data, Conservation of Arctic Flora and Fauna and Protection of the Arctic Marine Environment. Accessed 30 January, 2014, from http://caff.is/images/ArcticData_-_accessibility_to_Arctic_Council_data_—_CAFF_and_PAME.pdf

Carmody, T. (2013, January 10). Google and CDC show US flu epidemic among worst in a decade. *The Verge*. http://www.theverge.com/2013/1/10/3861538/google-cdc-show-us-flu-epidemic-among-worst-in-decade. Accessed January 30, 2014.

Cate, F. H. (2008). Government data mining: The need for a legal framework. *Harvard Civil Rights—Civil Liberties Law Review, 43*(2), 435–489.

Catone, J. (2011, June 27). How much data will humans create & store this year? [infographic]. *Weblog Mashable*. http://mashable.com/2011/06/27/data-infographic/. Accessed January 30, 2014.

Christopher, T. (2008, June 29). Can weeds help solve the climate crisis? *New York Times*. http://www.nytimes.com/2008/06/29/magazine/29weeds-t.html?pagewanted=all. Accessed January 30, 2014.

Cukier, K., & Mayer-Schoenberger, V. (2013). The rise of big data: How it's changing the way we think about the world. *Foreign Affairs, May–June*, 28–40.

Data.gov. Washington. http://www.data.gov/about. Accessed January 30, 2014.

Hamby, S. (2011, November 22). Top three technologies to tame the big data beast. *Weblog Huffington Post*. http://www.huffingtonpost.com/steve-hamby/top-three-technologies-to_b_1105022.html. Accessed January 30, 2014.

Helbing, D. (2013). Globally networked risks and how to respond. *Nature, 497*(7447), 51–59.

Higginbotham, J., Charron, A., & Manicom, J. (2012). Canada-US Arctic Marine Corridors And Resource Development. *Centre for International Governance Innovation, 24*(November), 1–8.

ICSU. (2010). Universality of science in the polar regions. *International Council for Science*. http://www.icsu.org/publications/icsu-position-statements/universality-polar-regions/download-statement. Accessed January 30, 2014.

IPY. (2012). Accessing, sharing and preserving data as a legacy of IPY. In *From knowledge to action*. http://132.246.11.198/2012-ipy/Abstracts_On_the_Web/by_theme.html. Accessed January 30, 2014.

Kamenetska, S. (2013, January 16). APIS change business. *Weblog Mashery*. http://www.mashery.com/blog/apis-change-business. Accessed January 30, 2014.

Kelley, M. (2012, January 16). The new normal: Annual library budgets survey 2012. *Library Journal*. http://lj.libraryjournal.com/2012/01/funding/the-new-normal-annual-library-budgets-survey-2012/#_. Accessed January 30, 2014.

Lenihan, D. (2013, February 26). Accountable first nations governance – minus the paternalism. *IPolitics*. http://www.ipolitics.ca/2013/02/26/accountable-first-nations-governance-minus-the-paternalism/. Accessed January 30, 2014.

Maguire, S. (2011). Can data deliver better government. *The Political Quarterly., 82*(4), 522–525.

Marks, J. (2012, May 2). 'Big data' could remake science – and government. *National Journal*. http://www.nationaljournal.com/tech/-big-data-could-remake-science-and-government-20120502. Accessed January 30, 2014.

McKendrick, J. (2010). *Keeping up with ever-expanding enterprise data—2010 IOUG database growth survey*. Unisphere Research, a division of Information Today, Inc. New Jersey: New Providence.

McKendrick, J. (2013, August 4). Why no one can be a 'passive consumer' in today's API economy. *Weblog Smartplanet*. http://www.smartplanet.com/blog/bulletin/why-no-one-can-be-a-passive-consumer-in-todays-api-economy/. Accessed January 30, 2014.

Munson, J. (2013, January 4). Russia set to dominate Arctic shipping unless Canada, US step up. *IPolitics*. http://www.ipolitics.ca/2013/01/04/russia-set-to-dominate-arctic-shipping-without-canada-us-co-operation/. Accessed January 30, 2014.

Murayama, Y. (2012). *The ICSU World Data System tackles IPY data challenges*. http://132.246.11.198/2012-ipy/Abstracts_On_the_Web/pdf/ipy2012arAbstract01111.pdf. Accessed January 30, 2014.

Nettler, J. (2013, March 3). The newest climate challenge: fatigue. *Planetizen*. http://www.planetizen.com/node/61023. Accessed January 30, 2014.

Radford, T. (2011). Of course scientists can communicate. *Nature, 469*(445). http://www.nature.com/news/2011/110126/full/469445a.html. Accessed January 30, 2014.

Seidler, C. (2013, February 20). Interview. *The Arctic Institute* podcast radio programme. http://www.thearcticinstitute.org/search?q=seidler. Accessed January 30, 2014.

Siegler, M. G. (2010, August 4). Eric Schmidt: Every 2 days we create as much information as we did up to 2003. *Weblog TechCrunch*. http://techcrunch.com/2010/08/04/schmidt-data/. Accessed January 30, 2014.

Subramony, D. P. (2010). Understanding the complex dimensions of the digital divide: Lessons learned in the Alaskan Arctic. *The Journal of Negro Education, 76*(1), 57–67.

The White House. (2012, March 29). Fact sheet: Big data across the Federal Government. *Executive Office of the President, Washington*. http://www.whitehouse.gov/sites/default/files/microsites/ostp/big_data_fact_sheet_3_29_2012.pdf. Accessed January 30, 2014.

Trapani, G. (2008, September 22). Information overload is filter failure, says Shirky. *Lifehacker*. http://lifehacker.com/5052851/information-overload-is-filter-failure-says-shirky. Accessed January 30, 2014.

United States Department of Agriculture. *What's new*. http://planthardiness.ars.usda.gov/phzmweb/AboutWhatsNew.aspx. Accessed January 30, 2014.
Varki, A. (2009). Human uniqueness and the denial of death. *Nature, 460*(7256), 684.
World Wildlife Fund. (2012). Important marine areas in the Arctic. *World Wildlife Fund*. http://awsassets.panda.org/downloads/arcticatlas_letter_v2g_web.pdf. Accessed January 30, 2014.
Yiu, C. (2012). The big data opportunity: Making government faster, smarter and more personal. *Policy Exchange* 1–36.

Global Diabetes Encounters: Accounting for Health Technologies in the Global

Annegrete Juul Nielsen and Henriette Langstrup

Abstract As already recognized, though little theorized within International Relations, the capacity of technology to ensure the achievement of preset policy goals is often grossly overrated. Drawing on Science and Technology Studies, this chapter proposes a lens to investigate global encounters, which take the technologies, involved in such encounters seriously both analytically and methodologically. This approach provides the opportunity not only to see how technology plays an important part in shaping global encounters and gets shaped itself in the process. It also allows us to appreciate the complex and productive nature of the friction these encounters produce. The proposed lens is used to analyze a specific global encounter illustrative of the way in which technologies increasingly become central to global encounters concerning health. In this encounter Danish and Indonesian partners are engaged in implementing a Danish developed quality assurance software for diabetes management in Indonesian primary care. We analyze what is at stake for the partners and what role the technology plays in shaping these stakes. The analysis points to the ambiguous and often contentious nature of technology in global encounters. The challenges met as the technology moves from Danish health care to an Indonesian context cannot be explained with reference to generalizing notions such as structural barriers, local culture or stakeholder interests. Instead attention to the frictions of global encounters provides insights to the conditions for and effects of movement of health technologies across geographical borders and thus to research at the intersection of Global Health and International Relations.

Keywords Globalization • Science and Technology Studies • Partnerships • Global health

A.J. Nielsen (✉) • H. Langstrup
Department of Public Health, University of Copenhagen, Center for Medical Science and Technology Studies, Copenhagen, Denmark
e-mail: anjn@sund.ku.dk; helan@sund.ku.dk

M. Mayer et al. (eds.), *The Global Politics of Science and Technology - Vol. 2*, Global Power Shift, DOI 10.1007/978-3-642-55010-2_14,

1 Introduction

This chapter takes as its point of departure the observation that global health initiatives increasingly revolve around promoting, developing, implementing and evaluating technologies that may alleviate the burdens of disease for citizens and nations (Lewis et al. 2012). As recognized in both global health and International Relations today most health problems have a global political and economic impact as they transcend national borders and the perspectives and concerns of individual nations (Kickbusch 2011; McInnes and Lee 2012). The belief among states, NGOs, scientists and other actors engaged in global health is that technological solutions are central to the management of global health problems. The aim of an increasing amount of actors is therefore to see to that technologies, which assist better uptake of drugs, vaccines, and diagnostics; improve delivery of health services; and help promote healthy behavior at the community and individual levels are disseminated and used on a global scale (McCannon et al. 2007; Kaplan 2006; Gardner et al. 2007). Like other modernizing or globalizing projects this generates enthusiasm because it "help[s] us imagine interconnection, travel and sudden transformation" (Tsing 2000: 330). However, as already recognized though little theorized within International Relations, the capacity of technology to ensure the achievement of preset policy goals (of economic development, security, transparent governance and more) is often grossly overrated. Often such expectations gloss over more complex conditions for and effects of technology in the global arena (Fritsch 2011; Ancarani 1995). So while the International Relation literature has acknowledged the need to promote global health issues as an important area of research for the field (Davies 2010; Garrett 2007), little attention has been given to the question of how such issues should be studied in a way that may acknowledge these more complex conditions for and effects of technology.

In this chapter we will propose a lens to investigate global encounters, which take the technologies involved in such encounters seriously methodologically and analytically . The lens we propose draws on Science and Technology Studies and—we will argue—provides us with the opportunity not only to see how technology plays an important part in shaping global encounters, but also to appreciate the complex and productive nature of the friction these encounters produce. The proposed lens is subsequently used to analyze a specific illustrative global health encounter revolving around technology. This encounter involve Danish and Indonesian partners jointly engaged in implementing a Danish quality monitoring software for diabetes management—the so-called '*Diabetes Quality Monitoring System*' *(DQMS)*[1]—in Indonesian primary care. We will ask what happens when capacity building partners with diabetes as a shared concern meet with the collective aim of implementing a specific tool—what is at stake for the partners and what role does the technology play in shaping these stakes? By describing this particular global diabetes encounter in ethnographic detail, rather than providing a formalized

[1] The DQMS is not the real name of the technology, but one given by the authors of this paper.

overview of the partners, program and outcomes, we argue that we may better appreciate the complexities of such encounters (Shenkar et al. 2008) and the role technology plays in these.

2 Global Health Technologies and IR

The field of Global Health, primarily positioned within a biomedical scientific tradition, has recently started addressing the question of technology as a lever for global health in particular in developing nations (WHO 2010). Significant hopes and expectations are attached to new Information and Communication Technologies (ICTs), such as computers, the Internet, and mobile phones, in addressing contemporary global health problems. McCannon and colleagues goes as far as arguing that "Scaling up [technologies] should become a major and sustained enterprise in the global health community" (McCannon et al. 2007: 1937). Development and dissemination of technological solutions for today's global health challenges are thus deemed one of the most important tasks by those teaching, researching, planning, and practicing global health across the world (Gerber et al. 2010). The belief is that technologies like ICTs can help health care organizations envision and implement new and more efficient ways of providing care (WHO 2010). For example ICTs are used in telemedicine services to help overcome geographical barriers and increase access to health care services in rural and underserved communities in developing countries (Clifford et al. 2008).

According to proponents for increased use of ICT in global health, the main problem is very slow dissemination—or diffusion—of new health technologies (Berwick 2003). Even though e.g. low-cost telemedicine applications have proven to be feasible, clinically useful, sustainable, and scalable in developing settings the applications are not being widely adopted allegedly due to a variety of barriers (WHO 2010). In an analysis of telemedicine support for the developing world Wootton points to practical and cultural barriers found in the local health services as well as more political issues hindering a fast spread of otherwise clinically useful and safe systems (Wootton 2008). For instance, Wootton discusses the possibility of perceived loss of control, which some developing countries might experience, when implementation of telemedicine requires expertise provided from outside the country. Clifford and colleagues in their analysis of the potential of medical information systems in developing countries argue that successful technology implementation and maintenance require what they term a multi-factorial approach. "Among some of the factors that must be taken into account are corruption, inequalities within the country, imposition of sub-optimal policies or technologies by authorities, and the lack of or incorrect information." (Clifford et al. 2008: 4).

These accounts of what makes technologies for global health diffuse and what hinders the flow share some common characteristics and related problems: Wootton and Clifford and others from the field of Global Health identify and recognize the need to take into account the social and political dimensions of ensuring the spread of innovations. Diverging interests and cultures of international partners, the local policies that may stand in the way of global solutions and the need of negotiations and agreements—these elements primarily concerning governance are put to the fore. However, the technologies themselves seem to be given a very limited explanatory value when wanting to understand how global health technologies move and with what effects.

Understanding the social and cultural issues and centrality of partnerships in the governance of global issues is at the heart of International Relations literature. Recent IR literature on global health has argued that the complexity of this global arena points to the need for IR to encompass a much more varied set of actors, when analyzing how global health issues are shaped and resolved (Davies 2010). Apart from the overall argument already referred to, that health issues increasingly transgress national boundaries and have become accepted as legitimate part of international politics, what is repeatedly emphasized in the IR literature is the escalation of the number and variability of actors involved in the arena of global health and the sheer complexity that this entails (Hill 2011; Feldbaum et al. 2010). As Davies, for instance, argues: "A range of actors, such as foreign governments, NGOs, pharmaceutical companies, private donors and international organizations drive a variety of different health agendas that influence priorities within individual states and affect the resources that are available to individual health workers and opportunities for patients" (Davies 2010: 1169). As such the IR literature confirms the point made in the literature on diffusion of global health technologies, that it is through analysis of the—albeit complex—cultural and social dynamics including negotiations, partnerships, and power struggles between human actors, that global health issues can be understood. The specificities of the actual encounters where actors engage in such negotiation are mostly left unattended in these strands of literature. Most notably, the technologies of global health remain unanalyzed and thus assume it to be stabile and neutral as they spread with difficulty or ease depending on the outcome of negotiations of or struggles between stakeholders. The frictions occasioned by the global health encounters are not recognized as something which may equally be occasioned by the technologies themselves, only by the cultural perceptions and predefined interests of the receiving actors. In the following we will outline how some basic notions taken from science and technology studies can provide new routes of analytical and empirical engagements with global encounters involving health technology.

3 STS: Studying Technology in Global Encounters

What we term global encounters in this chapter are concrete and potentially frictious contacts between specific human and non-human or technological entities meeting and creating relations that transgress national boundaries. These meetings are often occasioned by shared concerns that are already among the actors defined as 'global' in nature (see also Hathaway 2010 for a similar use of the notion).

Therefore talking about encounters points first and foremost to a certain methodological orientation toward what we consider to be an empirical phenomenon that should be studied in the context in which it unfolds. This is a very basic ethnographic notion—rather than generalizing phenomena such as "collaborations" or "technologies" in advance the researcher sets out to study these in specific contexts as they form part of everyday activities and experiences (for the use of ethnography in IR see de Volo and Schatz 2004; Vrasti 2008). This ethnographic frame of mind is at the heart of Science and Technology Studies (STS)—approaches on which we draw (Latour and Woolgar 1979; Morita 2013) in order to study the increasingly prominent role of technology in the global health arena. The Global Health literature presented above conceptualized global health technologies as neutral and discrete objects circulating unaltered through unspecified global flows potentially supported or hindered by policy and culture. In contrast and drawing on concepts and methodological principles from STS, we suggest that what a global health technology is and what it does is first and foremost an empirical question to be addressed by actually following specific entities as they move through various contexts and form part of various relationships. The analytical framework and methodological attunement we suggest depart from two basic notions generated in early and more recent science and technology studies respectively: the role of technology in shaping social order and global encounters as a process involving friction. Both these notions are the products of and calls for further ethnographic engagements with the specific contexts that make up 'the global'.

3.1 Technological Encounters

In the terminology of STS and more specifically Actor-Network Theory technologies are actors rather than passive objects to be disseminated. What in the social sciences generally has been considered to be phenomenons consisting primarily of social actors (society, partnerships, institutions) are in the vocabulary of ANT seen as heterogeneous networks consisting of relations between human and non-human actors. The agency of technology thus rests on its ability to connect to other entities in the network (Latour 2005). The technologies promoted vigorously in global health discourses are thus by definition dependent on their ability to engage in and connect to wider networks of actors (Mol and de Laet 2000). A telemedical device transferred to, say, rural Indonesia is in itself not enough to help overcome geographical barriers and increase access to health care services. State actors need to give consent, professional partners need to engage, health professionals need to know how to operate the device, patients need to show up and make use of the advice given, resources have to be put aside for repairs etc. All these actors are not seen as exogenous to the technology, but rather exactly what the technology is made up of together with plugs, circuits, processors and metal. This also means that a technology is contingent on the relations in which it is embedded and will be altered as it connects to new actors or new networks. This transformation is referred to as *translation* (Latour 1991) and concerns equally the transformation of the identity of the technology as it connects to new actors AND the transformation of the actors who become part of the network of the technology (Latour 1987). The dynamics between the durability of the material found in technology and its relationality and contingency may be captured with the concept of script: According to Akrich (1992) a script is the particular vision of the world in which the technology is expected to work, which is being built into the technology by its initial designers. The actual health technologies that Wootton and others cursorily relate to under singular headings may have a variety of embedded scripts, each suggesting particular roles for putative users, particular norms of action and particular infrastructural arrangements. A script may be powerful exactly because it is not readily visible in the technology. And, as the term translation suggests, the user will find his or her own visions more or less altered by engaging with the technology (Latour 1991). African women may come to regard their pregnancy differently, seeking help for complications more quickly after having been given a mobile phone with a special pregnancy-care app (Lund et al. 2012). Still, and this is central to Akrich's argument, no script will be able to predict or control the reality in which the technology is to work: "the fate of facts and machine is in the hands of later users" (Latour 1987: 59). The technologies themselves can be obeyed, dismantled, ignored or adapted as the script of the technology meets the actual contexts of use. We cannot expect global health technologies to harmonize healthcare to one universal standard. Rather we can explore how technologies, individuals, interests, knowledge and practice are being continuously reordered in somewhat unpredictable ways as technologies travel (Nielsen and Jensen 2013). These very basic notions from ANT point to the relevance of accounting ethnographically for the role that technologies play as specific actors in the complex global diabetes encounter, we are studying. Technologies are not neutral tools that will meet social systems and cultural barriers as they diffuse. They are always already embedded in social and material relations and inscribed with certain visions of the world and not others. As they are displaced they may have powerful effects, but only so if managing to relate to new actors, places, infrastructures and programs of action. As we will demonstrate, this displacement and establishment of relations across spatial, national, cultural boundaries is by no means simple or neutral. This leads to the second notion, on which our analytical framework rests, which relates to how we can understand global encounters and in turn globalization.

3.2 Frictionous Encounters

The understanding of globalization implicit in much Global Health literature on technology rests on assumptions of increased homogenization and potential "clashes" of cultures. Challenging such assumptions anthropologist Anna Tsing (2000, 2005) has developed the notion of friction as a metaphor for the diverse and conflicting social interactions that make up global encounters. According to Tsing globalization cannot be sufficiently understood as cultural oppositions or as cultural diversity being subjected to a time-space compression. Using the notion of friction, Tsing shows how curious and creative cultural differences are in the grip of global encounters, and how much is overlooked in contemporary theories of the global. Drawing on Tsing's work (2005) we suggest the metaphor of friction as a way of understanding the character and outcome of the concrete contacts that involve health technologies and the human and organizational actors engaging with them. Metaphors of distance—between cultures, interest or people—found in analysis of international phenomenon such as international management and technology transfer (Rogers 2002) are mostly based on abstract definitions of differences (Shenkar et al. 2008). In contrast, the metaphor of friction invites the analyst to look closer at the actual establishment of points of contact between the entities involved, and to acknowledge the tensions and discords which are inherently part of such encounters (Tsing 2005; Shenkar et al. 2008). In relation to the concepts of script and translation just presented, the concept of friction may be said to be much less concerned with the role of technology in particular, and more oriented toward the changes per se afforded by encounters, that always involve heterogeneous elements including technology. At the same time the concept of friction shares with the notion of translation an ethnographic aspiration to specify the transformations taking place as entities met and form relations. Tsing asks us to take more notice of the differences that may equally be produced through translation. Frictions do not necessarily entail conflict, but rather "difference-based interactions" (Tsing 2012: 707) that shape the arrangements in and through which we live. Tsing has also called friction "a metaphor for connection *with* difference" (Tsing 2009: 351). The groups that participate in the global encounters we are exploring, we do not see as holding fixed interests and having immutable characteristics that can explain the outcome of their encounters. Rather these very characteristics, interests and agendas may be partly explained by the very interactions they engage in (Latour 1991; Tsing 2000). In ethnographic studies of global environmental encounters scholars such as Tsing (2005), Hathaway (2010), and Jensen and Winthereik (2012) have pointed to the emergent and situated nature of global encounters and the fleeting nature of convergence between different actors. In the following analysis we will pay attention to both the individuals and groups engaging in an international partnership to intervene in Indonesian physicians' practices of diabetes treatment AND to the technology around which this partnership is established.

3.3 Ethnographic Encounters: Our Empirical Material

The analysis of the specific global encounter that we lay out in this chapter is based on empirical material collected in Copenhagen, Denmark and Jakarta, Indonesia over a period of 2 years starting primo 2011. Using an ethnographic approach we have interviewed and observed the work of the so-called DQMS capacity building partners both in Denmark and in Jakarta, Indonesia. The capacity building partners consists of the Danish developers and representatives of the DQMS, the Indonesian endochrinological medical association, and the Indonesian subsidiary company of a Danish pharmaceutical company. The partners collaborate on raising awareness of and skills in diabetes treatment among Indonesian internists and general practitioners and use the DQMS as part of this strategy. The analysis in the present paper is primarily based on interviews and observations carried out during 1 week-long visit in December 2012, where the capacity building partners met in Jakarta with the collective aim of planning the introduction of the DQMS to a selected number of general practitioners residing in Jakarta. Besides a number of board meetings, an introductory DQMS workshop was held for 25 general practitioners, who were to be the first to use the DQMS in an Indonesian context. In the following we introduce the details of the technology, the DQMS, relying on interviews and observations done among the Danish partners and developers of the technology.

4 The 'Diabetes Quality Monitoring System': Globalizing Danish Diabetes Expertise

Diabetes and other Non-Communicable Diseases (NCDs) only just recently became a legitimate part of the global health agenda. Leading stakeholders within the diabetes arena have argued that diabetes despite the size and toll of the epidemic for years has been marginalized on the global health agenda. However, diabetes is now a rising concern worldwide, and also increasingly seen as an opportunity for international partnerships and for expanding markets for pharmaceuticals, health services, and health technologies (The Danish Government 2013). Since 2012 a Danish outpatient clinic specialized in diabetes has collaborated with Indonesian endocrinologists and various Indonesian state actors. The collaboration has as its aim to expand the capacity for treating diabetes in Indonesia through education and dissemination of the Danish developed IT-assisted quality monitoring system—the '*Diabetes Quality Monitoring System*' (DQMS).

Indonesia is one of the countries in the Global South where the diabetes prevalence is rising epidemically and is expected to continue to do so in the coming years as the country experience rapid economic growth and improved living conditions (Soewondo and Pramono 2011). Political attention is given to the issue as it is expected to pose further challenges on the developing health system and ongoing reform process. Thus Indonesian health care providers, professionals, and patients are looking for innovative low-cost solutions to the increase in diabetes. Presently in Indonesia, various initiatives for reforming diabetes treatment are launched by public, private, and civil actors from within and outside of Indonesia. Multinational pharmaceutical companies (e.g. Novo Nordisk, Sanofi-Aventis and Eli-Lily) and international NGOs working with diabetes (e.g. the World Diabetes Foundation, the American Diabetes Association, Juvenile Diabetes Research Foundation) to a very large extend initiate and participate in such projects.

The Danish outpatient clinic is one of the leading clinics on diabetes care and research in Denmark. The clinic has in Denmark pioneered the work with developing and incorporating IT-systems as part of a patient-centered and evidence-based diabetes treatment. For years the clinic has been involved in diabetes education to Danish and foreign health professionals, both at its facilities in Denmark and abroad. The clinic has focused much of its activities in Asia, where the diabetes prevalence is rapidly increasing. It is in order to support an international need for diabetes management and further develop its position as a globally recognized expert on diabetes and to support that the Danish outpatient clinic has developed the DQMS.

The DQMS is a stand-alone self-surveillance software, which rest on the idea of the PDSA circle (Plan-Do-Study-Act).[2] The DQMS is designed to provide the individual physician with a tool to collect and monitor a range of diabetes related patient data. Analyses of the aggregated self-reported data are to provide input for how to improve the quality of the clinical work. The individual physician is thus to be assisted by the DQMS to engage in a circular temporal order of continuous data collection, analysis, and organizational change. In the excerpt below, the physician, who initiated the development of the DQMS, describes the considerations that guided the design of the system:

> We wanted to make a piece of software that enables the individual, say Chinese physician, to evaluate whether his clinical performance is good enough according to his own standards. We wanted to learn them to collect data. The software had to take into account that physicians in these countries are very busy and they have to make a living of it, so typing data should not take more than two minutes per patient. Also it should not be an electronic patient record, because then you run into juridical trouble.. [..] So we agreed on a maximum of 20 clinical indicators as that is what a trained person is able to type in two minutes. [..] We have tried to design the system so it can be used globally, because in fact it is just a simple database. The indicators chosen are the global ones and not ones that fit local concerns.

In short the DQMS is designed as a modest tool in the sense that the program only operates with 20 evidence-based clinical indicators, such as different measures of blood glucose levels, blood pressure, cholesterol levels etc. The system is also modest in the sense that it does not make use of advanced statistical analysis of the collected data, but only provides simple graphs and tables. By the designers this modesty is deemed a prerequisite for the system to be able to include as many users in as many diverse settings as possible. Finally the modesty is, as referred in the quote, also a question of avoiding conflicts with local policy as a medical patient record has a number of juridical implications concerning ownership, privacy, and data security among other things.[3]

---

[2] The PDSA circle was developed in industrial Total Quality Management, where the model has been fashionable since the early 1980s (Giroux 2006; Abrahamson and Fairchild 1999).

[3] However, the DQMS does allow viewing individual patient data anonymized and share it with patients.

On the face of it, the DQMS is thus designed to travel globally as the system only seems to require a PC and a person with 2 minutes on hand to type 20 different clinical indicators. However, as we will illustrate in the following analysis, the modest and universal aspirations of the DQMS are quickly challenged upon arrival in Jakarta.

5 Translating the Script: Redefining the Global

The following excerpt of field notes is from a meeting in Jakarta between the capacity building partners one day before the introductory workshop on DQMS begins. The Danish partners are represented by Dr. Pia[4] and her colleague Dr. Adam and the Indonesian endocrinologists are represented by Dr. Atin and Dr. Yutan. Preceding this encounter there has been a year-long process of interactions and negotiation between various actors leading up to the specific partnership. In this process the capacity building partners have in particular discussed and struggled with recruiting GPs.

Sitting in front of a laptop the Danish endocrinologist and responsible for the DQMS introduction, Dr. Pia, is going over some slides with the local partner from the Indonesian endocrinologist association, Dr. Atin. The slides are for the upcoming weekend's teaching program introducing DQMS first to a small group of Indonesian endocrinologist and then on the following day to a larger group of general practitioners working primarily within the public health system. Dr. Pia and Dr. Atin are both visibly excited and somewhat anxious about the upcoming event. Dr. Pia goes over the part of the program where she and her Danish colleagues will show the Indonesian physicians how to enter quality data collected in their own practices into the system and subsequently monitor and set goals for their clinical quality. Dr. Pia describes that the GPs should enter data on at least 200 patients during the 3-month data entry period. Dr. Atin seems surprised and responds that this is not possible. She says, that most of the GPs will probably only see around 20–30 different diabetes patients over a period of 3 months. Dr. Pia seems rather stunned. She says that it is important that they type at least 200: "If it is only 20–30 the statistical power will be skewed! A 100 would be all right, 200 are statistically preferable. It should be at least 100". Dr. Atin seems surprised: "In each clinic? There will be several GPs from the same clinic. They will be using the same computer". Everyone around the table seems to be speculating franticly if this perhaps will bring the number up the level sought by Dr. Pia. Dr. Atin says that from what they know about the clinics they might not have that many diabetes patients. Dr. Pia repeats that preferably there should be at least 100 patients in each clinic. Dr. Atin says that the GPs working in the public clinic should have 100. But in Jakarta the private GPs do not have that many patients. In other parts of the country

[4] The names of partners have been substituted in order to insure anonymity.

they might. Dr. Atin looks rather dejected: "Don't tell me you are going to cancel!?". Dr. Pia although still looking bewildered responds promptly: "Of course we are not going to cancel. Hopefully they will be able to type in more patients than 20".

The excerpt is illustrative of a number of analytical observations concerning the active role that technologies can play in shaping global encounters—and by this the very definitions of global. As described earlier, the system is designed to make its users collect and analyze data. They are to become quality monitoring doctors. Implicit in this is, however, also an adherence to the scientific—and thus 'global'—standards on which the system builds. In the excerpt Dr. Pia pursues this element of the technology's script as she explains the importance of the size of the database. Concretely this means that each dataset collected during the 3-month data entry period should contain at least 200 entries or patients as the statistical power otherwise will be too low or even worse invalid. This will have ramifications for the results generated for the GPs' self monitoring, but also for the prospect of producing scientific articles on the grounds of these data. Accessing and using these quality data for scientific publication has at steering committee meetings been stated as an interest of several of the partners in the program.[5] During the meeting it becomes clear that this part of the script does not fit well with the size of the Jakartan public primary care clinics and the patient population they serve. In the Danish outpatient clinic—on which practice the DQMS is modeled—all of the patients are diabetics and most of them go for a check-up once every third month. The Jakartan public primary care clinics do not only provide services for diabetics, but all kinds of patients. Furthermore, as the doctors are only allowed to prescribe medicine for 10 days at a time, the diabetic patients have to go to the clinic once every 10th day. In the Jakartan clinics the patients are therefore seen far more often. This means that although the Jakartan doctors have far more patient visits during a 3-month period than their Danish colleagues, the total number of patients with diabetes can still be as low as 20. Friction emerges as these differences between the assumptions of the global made by Danish designers and the practicalities and infrastructures of Jakartan primary care clinics surface in the meeting and rub against each other. The DQMS plays an important part in making these differences come to light, not only for the Danish partners, but also for the local endocrinologists. Previous to their engagement in enrolling GPs in the project these endocrinologists, working at hospital facilities, did not know much about the daily organization and patient population of primary care clinics in Indonesia. It is the technology's script in terms of seeking to ensure scientifically valid data, which shapes the interaction and produces this shared concern with the number of patient per clinic. In short the technology plays an active part in shaping the encounter and the friction that emerges.

---

[5] This is yet another version of the global being mobilized through the partnership: the global as constituted through collaborative work done across national boundaries.

A Global Health inspired analysis would analyze the friction as the surfacing of a 'barrier' in terms of a misfit between the technology and its putative users. However, this analysis would involve judging the whole of the primary health care structure of Jakarta to be a barrier for diabetes quality monitoring systems. Or it would involve judging the DQMS as being a badly designed quality assurance tool. Neither of these explanations would acknowledge nor be able to account for the definitional negotiation and translations about the very purpose of the global diabetes encounter, which unfolds as a consequence of the friction produced. As the Jakartan clinics do not fit the definition of the script, the purpose of the encounter can no longer be the securing of scientifically valid data for doing collaborative research on the results of the clinics. Still, the friction does not lead the capacity building partners to write off the DQMS or label Jakartan primary health care as the wrong site to introduce the DQMS. Instead the partial and contingent solution comes in the form of a translation of the technology's script: The inclusion of clinics no longer depends on their ability to produce and deliver scientifically valid data from their monitoring—at least not in the initial, statistically defined sense. The version of global concerning local adaptability gets presidency over the version of global concerning scientifically defined standards. Changes are thus introduced both to the script and to the world in which the script is to work.

6 Ends Shaped by Means: Becoming Technologically Engaged Endocrinologists

For the Indonesian endocrinology association, the DQMS initially was seen as a means of promoting more standardized diabetes treatment in primary care. Not 'just' for the sake of raising quality in the individual clinics, but also by freeing up time and resources in their own daily practices acutely overloaded by uncomplicated patients, many of whom should be treated in primary care. In the words of the international liaison of the endocrinologist association:

"If we can make the GPs to get more quality, more knowledge, then I believe we can do that [make the referral system work, ed.]. That the patient will get more trust in the GPs. And then we [the endocrinologists, ed.] will get fewer patients. It is not that we don't like to have many patients, it is just not right. It is not what we are supposed to do."

However, by pursuing their goal of promoting a more sustainable referral system through the DQMS, the endocrinologists got a very different and more involved role, as mentors of GPs, rather than as teachers and representatives working with GPs only at arm's length. This also involved the possibility of having to support tasks concerning system support for GP users, demanding that the endocrinologists themselves accepted to become active users of the DQMS.

At the meeting the plan for the introduction of DQMS to the GPs is discussed. Dr. Atin, the Indonesian endocrinologist, asks Dr. Pia, the Danish partner, why the endocrinologists have to be there when the GPs are introduced to DQMS. Dr. Pia says, that the endocrinologists will sit together with 'their' GPs—that is GPs who work in the same geographical area as the endocrinologist and for whom the specific endocrinologist will be a mentor in using DQMS. The idea is that the GPs will be able to contact the endocrinologists if they have questions later on concerning the use of the system. Dr. Atin is sceptical about this assumption. She says that in Indonesia there is very little if any contact at all between specialist doctors and GPs. As the GPs, Dr. Atin explains, have less training than a specialist, they are too shy to ask an endo (endocrinologist) a question or ask them for help. Dr. Atin explains that she has given her phone number to a number of GPs she has met on the workshop in diabetes that she teaches, but they have never used the opportunity to contact her. But, she says, "I hope with this program it will be different". Dr. Pia says that she is positive that it will. She continues to say, that she thinks that they need to make a clear agreement on how the contact between GPs and endos is to be ensured. Dr. Pia says: "We need to make an agreement stating that you (the endos) will meet with them (GPs) in two or three months. If they contact our clinic [the Danish clinic] when they have questions, you (the endos) will lose the contact". Dr. Atin comments half-jokingly to the rest of participants at the meeting: "She just leaves it to us!"

The excerpt is illustrative of how DQMS and the interests, roles, and relations of some of the partners are actively being re-shaped. Originally the system was designed for the autonomous medical practitioner, who wants to monitor and analyze his own medical performance for the benefit of himself and his patients. However, in the encounter the technology is being enrolled in the political and professional aspirations of the endocrinologists seeking to reform Indonesian health care. The global and highly individualized physician "with two minutes on his hands" choosing to autonomously manage his clinical quality as imagined by the system designers turns out to be difficult to disentangle from the local and largely collective infrastructures in which he or she is embedded. The endocrinologists take on the task of participating in this disentanglement as they see it as an opportunity for restructuring these very infrastructures, some of which are working very poorly in Indonesia—the referral system as a case in point. But passing through the means of the DQMS also has implications for the endocrinologist. Not least through a changed relationship to the GPs. Dr. Pia refers to the endos and "their GPs"—using this wording she suggests that the involved endocrinologists have a direct responsibility for a group of GPs and for their continued use of the DQMS. The contact established should not be dependent on previous relations of authority and status, but a specified division of work: The endocrinologists as mentors and super users responsible for meeting with GPs in order to help them use the system correctly. The half-joking comment made by Dr. Atin at the end of the field note illustrates the ambivalence of the emerging role of the endocrinologists in the project—she would like for her colleagues to establish relations to the GPs and for them to use their expertise more, but if they are to have a relation primarily concerning the use of DQMS, then the task may be more difficult.

Still, at the workshop for the GPs, the endocrinologists do act as mentors for GPs from their own geographical area—a professional connection not previously established. Also the endocrinologists end up by ensuring the GPs that they will invite 'their' GPs to a meeting every 2–3 month to keep track of the DQMS activities, providing them with email addresses and mobile phone numbers. By the end of the day Dr. Atin says to the GPs: "We will get back to you as soon as possible. Please understand that we are also new with the software. We can support you on the clinical issues, but not on technical issues."

The encounter between Indonesian endocrinologists and DQMS entails a transformation of both. Despite considerations on a strategic level of the interest and stakes that they as a professional group have in getting involved with the Danish partners, the interests and stakes which in practices unfold are highly contingent on the specificities of the technology. GPs and their activities become *interesting* for the endocrinologists in a new way as their use, non-use or mis-use of DQMS could have implication for the endocrinologists and their work. This is both in the aspired way of supporting a more sustainable referral system and in a more mundane and potentially challenging way, if the GPs will call upon endocrinologists to get help for their DQMS-use. Dr. Atin tries to safeguard against being over-run with questions by the GPs by saying that they can only "support on the clinical issues, but not the technical issues". In practice this distinction is however difficult to uphold, as clinical quality is to be accomplished exactly by technological means. Thus engaging with the means changes the ends and the stakes involved for the endocrinologists. For DQMS on its part ends are equally changed. According to the script of the DQMS, the system is addressing and seeking to engage the individual physician. However, in practice the DQMS no longer serves individual and medical goals alone, but also larger collective and political aims, such as a sustainable referral system. How the system is to provide quality and what quality entails have become less connected to the scientific ideal of 'statistically valid data' and less to the vision of the autonomous physicians conducting self-monitoring for the benefit of their individual patient and own professional satisfaction. The system is now also connected to experimenting with establishing better relations between primary and specialist care and an overall vision of improving Indonesian healthcare.

7 Connections *with* Differences: Final Reflections

Our account of global diabetes encounters involving the dissemination of technology has first and foremost used dense ethnographic descriptions to provide insight to the ambiguous, contentious and potentially transformative nature of such encounters. The health technology, DQMS, supposed to travel from Denmark to Indonesia with ease due to its modest and 'global' design, turned out to be anything but a passive and stable entity, unchanged by its meeting with other actors at the point of arrival. A global health-perspective might have us discuss how certain 'barriers' or 'facilitators' become visible in the course of introducing the DQMS in Indonesia—in particular when the encounters are linked to unsmooth introductions of technology as we might interpret the above empirical stories. However, as we have already suggested, it would be a gross simplification to reduce the interaction to one either allowing unhindered flow of a fixed entity or providing—cultural, financial, professional—obstacles to this flow.

Rather than being 'global' in and of itself, this very inscription of the technology got translated through the specific encounters it occasioned. Here the very meaning of 'global' became a challenging and potentially contentious matter among capacity building partners: Was the technology's 'global' quality—justifying its travel across the globe—most strongly connected to its scientific nature or to its ability to 'localize'—that is its adaptability to local conditions and concerns? There is no simple answer to this question, but in the context of the encounter described, negotiations were made and solutions found that involved frictions between various versions of 'global' and between the involved actors and their emerging interest and roles in relation to the technology. Investigating global encounters with an approach, which take the technologies involved in such encounters seriously both analytically and methodologically, we have demonstrated how technology plays an important part in shaping global diabetes encounters—and importantly how the DQMS got shaped itself in the process. And how the friction that emerges in these processes cannot in advance be seen as either unproductive or leading to a specific outcome.

Frictions do not necessarily entail conflict, but rather "difference-based interactions" (Tsing 2012: 707) that shape the arrangements in and through which we live. Tsing has also called friction "a metaphor for connection *with* difference" (Tsing 2009: 351) highlighting the critique against a diffusionist understanding of technology for global health that we have tried to articulate in this chapter. If seeing frictions as unsuccessful attempts to transfer technology from the North to the South and as barriers to be identified and avoided, then we miss the chance to see the practicalities and creativity involved in worlds coming into being *through* such differences. The translation of DQMS came with certain trade-offs—e.g. the possibility of getting scientifically valid data from GP uses for scientific publication—but also making possible the inclusion of more visions, more actors and different realities. Also, circulation, diffusion or transmission—these concepts make us focus on the flows of technologies, not on the digging of canals making such flows possible or on the transformative nature of such work. As both Hathaway (2010) and Jensen and Winthereik (2012) have also pointed to, work and transformation is necessary to build partnerships. Getting the DQMS from Denmark and all the way to the daily practices of Indonesian physicians treating diabetes patients entails the collaborative establishments of links that were not there in advance, not only between Danish and Indonesian partners but also between Indonesian endocrinologists and Indonesian general practitioners. Highlighting how scripts are continuously translated and may occasion variable versions of a technology, we have aimed to show that technologies as discrete objects do very little of and by themselves, but may become part of and in turn shape interests, practices, and roles in somewhat unpredictable ways. For International Relation literature these analysis may inspire to a broadening of methodological and analytical avenues into the very pertinent questions arising from the increased internationalization of politics and practices of health and illness.

References

Abrahamson, E., & Fairchild, G. (1999). Management fashion: Lifecycles, triggers, and collective learning processes. *Administrative Science Quarterly, 44*(4), 708–740.

Akrich, M. (1992). The description of technical objects. In W. E. Bijker & J. Law (Eds.), *Shaping technology/building society: Studies in sociotechnical change* (pp. 205–224). Cambridge, MA: MIT Press.

Ancarani, V. (1995). Globalizing the world: Science and technology in international relations. In S. Jasanoff, G. E. Markle, J. C. Petersen, & T. J. Pinch (Eds.), *Handbook of science and technology studies* (Rev. ed.) (pp. 652–670). Thousand Oaks, CA: Sage.

Berwick, D. M. (2003). Disseminating innovations in health care. *The Journal of the American Medical Association, 289*(15), 1969–1975.

Clifford, G. D., Blaya, J. A., Hall-Clifford, R., & Fraser, H. S. F. (2008). Medical information systems: A foundation for healthcare technologies in developing countries. *BioMedical Engineering OnLine, 7*, 18.

Davies, S. E. (2010). What contribution can International Relations make to the evolving global health agenda? *International Affairs, 86*(5), 1167–1190.

de Volo, L., & Schatz, E. (2004). From the inside out: Ethnographic methods in political research. *Political Science and Politics, 37*(2), 267–271.

Feldbaum, H., Lee, K., & Michaud, J. (2010). Global health and foreign policy. *Epidemiologic reviews, 32*(1), 82–92.

Fritsch, S. (2011). Technology and global affairs. *International Studies Perspectives, 12*(1), 27–45.

Gardner, C. A., Acharya, T., & Yach, D. (2007). Technological and social innovation: A unifying new paradigm for global health. *Health Affairs, 26*(4), 1052–1061.

Garrett, L. (2007). The challenge of global health. *Foreign Affairs, 86*, 14–38.

Gerber, T., Veronica, O., Brown, K., & Pablos-Mendez, A. (2010). An agenda for action on global e-health. *Health Affairs, 29*(2), 233–236.

Giroux, H. (2006). 'It was such a handy term': Management fashions and pragmatic ambiguity. *Journal of Management Studies, 43*(6), 1227–1260.

Hathaway, M. J. (2010). Global environmental encounters in Southwest China: Fleeting intersections and "transnational work". *Journal of Asian Studies, 69*(2), 427–451.

Hill, P. S. (2011). Understanding global health governance as a complex adaptive system. *Global Public Health, 6*(6), 593–605.

Jensen, C. B., & Winthereik, B. R. (2012). Recursive partnerships in global development aid. In S. Venkatesan & T. Yarrow (Eds.), *Differentiating development: Beyond an anthropology of critique* (pp. 84–101). London: Berghahn.

Kaplan, W. A. (2006). Can the ubiquitous power of mobile phones be used to improve health outcomes in developing countries? *Globalization and Health, 2*, 9.

Kickbusch, I. (2011). Advancing the global health agenda. *UN Chronicle, XLVIII*, 4.

Latour, B. (1987). *Science in action: How to follow scientists and engineers through society.* Cambridge: Harvard University Press.

Latour, B. (1991). Technology is society made durable. In J. Law (Ed.), *A sociology of monsters: Essays on power, technology and domination* (pp. 103–131). London: Routledge.

Latour, B. (2005). *Reassembling the social. An introduction to actor-network theory*. Oxford: Oxford University Press.

Latour, B., & Woolgar, S. (1979). *Laboratory life: The social construction of scientific facts*. Princeton, NJ: Princeton University Press.

Lewis, T., Synowiec, C., Lagomarsino, G., & Schweitzer, J. (2012). E-health in low- and middle-income countries: Findings from the Center for Health Market Innovations. *Bulletin of the World Health Organization, 90*, 332–340.

Lund, S., Hemed, M., Nielsen, B. B., Said, A., Said, K., Makungu, M. H., et al. (2012). Mobile phones as a health communication tool to improve skilled attendance at delivery in Zanzibar: A cluster-randomised controlled trial. *BJOG: An International Journal of Obstetrics and Gynaecology, 119*(10), 1256–1264.

McCannon, C. J., Berwick, D. M., & Rashad Massoud, M. (2007). The science of large-scale change in global health. *JAMA: The Journal of the American Medical Association, 298*(16), 1937–1939.

McInnes, C., & Lee, K. (2012). *Global health and international relations*. Cambridge: Polity Press.

Mol, A., & de Laet, M. (2000). The Zimbabwe bush pump: Mechanics of a fluid technology. *Social Studies of Science, 30*(2), 225–263.

Morita, A. (2013). The ethnographic machine: Experimenting with context and comparison in Strathernian ethnography. *Science, Technology & Human Values, 39*, 214–235.

Nielsen, A. J., & Jensen, C. B. (2013). Travelling frictions: Global disease self-management, local comparisons and emergent patients. *Science and Technology Studies, 26*(2), 61–79.

Rogers, E. M. (2002). The nature of technology transfer. *Science Communication, 23*(3), 323–341.

Shenkar, O., Luo, Y., & Yeheskel, O. (2008). From "distance" to "friction": Substituting metaphors and redirecting intercultural research. *Academy of Management Review, 33*(4), 905–923.

Soewondo, P., & Pramono, L. A. (2011). Prevalence, characteristics, and predictors of pre-diabetes in Indonesia. *Medical Journal of Indonesia, 20*(4), 283–294.

The Danish Government (Regeringen). (2013). *Vækstmarkedsstrategi Indonesien – batik og business*. http://indonesien.um.dk/en/~/media/Indonesien/Documents/Vaekstmarkedsstrategi_Indonesien.PDF. Accessed January 3, 2014.

Tsing, A. L. (2000). The global situation. *Cultural Anthropology, 15*(3), 327–360.

Tsing, A. L. (2005). *Friction: An ethnography of global connection*. Princeton, NJ: Princeton University Press.

Tsing, A. L. (2009). Beyond economic and ecological standardisation. *The Australian Journal of Anthropology, 20*(3), 347–368.

Tsing, A. L. (2012). Frictions. In G. Ritzer (Ed.), *Wiley-Blackwell encyclopedia of globalization* (pp. 707–709). West Sussex: Wiley-Blackwell.

Vrasti, W. (2008). The strange case of ethnography and international relations. *Millennium-Journal of International Studies, 37*(2), 279–301.

Wootton, R. (2008). Telemedicine support for the developing world. *Journal of Telemedicine and Telecare, 14*(3), 109–114.

World Health Organization. (2010). *Telemedicine: Opportunities and developments in Member States: Report on the second global survey on eHealth*. Geneva: World Health Organization.

Chukudu Politics: The Relevance of Infrastructural Absences for Governmental Power and State Failure in the Post-Colonial world

Peer Schouten

Abstract This chapter explores an analytical gap emerging at the intersection of political science and science and technology studies when their gaze is turned towards postcolonial Africa. Where political science has been bad at dealing with the importance of technological infrastructures for the constitution of governmental power, science and technology studies have hitherto been less attentive to situations in which such 'infrastructures of rule' are *absent*. This chapter aims at tentatively exploring this gap by pointing to the relevance of technological infrastructures in postcolonial Sub-Saharan Africa. It does so both through a cursory overview of literatures that might help articulate this problematique and by discussing the role of absences in technological infrastructures—such as roads, statistical tools, and government offices—in the historical unfolding of state formation of the Democratic Republic of Congo. The example of the *chukudu*—an improvised wooden transport utility that allows Congolese to go about despite decrepit infrastructures—is used to unpack the significance for infrastructural absences for social processes. Generalizing beyond the example of the *chukudu*, this chapter offers novel insights in the importance of technological asymmetries in global power-relations in postcolonial contexts.

Keywords Infrastructure • Postcolonial state formation • State failure • Governmental power • Underdevelopment • Sub-Sahara Africa

---

P. Schouten (✉)
School of Global Studies, University of Gothenburg, Gothenburg, Sweden
e-mail: peer.schouten@globalstudies.gu.se

M. Mayer et al. (eds.), *The Global Politics of Science and Technology - Vol. 2*, Global Power Shift, DOI 10.1007/978-3-642-55010-2_15,

1 Introduction

The *chukudu* (Fig. 1) is unique to Eastern Congo. It is a crossover between a wheelbarrow and a bicycle only found here; its robustness and practical simplicity make it the means of transport of choice for charcoal, bananas, construction materials, and other heavy things. The chukudu, now ubiquitous in Goma, the provincial capital of North Kivu, was allegedly invented in the 1970s when the politico-economic apparatus of the Mobutu state began to crumble, and Congo's infrastructure with it. After the volcano eruption in 2002, Goma's roads no longer permitted automated vehicles at all, and until the UN started repairing some of the main roads, chukudus were the only way to transport stuff. It reminds of a Flintstone mobile, and its simple outline in fact makes it particularly apt to navigate the choreography of potholes and puddles that Goma has become. The chukudu has become a viable source of employment for poor young men, which can be seen sweating while pushing a chukudu buried under huge bags, adorned with fading UN logos, of onions or other produce. At the center of one of Goma's roundabouts, a golden statue of a boy with a chukudu can be found. President Joseph Kabila erected it as a monument for the hard work of the people of North Kivu. Yet in another way, not born out by Kabila's intentions, it symbolizes the creative ways in which the Congolese have adapted to the conditions that places their country at the bottom of most developmental rankings. Congolese have become exceptionally creative in surviving in the 'debris of the state' (Kabamba 2008). And the chukudu seems a resolute proof of the capacity to navigate a country the size of Western Europe, yet without the infrastructure that makes Europe so small. The chukudu was invented about the time when *Maréchal* Mobutu pronounced—in what was still Zaire—the famous words *debrouillez-vous* ('fend for yourselves'). The chukudu seems to attest that the Zairois took this to heart.

While the chukudu stands symbol for human creativity under conditions of extreme suffering, I want to propose that it can tell much more about Congo's predicament and the role of technological infrastructure, or rather the disintegration thereof, than contemporary studies of state failure allow for. The discursively expressed unwillingness of the Zairois state to provide for its citizens in even the most basic services still haunts the specter through which donors and the humanitarian community understand Congo's contemporary predicament: its state failure is a social ill, to be remediated through programs targeting 'governance'. In their search to hold human actors accountable for conflict and underdevelopment in situations such as in Congo, studies of state failure have overlooked the importance of infrastructure. Loyal to the 'social' predicate in social sciences, Congo's predicament is dominantly explained in terms of a 'broken social contract' between the state and its citizens, attributing Congo's state failure to the absence of viable 'institutions' (Leonard et al. 2011; Kabamba 2012). This understanding translates into international efforts geared towards fixing the country's broken institutions that

Fig. 1 Congolese with a *chukudu* in Goma (picture by author, November 2010)

are to be responsible for re-establishing the social contract.[1] Indeed, many of the western donor programs concern efforts at changing governance patterns by educating Congolese government officials or security forces in such western values as accountability, equality, and justice that are considered to underpin well-functioning democratic states. Thus, the mainstay of external state building efforts in the DRC consists of capacity building programs and other forms of training targeting a change in behavior on part of the representatives of the formal Congolese state (Moshonas 2013). The most vocal critics of external reform efforts depict it as a masquerade, a discursive façade behind which donors and Congolese elites informally accommodate each other to not lose face (Trefon 2011).

Yet the specter of chukudu politics as a way of creatively negotiating life under conditions of the absence of viable transport and communication infrastructures raises a series of questions regarding the aptitude of state building efforts premised on a thoroughly 'social' understanding of state failure. Beyond the particulars of the DRC, this example allows introducing a phenomenon that more broadly marks the 'postcolonial condition' of much of Sub-Saharan Africa, but goes unacknowledged in various analytical approaches deployed to understand this very condition. This chapter explores the relative technological and infrastructural disassembly that is found in much of the post-colonial world particularly with reference to the disciplines of International Relations and Science and Technology Studies. The chapter

[1] The pervasiveness of social contract theory for the framing of state building efforts in policy circles is illustrated, for instance, by the fact that UNDP explicitly subtitles its 2012 policy framework report 'securing the social contract' (UNDP 2012).

focuses particularly on the political implications, most acutely articulated in the Democratic Republic of Congo, of infrastructural absences and disintegration for 'governmental power' or the capacity to effectively govern. This terrain remains hitherto unexplored within IR because of its bias towards 'social' explanations of differences in governmental capacity. In locating the source of Congo's predicament—and state failure more generally—in social processes or the absence thereof, mainstream approaches to IR present a distorted view of what state failure is about. Rather than a purely 'social' evil, Congo's state failure also needs to be understood in 'material' terms, that is, as a consequence of the absence of physical infrastructure that is constitutive of modern government. Conversely, while STS has focused largely on how technology is weaved through social and political relations in the 'developed world', the field has remained remarkably silent about places where technological infrastructure is not pervasive. What is specific about contexts such as Congo is not the ubiquity of technological infrastructures, but rather the relative absence, and progressive disintegration, thereof (Mann 2008: 361ff; Rotberg 2002). Thus, in the two fields most concerned with, respectively, questions of governmental power and science and technology, we cannot find any explicit discussion of the extent to which the lack of basic infrastructures might play a role in explaining conditions of underdevelopment, volatility, and outright state failure.

This chapter aims at addressing this imbalance, by posing to studies of state failure, the urgent question that Bruno Latour has posed to social science in general: where are the missing masses (Latour 1992)? However, whereas Latour posed his question rhetorically to ask why ubiquitous technologies that infuse social life do not play a role in social sciences accounts, the point of this chapter is that in many instances in the post-colonial world, those missing masses really are missing, or in utter decay. Indeed, according to technical expert literature, Congo is the most extreme example of infrastructural disassembly (Foster and Benitez 2011), yet this plays no more than an anecdotal role in social science studies of Congo's state failure. In contrast, it is argued here that we can better understand state failure and challenges to governmental power in the developing world by deploying insights from science and technology studies (STS).[2]

2 The Analytical Absence: Infrastructural Power

In mainstream political science and development studies, Sub-Saharan Africa's political and developmental problems are commonly presented as a 'governance' issue, which refers to the ensemble of institutionalized social practices by Africa's governmental elites (cf. Bayart 2009; Daloz 2003). Within this understanding, governmental power—the capacity to efficiently govern processes within a given

[2] cf. Barad (2003), Haraway (1991), Jasanoff (2004), Latour (1993), Law and Mol (1995), Mol (1998), J. Murdoch (1997).

political territory—is thus a thoroughly social problem, placing it firmly within the realm of social sciences. This reflects how international institutions and bilateral donors have construed their development efforts since roughly the end of the Cold War (Harrison 2004; World Bank 2008). In contrast to this problem-solving approach, critical approaches to development efforts and intervention focus on discursively silenced power disparities that condition the relations between intervener and intervened, often in light of liberal efforts to maintain an international status quo (Barnett 2010; Duffield 2007; Englebert and Tull 2008). Here, governmental power is often equated explicitly with Foucaultian 'knowledge-power' or concerns the social construction of political identities and institutions more generally (Chopra 2002; Hendrie 1997; Mosse and Lewis 2005).

What both mainstream and critical approaches to postcolonial politics and development share is a thoroughly *social* understanding of what political and economic underdevelopment in Sub-Saharan Africa is about. The 'social' bias of both is worth emphasizing because it makes these academic studies stand in marked contrast with the approaches taken to the same concerns in a range of 'technical' studies and audits of development programs, which tend to adopt a less 'social' language, and often invoke a range of infrastructural hindrances that stand in the way of completion of interventions and developmental efforts (see, for instance, African Union 2011; Bynens and Taylor 2012). 'Social' development efforts targeting governance processes founder when confronted with the absence of often basic infrastructures and technological conditions that are required social processes of governance possible. Reflecting similar insights by anthropologist Akhil Gupta (2012), most development efforts focus only on the bureaucratic process, while leaving implicit the infrastructure of rule that is needed to render these processes possible. Point of this discussion is, that the body of critical studies of development and intervention focus on the discursive production of the object to be intervened upon, and take for granted how disparities in the power to produce such pervasive or dominant framings is itself the product of disparities in technological infrastructures.

Drawing on insights from science and technology studies can redress this analytical imbalance. STS consistently analyzes the 'social' as an entanglement of social and material elements.[3] Such approaches allow for rendering visible the crucial importance of the absences and deficiencies of the 'infrastructure of rule' or 'political infrastructures' in Sub-Saharan Africa. These terms borrow from the notion, introduced by Mann (1984), of 'infrastructural power' as the capacity to effectively entangle governmental efforts with the processes that they target. However, Mann's understanding differs from the one advanced here for it tends to refer to infrastructure and technology as a passive 'thing' exogenous to political negotiations and struggles (cf. Hecht 2012: 2).

However, taking insights from STS to Sub-Saharan Africa is by no means self-evident. STS has largely been used to study complex socio-technical entanglements

[3] For a more elaborate discussion, see, for instance, Latour (2005) and Müller (2012).

such as scientific laboratories and other highly modern settings in Europe or the USA, where the infrastructure of rule is so pervasive that we usually fail to notice it. To a larger extent than in Africa, infrastructures in such settings were made to behave as passive, stable, entities, making it possible to leave them out in the study of social processes (Olwig 2002). STS's focus on how technology is weaved through social relations has contributed to a better understanding of the intricacies of power in the 'developed world', but STS has remained remarkably silent about its implications for politics and development in the developing world.[4] In that regard, Congo's predicament presents a critical case for STS. As we will see, the main difference between existing accounts and the STS-inspired approach presented here, resides in the role that 'things' such as technology and infrastructure, and their absences, play in accounting for state failure and disparities in governmental power in Sub-Saharan Africa.

How can we study the importance of infrastructures in postcolonial Sub-Saharan Africa when it has consistently been silenced? While technological infrastructures are by and large absent of studies of Congo's politics in particular and African politics more generally, several works discuss the relationship between technological infrastructures and development on the continent. Given that there is hardly any data directly available, we can start exploring this concern by looking at two bodies of scholarship addressing disparate issues. First are—mainly Francophone—studies of Sub-Saharan Africa that build on insights from geography to map the impact of colonial state-building efforts on postcolonial politics and social relations. Second is the still expanding body of technical policy documents that assesses contemporary Sub-Saharan African Infrastructural absences as a hindrance to, or steppingstone for, economic development.

What the first body of work, comprising geographical studies of colonial rule—often of an inclination not unsympathetic to Marxism—shares, is an appreciation for how the spaces of colonialism have survived—or not—to influence the 'postcolonial condition'. A striking insight of this body of work is the argument that colonial rule has consistently been draped around the edges of the African continent, only to penetrate the territory to the extent that exploitation of natural resources required it (Badibanga 1985, cf. Benton 2006; Boone 2003). For instance, a map of contemporary population distribution by Coquery-Vidrovitch (1988: 50) reveals that much of postcolonial urbanization is still is concentrated around colonial coastal outposts and along main infrastructural corridors carved out during colonial rule (Fig. 2).

As Badibanga (1985: 849) puts it: contemporary African towns are draped around what used to be centers of colonial activity—'around a port, border, or primary resource'. Postcolonial social developments are strongly conditioned by the infrastructural pathways violently carved out by colonial rule. Unless postcolonial regimes substantially intervene in infrastructure development (the megalomaniac infrastructural development schemes pushed by African leaders

[4] Save for a few exceptions, most notably Hecht (2012) and Rottenburg (2009a, b).

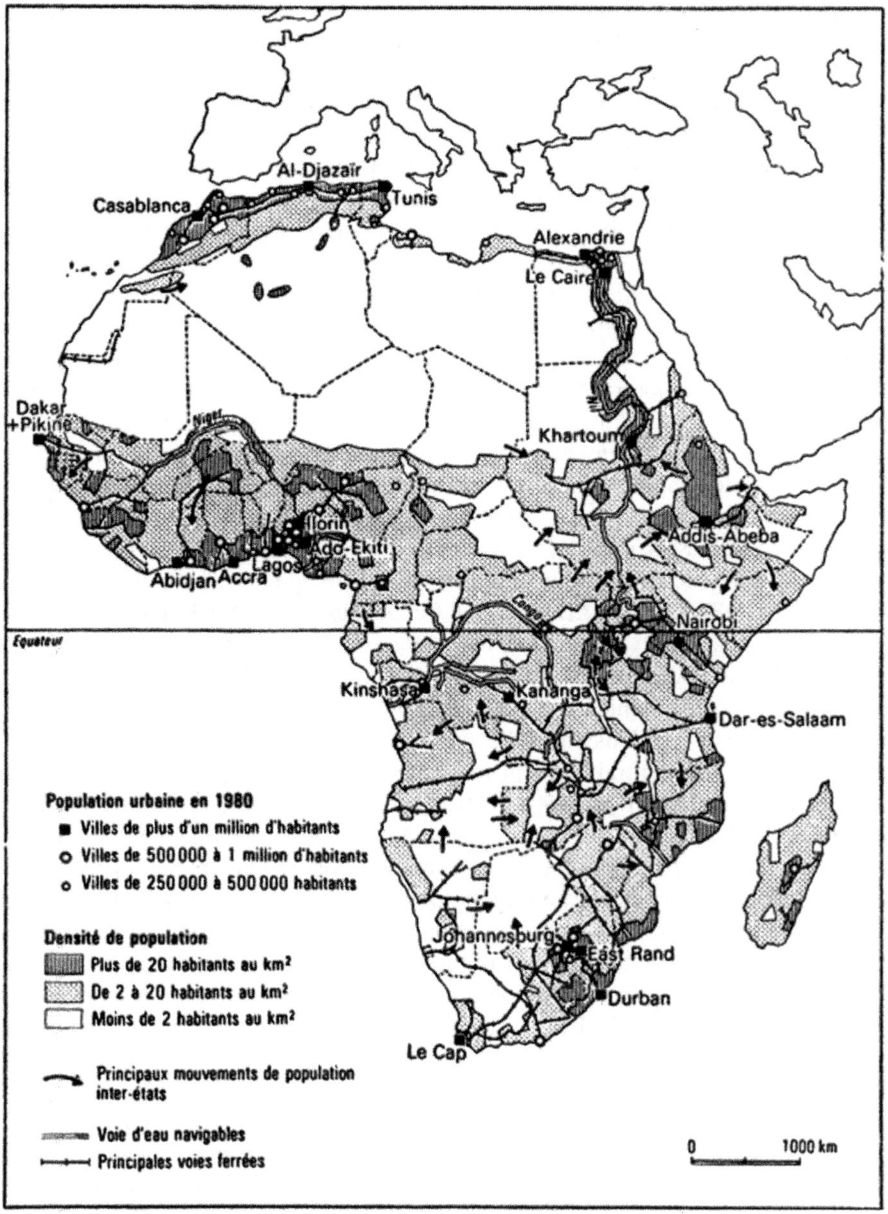

Fig. 2 Population distribution in postcolonial Africa

and undertaken by western engineers ('white elephants') have, for most part, fallen into disuse), postcolonial rule seems condemned to perpetuate the colonial gaze of the extraverted political economy (cf. Mann 2008: 364). And even if post-colonial regimes—like Mobutu's Zaire—extended infrastructures into previously unexplored areas, these were perceived locally as a problematic penetration of western-backed dictatorial power into society (Fairhead 1992; Ntoko 2011). While this line of research echoes a particular post-Marxist understanding of technologies that is heavily biased to seeing infrastructure as a passive conduit of governmental power (see, for instance, Feenberg 2005), it does raise the question to which degree the postcolonial condition is a matter of the path-dependences generated by selective infrastructure development under colonial rule. Arguably, such path-dependencies limit and direct contemporary processes of government in ways that differ from countries where infrastructure development followed a different logic. What they point out is essentially an unexplored territory for research that would incorporate insights from existing work on the interactions between science and technology and colonial or imperial governmental power[5] into the study of contemporary African politics and postcolonial state formation more generally.

The second body of literature is less concerned with the heritage of colonial infrastructure development for postcolonial social relations, and more with the state in which contemporary African technological infrastructures find themselves. Contemporary studies of African infrastructure and its implications for development are singularly technical reports commissioned by development agencies such as the World Bank, DFID and the African Bank for Development.[6] Such studies show that infrastructure has hardly been expanding in many African countries since the end of colonialism and if it did, it often concerns privately owned projects for extractive purposes, the costs of which are discounted by African governments in order to be inviting to investment. This raises the question whether it matters that 'technologies of government'—a central concept in poststructuralist critiques of governmental power—are singularly in *private* and often international hands in Africa rather than in the hands of democratically accountable African governments (Ferguson 2005). A much-overlooked study points to an important concern in this regard: by meticulously mapping how science and technology in Sub-Saharan Africa is structured, it reaches the dual conclusion that 'the areas of science and technology most vigorously pursued are increasingly determined by foreigners', and that the choice of these areas 'is based primarily upon the effects of such choice on the developed countries' (Enos 1995: vii). If, as Mann (1984) hypothesized, infrastructural power is key to understanding the autonomous power of the 'modern' state, then what does it mean for African politics that infrastructural power is *not* part of African state apparatuses, but rather of *private* apparatuses steered elsewhere? Africanist J.-F. Bayart famously characterized African political economy as extraverted (Bayart 2000), by which he meant that African postcolonial political and economic

[5] See, for instance, Carroll (2006), Guldi (2012), Hannah (2000), Headrick (1979), Mitchell (1988), Murdoch and Ward (1997).

[6] African Union (2011), Bynens and Taylor (2012), Foster and Briceño-Garmendia (2010), Hoeffler (1999), Reed and Miranda (2007), Schwartz et al. (2004), The African Development Bank Group (2010), Ulimwengu et al. (2009).

relations are largely directed towards, and conditioned by, external power sources. A cursory overview of literature dealing with political infrastructure indicates that this pattern also holds for the political dynamics related to technology and infrastructure. Yet counter-examples, where African governments appropriate transnational private-sector infrastructure programs—in the case of Rwanda and South Africa most notably, (see Booth and Golooba-Mutebi 2012; Schensul 2008)—caution against inferring simple causal relationships between technology and politics.

The similar extraverted bias pervades the contested design of technologies in Africa. Literature on 'tropical underdevelopment' raises the concern of affordances and propensities of technological infrastructures and the possibility of translating the use value of western technologies into different contexts. This issue has been addressed by a number of studies of the late 1990s which attested that there seems to be a little-discussed barrier constituted by the fact that technologies (including such basic infrastructures as paved roads) originally designed in and for use in the 'global north' quickly perish under conditions of tropical humidity and high temperatures, meaning that development efforts hinging on such technologies might work for a few years to quickly founder (Oyebisi 2000; Spencer 1996). Indeed, many of the colonial technological infrastructure schemes depended heavily on constant maintenance and repair operations, which perished first as African regimes gained independence and the engineers left (Pourtier 1991), raising the issue whether other types of infrastructural projects—not dependent on concrete and asphalt—would be more propitious. If one adds to that the scarcity of spare parts, it becomes easy to envision that current development models based on simple technology transfer miss the complete picture. Yet often, development workers attribute these failures to Africans' lack of ability to learn and use technologies or their willingness to engage in maintenance.[7] This not only holds for logistical infrastructure—the most concrete example of 'infrastructure of rule'—but also for bureaucratic apparatuses that were, during colonial times, distributed spatially between metropole and colonial administrative center.

A counterexample of technologies that are particularly suited to travel through tropical contexts is the 'flexible technology' of the Zimbabwean bush pump as brilliantly discussed by De Laet and Mol (2000). In similar vein, a group of STS scholars has recently produced a nuanced attempt to rethink knowledge infrastructures in Sub-Saharan Africa, in which they first and foremost emphasize a dearth of knowledge of African knowledge infrastructures (Jackson et al. 2008). This attests to the observation that we need to nuance sweeping generalizations regarding the importance of infrastructure for postcolonial politics on the basis of in-depth empirical analysis. The following presents a discussion of these dynamics for the case of Congo.

---

[7] Source: interviews conducted in Kinshasa, September 2012.

3 The Colonizing Effort as an Infrastructural Enterprise[8]

In historical accounts of the colonial enterprise, what is noticeable is that from the outset, the battle was 'directed mainly against the territory's forbidding landscape, not its people' (Hochschild 1998: 125). Absent road structures or railways, in 1890, journeys of a few miles could take weeks, and cost the lives of thousands of porters. Pivotal in colonizing schemes was infrastructure: both for purposes of extraction, but also to be able to in effect rule a territory that is 80 times the size of Belgium. Over the developing infrastructure, Belgians were exponentially able to add machinery to their relations to the Congolese soil and populations. In the 1950s, 'there was no good reason to believe that Congo would do otherwise than continue to assemble the components of a modern industrial economy' (Reno 2006: 45). At independence, the territory was mapped; translated into statistics that revealed its constituent parts as predictable and legible aggregate entities; and carefully maintained infrastructures allowed efficient governance at a distance. By the 1930, Congo was endowed with a state apparatus, consisting of an entanglement—albeit exploitative—of humans and non-humans, bodies and copper, concession companies, bureaucracies and infrastructures.

Yet the most striking aspect of the Belgian colonial state was precisely that it was most firmly entangled with those material entities that formed part of what the Belgians considered *l'Afrique utile*. The colonial state was an apparatus geographically concentrated around those spheres of activities of 'rational' interest to the regime (Pourtier 1989). Hochschild speaks of an 'apparatus of exploitation' (Hochschild 1998: 185), to highlight that the socio-material entanglements of Congo's political economy formed a transnational extractive assemblage that forcibly translated Congolese *natural* products into *economic* commodities through complex machinations composed of Congolese labor, machines, colonial law, shipping, and so forth. The Force Publique—'la police privée des colonisateurs'—was to 'insure tranquility and security where foreign nationals were found' (Shaw 1984: 17 and 57), which in practice coincided with the technical assemblages of extraction. In technological infrastructural terms, then, we'd see Congo as a rather flat geographical relief—with large parts of Congolese not intensely entangled with technological entities of governmental infrastructures—peaking in the more intense socio-technical entanglements of extractive assemblages which were draped around the country's edges. At independence (30th of June 1960), Congo was Africa's second most industrialized country (Fairhead 1992: 17); it boasted infrastructures that filled even South Africans with envy at the time (Fig. 3). Big part of Belgian's colonial legacy is thus material—both in terms of what was removed from Congo (the extraction of mineral resources and human bodies), but also in terms of what was put in place: it would be hard to deny colonial towns, prisons, garrisons, railways, and other infrastructures of extraction and rule their significance for governmental power.

[8] This section is an adaptation from Schouten (2013).

Fig. 3 Remains of the industrial plant that made mining tools for Belgian Congo's gold mining company SOKIMO in Nizi, Ituri. *Source*: picture by author, November 2011

4 The Materiality of State Failure

While Congo became politically independent during the political roundtable in February 1960, an economic roundtable 2 months later broke up the colonial extractive apparatus. In effect, it entailed the disentanglement of much of the economic infrastructure, mineral resources, and other sources of wealth (the whole of Congo had practically been carved up between concessionary companies) from the political apparatus—now an empty bureaucracy with significantly less funds to upkeep its infrastructure of rule (Depelchin 1992). The separation of the (now Congolese) state apparatus from the core of the Belgian economic apparatus meant that shortly after Congo's independence, with the assassination of Lumumba on January 16, 1961, 'the State apparatus began to crumble' (Peemans 1975: 154). This deliquescence was most notable in the transport sector, crucial infrastructure not only for the economy but for the bureaucracy to function as well (Pourtier 1993). In the 1970s, Mobutu, under the header of 'zairianization', nationalized Zaire's economic infrastructure and redistributed it amongst his peers, who, lacking any corporate management experience, sold off much of the material assets and lived on easy rents generated by steadily disassembling 'state' enterprises (Schatzberg 1980). Mobutu, very weary for challenges to his power, actively weakened the state security apparatus and eventually also conceived of economic infrastructure as a site of possible entanglements *against* him:

> Mobutu believed that investments in economic infrastructure, including those as simple as maintaining the network of roads left by the Belgian colonials, would pose a threat to his hold on political power by facilitating collective mobilization against his regime (Dunning 2005: 465).

Especially in the 1980s and 1990s, he increasingly let the state apparatus crumble; its bureaucratic infrastructure prey to unpaid civil servants, roads and the machines of the large mining firms of yore rusting in disuse while the economy informalized nearly completely. The two Congo Wars, between 1997 and 2003, only further destroyed much of what was left of the material infrastructure of the country. People have turned to walking with bicycles loaded with goods as the main means of logistics, on what used to be the country's main roads; for a Ministry of Mining official, a visit to a single mining site to levy taxes can take up to 5 days on foot through the jungle. According to a recent World Bank report on infrastructure in Congo,

> The Democratic Republic of Congo faces what is probably the most daunting infrastructure challenge on the African continent [. . .] Road and rail infrastructure are in dilapidated condition, and the rail network has fallen into disuse [. . .] The DRC's infrastructure backbones have yet to form a national network. (Foster and Benitez 2011: 1, 5)

As of 2012, Congo—two-thirds the size of Western Europe—only has around 3,000 km of paved roads (Bynens and Taylor 2012: 4). Social relations in Congo—including government—are to a larger degree *un*mediated by entities that would render them durable.

5 Chukudu Politics

Life in Congo is extremely complex. In response to the material disassembly of the state, the number of human intermediaries has proliferated, to the extent that in the 1990s, it was estimated that up to 1,000,000 state functionaries made up Congo's baroque bureaucracy (Meni 2000: 13). 'There is not the deficit of state but an excess of statehood practices: too many actors competing to perform as state' (Aretxaga 2003: 396). There are hardly infrastructures of rule in place to monitor and account for them; as a result, nobody knows how much individuals constitute this choreography of bureaucrats and security forces, nor would it be possible to keep track of them. If Wasinski's analysis of the importance of socio-technical mediators for military chains of command (Wasinski 2011) is correct, then much of the disciplinary problems of the Congolese armed forces (FARDC) could also be interpreted in terms of the absence of material conduits of hierarchy (Jackson and Rosberg 1982b: 10). More generally, if such categories as 'national population' and 'the economy' only exist as governable objects in virtue of statistical tools (see, for instance, Elden 2007), then the absence of aggregate national data in Congo means we hardly speak of such entities in Congo. The last comprehensive population census took place in 1984, but a biometric census of Congolese state functionaries, currently performed by international organizations, is to remedy the uncertainty over numbers of state representatives.

In face of the human over-presence of the state, 'être branché'—to be connected—has become the main tactic of survival, while wads of money are the central material lubricant for social relations. As a response to the material disassembly of the state apparatus, the Congolese state is thus a 'mediated state' (Raeymaekers et al. 2008), constantly negotiated and reproduced in individual interactions; it is thus not a state of atomic disentanglement, but rather one of intense human entanglement. As a Ugandan frequently doing business in Congo strikingly joked: 'In Congo, you don't need technical know-how, you need technical know-who'.[9] However, to say that governance in Congo is 'networked' and 'personalized' (Acemoglu et al. 2004; Jackson and Rosberg 1982a) is thus to state the obvious. Yet seeing this as fundamentally different and an aberration from our own ways of governing—often performed within imaginary institutional spaces—would be to mistake a lack of material conduits for qualities or the absence thereof inherent in Africans themselves. Any effort at governmental activity—or avoiding it—becomes an instance of chukudu politics, a creative process of assembling at one's best capacity given partially disassembled technologies and infrastructures in an effort to sustain social relations and economic activity.

6 Conclusion

The social bias of IR vis-à-vis governmental power in Africa is pervasive. State failure and underdevelopment are commonly explained in terms of a 'broken social contract', foregrounding *social* dynamics as the source of its regression into an anarchic state of nature. In this chapter, I have tried to take a different route to understand governmental power in Africa by deploying insights from science and technology studies. State failure and lack of governmental power in Africa should not be attributed to something inherent 'social' or 'human', but can be to a considerable extent accounted for by the absence of the technical infrastructure of rule that makes it possible to stabilize social relations. This sensitivity to socio-material relations rather than social relations alone offers an utterly different description of what matters in state failure and what accounts for differences in governmental power. This is because STS entails the recognition of the importance of material underpinnings of rule—technological and infrastructural power, which in Congo, as in colonial times, resides largely with external actors. Rather than mere technical concerns marginal to political power, such socio-technical systems constitute the critical infrastructure of political order and accumulation (see, for instance, Barry 2006; Collier and Lakoff 2009; Mitchell 2009).

[9] Interview with gold trader, Arua, April 2012.

More broadly, STS uncovers as profoundly political the 'technical' issues revolving around who owns, controls, and builds Congo's infrastructure, and sheds a different light on the concerns of development programs. The implications of attention for the role of technological infrastructures for our analysis of governmental power in Sub-Saharan Africa leads to a different assessment of extant development policy, raising questions regarding the focal points of such interventions. If the goal is to enhance governmental capacity—whether of state representatives or of 'civil society'—then perhaps more attention should be paid to absences and deficiencies in technological infrastructures that mediate efficient rule. If effective rule is inextricably linked to large infrastructural modernization schemes, then the controversial minerals-for-infrastructure deal between the Congolese government and the Chinese Sicomines, the largest arrangement of its kind, involving around $6 billion of investments, becomes a more viable policy option. In this scheme, Chinese investors would put up infrastructures in the DRC equal to the amount of minerals it would extract (mainly copper from the Gécamines concessions in Katanga). What makes the deal interesting for purposes of this discussion is that Chinese singularly aim at refurbishing currently decrepit colonial infrastructures geared towards exports via Tanzania in the east and Angola in the west. While it in principle extends Congolese infrastructures of rule, the Sicomines deal is thoroughly lodged in the extractive train of path-dependency set in motion by earlier colonial violence.

An alternative to large-scale modernization efforts 'from above' might be to investigate the material propensities of different ways of mediating social relations. Concretely, this means looking at water pumps, irrigation patterns, local expertise, and plant growth as all relevant elements that can be entangled in novel ways to produce more sustainable and equitable governance outcomes. Analytically, this would entail engaging with co-productionist approaches to the politics of science and technology that see the design of technologies and infrastructures as equally 'social' as, and not more privileged than, anthropological knowledges. For the broader discipline of IR this entails taking technological infrastructures not as exogenous to the study of such issues as governmental power, development, and governance more broadly, but rather as central elements that compose international relations. As Mann (1984, 2008) cautions, infrastructural power is profoundly ambiguous as infrastructural developments by one type of 'actor' can be appropriated creatively by others. The example of chukudu politics—understood as the creative entangling of tools in social relations under conditions of generalized technological disassembly—has attempted to elucidate how technology is not static but rather unstable, and subject to creative agency rather than fixed hegemonically from 'above' by the state, prevalent both in neo-Marxist interpretations of 'technopolitics' and determinist approaches to IR (e.g. Feenberg 2005; Fritsch 2011).

Conversely, while IR has little to say about the difference 'things' make to political power, taking STS to Congo shows how science and technology studies fall short in studying technology were the possession of even basic technology makes most difference to relations of power. While it is often heard that 'science and technology are everywhere' or that 'our everyday physical environment has become peppered with the tools and techniques of surveillance' (Aas et al. 2009: 258), in fact, radical disparities exist in the spatial distribution and concentration of complex infrastructures across the globe; disparities, that correlate with differences in governmental power. It is thus only selectively true that '[t]he social and cultural relations between individuals in late modern societies would seem to be, more than at any time in the past, mediated via material objects' (Dant 2006: 290). Yet STS, in its focus on the ubiquitous presence of materiality in social relations, has been extremely bad in dealing with the significance of *absences* of these very same material conduits of power (Fowles 2010: 31). STS appreciates the creative ways in which human subjects entangle with technological devices to produce new modes of governance, but fails to recognize the disparities in governmental power that result from the differences between entanglements figuring airplanes or chukudus. This chapter should be seen as a tentative invitation for STS to engage with these absences and asymmetries as politically meaningful.

References

Aas, K. F., Gundhus, H. O., & Lomell, H. M. (Eds.). (2009). *Technologies of in security: The surveillance of everyday life*. New York, NY: Routledge-Cavendish.

Acemoglu, D., Verdier, T., & Robinson, J. A. (2004). Kleptocracy and divide-and-rule: A model of personal rule. *Journal of the European Economic Association, 2*(2–3), 162–192.

African Union. (2011). *Exploiting natural resources for financing infrastructure development – policy options for Africa*. Addis Ababa: African Union.

Aretxaga, B. (2003). Maddening states. *Annual Review of Anthropology, 32*, 393–410.

Badibanga, A. (1985). L'urbanisation mimétique et l'extraversion des villes africaines. *Tiers-Monde, 26*(104), 849–859.

Barad, K. (2003). Posthumanist performativity: Toward an understanding of how matter comes to matter. *Signs: Journal of Women in Culture and Society, 28*(3), 801–831.

Barnett, M. N. (2010). *The international humanitarian order*. London: Routledge.

Barry, A. (2006). Technological zones. *European Journal of Social Theory, 9*(2), 239–253.

Bayart, J.-F. (2000). Africa in the world: A history of extraversion. *African Affairs, 99*(395), 217–267.

Bayart, J.-F. (2009). *The state in Africa: The politics of the belly* (2nd ed.). Cambridge: Polity.

Benton, L. (2006). Spatial histories of Empire. *Itinerario*, 30(3), 19–34.

Boone, C. (2003). *Political topographies of the African state: Territorial authority and institutional choice*. Cambridge: Cambridge University Press.

Booth, D., & Golooba-Mutebi, F. (2012). Developmental patrimonialism? The case of Rwanda. *African Affairs, 111*(444), 379–403.

Bynens, E., & Taylor, G. (2012). *Supporting infrastructure development in fragile and conflicted-affected states. Case study – DR Congo*. London: DFID.

Carroll, P. (2006). *Science, culture, and modern state formation*. Berkeley, CA: University of California Press.

Chopra, J. (2002). Building state failure in East Timor. *Development and Change, 33*(5), 979–1000.

Collier, S. J., & Lakoff, A. (2009, February). On vital systems security. *International Affairs Working Paper* 2009-01.

Coquery-Vidrovitch, C. (1988). Vlles coloniales et histoire des Africains. Vingtième Siècle. *Revue d'histoire, 20*, 49–73.
Daloz, J.-P. (2003). "Big Men" in Sub-Saharan Africa: How elites accumulate positions and resources. *Comparative Sociology, 2*, 271–285.
Dant, T. (2006). Material civilization: Things and society. *The British Journal of Sociology, 57*(2), 289–308.
de Laet, M., & Mol, A. (2000). The Zimbabwe bush pump: Mechanics of a fluid technology. *Social Studies of Science, 30*(2), 225–263.
Depelchin, J. (1992). *From the Congo free state to Zaire: How Belgium privatized the economy: A history of Belgian stock companies in Congo-Zaïre from 1885 to 1974*. Dakar: CODESRIA.
Duffield, M. R. (2007). *Development, security and unending war: Governing the world of peoples*. Cambridge: Polity.
Dunning, T. (2005). Resource dependence, economic performance, and political stability. *Journal of Conflict Resolution, 49*(4), 451–482.
Elden, S. (2007). Governmentality, calculation, territory. *Environment and Planning D: Society and Space, 25*(3), 562–580.
Englebert, P., & Tull, D. M. (2008). Postconflict reconstruction in Africa: Flawed ideas about failed states. *International Security, 32*(4), 106–139.
Enos, J. L. (1995). *In pursuit of science and technology in sub-Saharan Africa: The impact of structural adjustment programmes*. New York, NY: Routledge.
Fairhead, J. (1992). Paths of authority: Roads, the state and the market in Eastern Zaire. *The European Journal of Development Research, 4*(2), 17–35.
Feenberg, A. (2005). Critical theory of technology: An overview. *Tailoring Biotechnologies, 1*(1), 47–64.
Ferguson, J. (2005). Seeing like an oil company: Space, security, and global capital in neoliberal Africa. *American Anthropologist, 107*(3), 377–382.
Foster, V., & Benitez, D. A. (2011). *The Democratic Republic of Congo's infrastructure – a continental perspective*. New York, NY: The World Bank.
Foster, V., & Briceño-Garmendia, C. (Eds.). (2010). *Africa's infrastructure – a time for transformation*. Washington, DC: World Bank.
Fowles, S. (2010). People without things. In M. Bille, F. Hastrup, & T. F. Sørensen (Eds.), *The anthropology of absence: Materialisations of transcendence and loss* (pp. 23–41). New York, NY: Springer.
Fritsch, S. (2011). Technology and global affairs. *International Studies Perspectives, 12*(1), 27–45.
Guldi, J. (2012). *Roads to power: Britain invents the infrastructure state*. Cambridge, MA: Harvard University Press.
Gupta, A. (2012). *Red tape: Bureaucracy, structural violence, and poverty in India*. Durham: Duke University Press.
Hannah, M. G. (2000). *Governmentality and the mastery of territory in nineteenth-century America*. Cambridge: Cambridge University Press.
Haraway, D. J. (1991). *Simians, cyborgs, and women: The reinvention of nature*. New York, NY: Routledge.
Harrison, G. (2004). *The World Bank and Africa: The construction of governance states*. London: Routledge.
Headrick, D. R. (1979). The tools of imperialism: Technology and the expansion of European colonial empires in the nineteenth century. *The Journal of Modern History, 51*(2), 231–263.
Hecht, G. (2012). *Being nuclear: Africans and the global uranium trade*. Cambridge, MA: MIT Press.
Hendrie, B. (1997). Knowledge and power: A critique of an international relief operation. *Disasters, 21*(1), 57–76.
Hochschild, A. (1998). *King Leopold's ghost*. New York, NY: Houghton Mifflin.

Hoeffler, A. (1999). *Challenges of infrastructure rehabilitation and reconstruction in war-affected economies*. Oxford: University of Oxford.

Jackson, S. J., Batcheller, A. L., Edwards, P. N., Bowker, G. C., Cisler, S., & Star, S. L. (2008). *Extending African knowledge infrastructures: Sharing, creating, maintaining*. Washington, DC: World Bank.

Jackson, R. H., & Rosberg, C. G. (1982a). *Personal rule in Black Africa: Prince, autocrat, prophet, tyrant*. Berkeley, CA: University of California Press.

Jackson, R. H., & Rosberg, C. G. (1982b). Why Africa's weak states persist: The empirical and the juridical in statehood. *World Politics, 35*(1), 1–24.

Jasanoff, S. (2004). *States of knowledge: The co-production of science and social order*. London: Routledge.

Kabamba, P. S. (2008). *Trading on war: New forms of life in the debris of the state*. Unpublished PhD thesis, Columbia University, New York.

Kabamba, P. S. (2012). *In and out of the state: Working the boundaries of power in the Democratic Republic of the Congo*. Open Anthropology Cooperative Press Working Paper Series (15).

Latour, B. (1992). Where are the missing masses? The sociology of a few mundane artifacts. In W. E. Bijker & J. Law (Eds.), *Shaping technology/building society: Studies in sociotechnical change* (pp. 225–258). Cambridge, MA: MIT Press.

Latour, B. (1993). *We have never been modern*. Cambridge: Harvard University Press.

Latour, B. (2005). *Reassembling the social: An introduction to actor-network-theory*. Oxford: Oxford University Press.

Law, J., & Mol, A. (1995). Notes on materiality and sociality. *The Sociological Review, 43*, 274–294.

Leonard, D. K., Mushi, F. M., Samantar, M., & Vincent, J. (2011) *Social contracts and security in Sub-Saharan African conflict states: The Democratic Republic of Congo, Sierra Leone and Somalia*. Paper presented at the African Studies Association Conference.

Mann, M. (1984). The autonomous power of the state: Its origins, mechanisms and results. *European Journal of Sociology, 25*(2), 185–213.

Mann, M. (2008). Infrastructural power revisited. *Studies in Comparative International Development, 43*(3–4), 355–365.

Meni, M. (2000). L'importance du secteur informel en RDC. *Bulletin de l'ANSD, 1*, 21–40.

Mitchell, T. (1988). *Colonising Egypt*. Berkeley, CA: University of California Press.

Mitchell, T. (2009). Carbon democracy. *Economy and Society, 38*(3), 399–432.

Mol, A. (1998). Ontological politics. A word and some questions. *Sociological Review, 46*(S), 74–89.

Moshonas, S. (2013). Looking beyond reform failure in the Democratic Republic of Congo. *Review of African Political Economy, 40*(135), 132–140.

Mosse, D., & Lewis, D. (Eds.). (2005). *The aid effect: Giving and governing in international development*. London: Pluto Press.

Müller, M. (2012). Opening the black box of the organization: Socio-material practices of geopolitical ordering. *Political Geography, 31*(6), 379–388.

Murdoch, J. (1997). Inhuman/nonhuman/human: Actor-network theory and the prospects for a nondualistic and symmetrical perspective on nature and society. *Environment and Planning D: Society and Space, 15*(6), 731–756.

Murdoch, J., & Ward, N. (1997). Governmentality and territoriality: The statistical manufacture of Britain's 'national farm'. *Political Geography, 16*(4), 307–324.

Ntoko, K. D. (2011). *Roads and development under dictatorship: A study of the Democratic Republic of Congo*. Unpublished PhD thesis, New School University, New York.

Olwig, K. (2002). Landscape, place, and the state of progress. In R. D. Sack (Ed.), *Progress: Geographical essays* (pp. 22–60). Baltimore, MD: John Hopkins University Press.

Oyebisi, T. O. (2000). On reliability and maintenance management of electronic equipment in the tropics. *Technovation, 20*(9), 517–522.

Peemans, J.-P. (1975). The social and economic development of Zaire since independence: An historical outline. *African Affairs, 74*(295), 148–179.
Pourtier, R. (1989). Les États et le contrôle territorial en Afrique centrale: principes et pratiques. *Annales de Géographie, 98*, 286–301.
Pourtier, R. (1991). L'inéluctable défi des transports. *Politique Africaine, 41*, 22–31.
Pourtier, R. (1993). Désorganisation des transports et spirales du sous-développement au Zaïre. In C. Blanc-Pamard (Ed.), *Dynamique des systèmes agraires: politiques agricoles et initiatives locales: adversaires ou partenaires* (pp. 49–69). Paris: ORSTOM.
Raeymaekers, T., Menkhaus, K., & Vlassenroot, K. (2008). State and non-state regulation in African protracted crises: Governance without government? *Afrika Focus, 11*(2), 7–21.
Reed, E., & Miranda, M. (2007). *Assessment of the mining sector and infrastructure development in the Congo Basin Region*. Washington, DC: WWF.
Reno, W. (2006). Congo: From state collapse to 'absolutism', to state failure. *Third World Quarterly, 27*(1), 43–56.
Rotberg, R. I. (2002). The new nature of nation-state failure. *The Washington Quarterly*, 25(3), 83–96.
Rottenburg, R. (2009a). *Far-fetched facts: A parable of development aid*. Cambridge, MA: MIT Press.
Rottenburg, R. (2009b). Social and public experiments and new figurations of science and politics in postcolonial Africa. *Postcolonial Studies: Culture, Politics, Economy, 12*(4), 423–440.
Schatzberg, M. G. (1980). The state and the economy: The "radicalization of the revolution" in Mobutu's Zaire. *Canadian Journal of African Studies/Revue Canadienne des Études Africaines, 14*(2), 239–257.
Schensul, D. (2008). From resources to power: The state and spatial change in post-apartheid Durban, South Africa. *Studies in Comparative International Development, 43*(3–4), 290–313.
Schouten, P. (2013). The materiality of state failure: Social contract theory, infrastructure and governmental power in Congo. *Millennium – Journal of International Studies, 41*(3), 553–574.
Schwartz, J., Hahn, S., & Bannon, I. (2004). *The private sector's role in the provision of infrastructure in post-conflict countries*. Washington, DC: World Bank.
Shaw, B. P. (1984). *Force publique, force unique: The military in the Belgian Congo, 1914–1939*. Unpublished PhD thesis, The University of Wisconsin, Madison, Wisconsin.
Spencer, D. S. (1996). Infrastructure and technology constraints to agricultural development in the humid and subhumid tropics of Africa. *African Development Review, 8*(2), 68–93.
The African Development Bank Group. (2010). Infrastructure Deficit and Opportunities in Africa. *Economic Brief, 1*(September), 1–15.
Trefon, T. (2011). *Congo masquerade*. London: Zed Books.
Ulimwengu, J., Funes, J., Headey, D., & You, L. (2009). *Paving the way for development: The impact of road infrastructure on agricultural production and household wealth in the Democratic Republic of Congo*. Washington, DC: International Food Policy Research Institute.
UNDP. (2012). *Governance for peace: Securing the social contract*. New York, NY: UNDP.
Wasinski, C. (2011). On making war possible. *Security Dialogue, 42*(1), 57–76.
World Bank. (2008). *Democratic Republic of Congo growth with governance in the mining sector*. New York, NY: World Bank.

The Power of Large-Scale Interactions Through Information Technologies and Changes in Cultural Identity Politics

J.P. Singh

Abstract Large-scale and dense communicative interactions among multiple actors are creating new and shared understandings in global politics that require re-conceptualizing power. The existing 'instrumental' conceptions of power explain the constraining or expanding influence of technologies on global actors, but underestimate the role of communication in these politics. However, information and communication technologies are both products of human agency to communicate and, in turn, intensify these communications. The current debates on cultural identity are taken as an exemplar of the new meanings that arise in global politics as a result of the intense and large-scale communicative interactions. National identity is not replaced but supplemented with other forms of cultural identity in a networked world. The concept of meta-power captures the emergence of these new meanings that information and communication technologies facilitate.

Keywords Power • Meta-power • Communication • Transformation • Interactivity • Cultural identity • Global actors

Information networks facilitate communicative interactions among new and old global actors at a scale that was perhaps unimaginable when undersea telegraph cables started connecting distant continents in the mid-nineteenth century. These interactions transform the way global actors communicate with each other and understand their identity in the global environment. The ability of interactive communication processes to propose new meanings among social actors is termed meta-power in this essay. Just as powerful institutions, be they nation-states or religious organizations, struggle to control interactions over information networks to reinforce existing cultural identities, these networks erase the differences among old identities and propose new meanings. National identity is supplemented with

---

J.P. Singh (✉)
George Mason University, Fairfax, VA, USA
e-mail: jsingh19@gmu.edu

M. Mayer et al. (eds.), *The Global Politics of Science and Technology - Vol. 2*, Global Power Shift, DOI 10.1007/978-3-642-55010-2_16,

other forms of cultural identity and the language of universal rights with cultural rights.

Communication networks, encompassing artifacts and human agency, involve multiple actors negotiating multiple meanings, which are different from the power-as-usual approach in which the meaning of politics has been defined and the actors merely utilize communicative 'instruments' to enhance or limit the attainment of pre-specified goals. In short, the meta-power of a communication network specifies the transformative dimension of global politics: it literally 'identifies' actors and their issues in helping us to understand collectively shared meanings among them. Communicative interactions both shape networks and are shaped by them. The network architecture of Facebook has, for example, allowed for new forms of collective identity formation while also being shaped by the communicative concerns that are brought up by its nearly one billion members. The collective outcomes of networked interactions, in general, continue to challenge how we collectively imagine ourselves—in terms of our individual identities, and the nature of our social ties with each other.

This essay first describes a few large-scale interactions before turning to the need for a new understanding of power. It then prefaces currently underway debates on cultural identity for empirical substantiation. A subsidiary argument throughout the essay, picking up on major strands of thinking on the social construction of technology, is the mutual constitution of information technology artifacts and social meanings (Hughes 2012; Pinch and Bijker 2012). I conclude that nation-state and national identities are not replaced but supplemented with new political formations and identities best described in networked terms. As these networks proliferate, nation-states sit bestride the politics of increasing global consciousness and local or transnational cultural identity movements.

1 Large-Scale Interactions

Interactions across social groups are the basis of global politics and have existed throughout history. However, the scale of these interactions has increased exponentially and, therefore, the ability of various groups, as opposed to only dominant groups such as nation-states or global firms, to contribute to the constitution of identities of issues and actors that flow from these interactions.

There are two predominant ways these interactions are conceptualized: one traces the influence of interactions through existing actors and their collective imaginaries, which remain essentially unchanged through these interactions; the second way is to trace them via new forms of organizing and organizations. Buzan and Little (2000: 168–169) note that the interactional capacity of the international system, dating back to the Sumerian city-states of 3500 BC, has greatly increased with communication technologies. However, the identification of the units of this system (variations of the nation-state) and the definition of their activities (security, commerce) remain unchanged in their account. Similarly, Fung et al. (2013)

provide a nuanced conception of how collective action is enabled through existing actors and institutions to enable various forms of advocacy, accountability, and deliberation throughout the world. Fung et al. balances pessimistic scenarios that assign no agency to actors stemming from technology against deterministic ones that overestimate technology's effects. Nevertheless, the identity dimensions of actors and their institutions remain essentially unchanged. The proliferating literature on electronic government also offers a similar conclusion: the varying levels of engagement among citizens and governments in the provision of services are novel, but not so transformative (Livingston and Walter-Drop 2014; Gil-Garcia 2012). We must also keep in mind that nation-states themselves understand that information networks weaken the legitimacy of borders and strive to maintain existing meanings, ranging from authoritarian measures to censor information to the practice of subtler forms of national identity politics which, a few might argue, includes the dominant role of the United States in Internet governance or control of the root servers (Deibert et al. 2008). The current soft power politics and the use of the social media in public diplomacy to improve the image of nation-states can also be analyzed in similar ways (Nye 2004; Cull 2013).

Information networks also enable new forms of organizing and organizations. Clay Shirky's (2008) thesis on organizing without organizations with the falling transaction costs of joining networks is well known in speaking to the decentralization of hierarchies and new ways of organizing and imagining politics. These new political organizations are perhaps best understood with the language of networks (Slaughter 2004; Grewal 2008). Choucri (2012) is inching toward a similar perspective in noting a "diffusion of power, different types of asymmetries, and a relatively weak hierarchy" (Choucri 2012: 223). The new subjectivities and meanings in global politics being proposed should not surprise us. After all, Marxian identification of class systems or identities also followed modes of production with feudal or industrial technologies. Anderson (1983), Castells (1997), and Rosenau and Singh (2002) propose similar theses.

Technologies facilitate social membership. Communication scholars have long argued that the media they study are central to social and political imaginations (Innis 1950; McLuhan 1964). The incredible growth of various forms of social media provides the best-known example of new forms of large-scale social membership, which I will argue later, propose new cultural meanings. Such large-scale interactions were imaginable only for the format of the nation-state merely a decade ago. There are nearly one billion Facebook users (CheckFacebook 2013), or one-seventh of the world's population, and nearly half of the population of United States utilize Facebook with 159 million users. The top ten countries on Facebook include developing countries such as Brazil (72 million), India (64 million), Indonesia (48 million), Mexico (43 million), Turkey (33 million) and the Philippines (30 million). An estimated 555 million twitter users worldwide send an average of 58 million tweets each day (Statistic Brain 2013). Meanwhile one billion unique visitors go to YouTube each month and watch over four billion hours of YouTube video (YouTube 2013). There are also regional platforms: the Chinese micro-blogging platform Sina Weibo had over 500 million users within 3 years of its launch in 2009 (Wikipedia 2013).

Yet, numbers by themselves, awe inspiring as they might be, do not reveal the structuring influences of human agency and the constitution of artifacts themselves. The most important co-constitution involves story telling. In terms of social membership a dominant story about individuals remains their national citizenship. Even the Facebook statistics described above corroborate that claim. However, national belonging is one among many stories about social membership being narrated on social media, which has facilitated multiple forms of global collective action and the ability for individuals to imagine themselves as members of multiple and overlapping groups. Analyses of Facebook connections range from the degrees of separation among individuals to the language networks that bring people together.[1] While uniting members across organizations (Karpf 2012), this has also resulted in polarizing, stove-piping, or homophily effects (McPherson et al. 2001; Centola et al. 2007; Farrell 2012). Nevertheless the rise of global social movements whether it concerns indigenous people's rights or that of lesbian, gay, bisexual, and transgender (LGBT) populations has benefitted from the ability of members of these groups to narrate their stories—collectively or at near epistemic distances—online. Kollman (2007), for example, shows that the perceived legitimacy of LGBT rights through international networks has led to the proliferation of same-sex unions in established democracies, along extant (heterosexual) norms of civil union or marriage. Similarly, Keck and Sikkink (1998) notes that networks are about persuasion and socialization and human rights activists adopt frames that appeal to a wide variety of groups.

The architecture of networks has responded in various ways to facilitate narration of stories. The artifacts are 'user-friendly.' Unlike the classic stories of the Video Cassette Recorder that remained opaque to users even with detailed manuals (or, because of them), most social media do not need a lengthy primer beyond minimum forms of digital literacy. YouTube and Facebook are, therefore, continually updating their features to allow for social commentaries and digital content. Even traditional journalistic story-telling has changed and most online newspapers now enable reader comments and critique. However, the corporate social media owners have also responded in structurally controlling ways to problems brought up by users on the loss of privacy. For example, firms continually change the architecture to seemingly provide more transparency while simultaneously obfuscating new design features that make users' information publicly available especially for marketing purposes. Similarly, as detailed later, nation-states themselves try to control networks and the information or the meanings that flow across them. Nevertheless the evolution of large-scale interactions over social media meets the description of technological development that Bijker et al. (2012: 7) provide as "a

---

[1] One Facebook data analysis in November 2011 examined 721 million users with 69 billion friendships among them to find that there were only three to four degrees of separation—or "hops" among users—for most people (Backstrom, November 21, 2011).

non-determined, multidirectional flux that involves constant negotiation and renegotiation among and between groups shaping the technology."

2 Meta-power[2]

The ability of information technologies to foster interactions that alter the identity of actors and the meanings of issues in global politics is termed "meta-power" in this essay. This conceptual understanding, however, does not propose yet another form of technological determinism. Rather, information technologies can be understood as *affordances*, which Earl and Kimport (2011: 32) describe as "actions and uses that technology makes qualitatively easier or possible when compared to prior like technologies." These affordances arise from lowered costs and facility of large-scale and diverse interactions, encompassing many actors and engendering new meanings. Meta-power results from these affordances, but has to be understood as a social *process* rather than a determinative *resource*. Anthony Giddens (1984), similarly, notes that power must be understood through social relationships that sustain or transform particular practices.

Meta-power proposes meanings for global actors and their issues through communicative interactions.[3] Conceptually, therefore, *meta-power* is antecedent to instrumental notions of power: before actors exercise power in the traditional sense, meta-power specifies the meaning of their collectively understood identity and interests. When Platonic subjects gaze at the shadows on the wall in front of them, the meaning has been created for them through prior interactions. When nation-states fight territorial wars, similarly the meaning of a nation-state or security understood in territorial terms has already been *imagined*.[4] In practice, however, meta-power processes overlap traditional ways of exercising power. Actors may continue to do what they do, especially in a hierarchical environment, while decentralized relations elsewhere foster meta-power processes that create new meanings. Figure 1 illustrates the *conceptual sequence* in which information technologies lead to meta-power or the ability of interactions to change the identity of the actors and the meanings of issues in global politics. A feedback loop between outcomes and information networks would address the way prior interactions affix identity (Fig. 1).

It is important to distinguish meta-power from two forms of power that are often discussed: instrumental power and productive power. Technologies in the

[2] This section is adapted from Singh (2013a).

[3] The rise of cultural identity politics in the last few decades can be understood both as a new issue in global politics with new global actors who now make cultural identity claims.

[4] In his seminal monograph, Anderson (1983) shows how information technologies of a prior era, namely the printing press, enabled the formation of the European nation-state around linguistic lines as printing proliferated in the vernacular, rather than Latin.

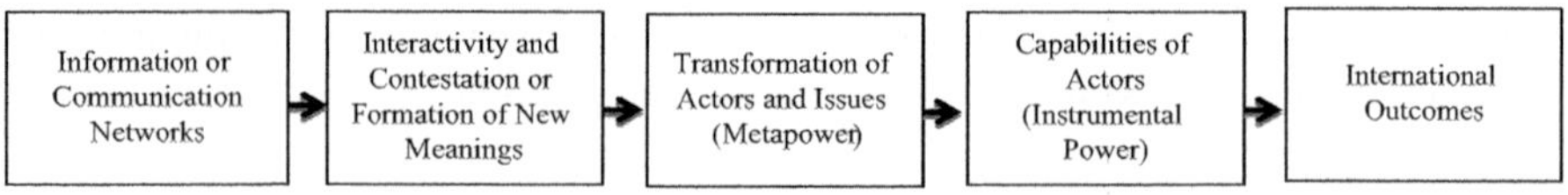

Fig. 1 Information networks and power

instrumental or traditional power equations enhance or diminish the capabilities of actors. Instrumental notions reign supreme in international relations in the outcomes of realist power politics, in fostering the interdependence in liberal institutionalist analyses, and in the class and ideological warfare of critical analyses (see Fritsch 2014). Many scholars, in these traditions, who had expected new meanings in our politics from information technologies, have now retreated to re-specify the old ones. For instance, a lively forum on "Who Controls the Internet?" in the *International Studies Review* (Eriksson and Giacomello 2009: 205–230) mostly responded by noting that existing state actors and ideologies control the Internet. Other scholarship on Internet governance has reached similar conclusions. Those lining up behind state actors include Goldsmith and Wu (2006), and Drezner (2007).

Scholarship on instrumental power is sophisticated in showing how states, firms, or classes reassert their rules or ideologies through new conditions of multilateralism or technological uncertainty. Nevertheless, when these scholars accord attention to information technologies, these technologies seem to be no different from nuclear bombs, medicines, or agricultural tractors. They miss the role of communication in meaning creation and cultural alteration because most dominant paradigms in international relations assume that human communicative interactions produce no new meanings although theories based in social constructivism now challenge these 'old' paradigms.

In contrast, conceptualizations of power as social processes that constitute the identities and interests of actors provide a useful point of entry for this essay.[5] Barnett and Duvall's (2005) concept of productive power is a variant of power in a constitutive sense, in explaining social meanings. Productive power is imagined as a discourse, that is "the social processes and the systems of knowledge through which meaning is produced, fixed, lived, experienced, and transformed" (Barnett and Duvall 2005: 55). However, meta-power as conceptualized here is different from productive power because it is antecedent to the exercise of other forms of power, follows from intensive and extensive communicative interactions, and has a normatively positive and transformative dimension. For example, the idea of interstate commerce might be changing in an era of networked firms and intense information exchanges where intra-firm transactions and value-added across several national frontiers may be more important than what one country exports. For example, are Apple products manufactured in China really American? Or, to take

[5] Sociological understandings of power can be traced back to Weber and Durkheim and elaborated in current contexts through Foucault and Bourdieu.

an example explained in the next section, can personal identity be understood in national terms when individuals inhabit several forms of cultural identity online? Similarly, the meaning of 'digital property', broadly defined as ownership of particular forms of knowledge and digital content, is being redefined through intense interactions among nation-states, firms, civil-society and advocacy groups, individuals, and academics and think-tanks. These struggles underscore social struggles over the meaning of property and a broadening of this definition to accommodate both private ownership and new forms of global sharing such as the creative commons or Wikipedia.[6]

Finally, there are intellectual antecedents to the term meta-power. Meta-power conceptualizations share with Baumgartner, Buckley and Burns (1975), and Burns and Hall (2013) the sociological understanding of what they also term meta-power to note that relationships among international actors enable organizational rules and world orders. However, these understandings do not involve the ascription of new meanings. Krasner (1985) had employed the concept of meta-power in a similar sociological fashion to speak to the ways in which the Third World sought to change the rules of the game through its advocacy in the 1970s and 1980s. Krasner concluded that the Third World lacked resources to change the meta-rules in its favor. This is not a surprise. Krasner's meta-power seems to be arrested in a structural straightjacket of a hierarchical ordering of actors. It ignored, like most social science does, the communicative stimuli of social interactions. Particularly, these precursors of meta-power fell short of coping with the cultural identity politics detailed in the following section.

3 The Rise of Cultural Identity Politics[7]

This section discusses two identity outcomes that result from interactive global cultural politics: multiple discourses of cultural identities circulating in large-scale networked environments, and the concurrent overlapping languages of 'universal' and 'limited' cultural rights in human rights discourses. In networked environments, what it means to be a nation-state or to command due obedience through traditional means comes undone. This is equally valid for civil society and individuals who search and negotiate for meanings of their identity and of the issues such as human security or human rights in which they indulge.

In an age of social media and content-based story-telling, the global governance contestations around information network and cultural content are especially interesting. The proliferation and liberalization of these networks offers the first instance of the challenges to cultural identity politics, which are then seconded with moves

[6] See, creative commons at http://creativecommons.org/ and Wikipedia at http://www.wikipedia.org/. Accessed October 2, 2013.

[7] This section is adapted from Singh (2008, 2011).

to encourage or thwart further debates through global governance institutions. Until recently, telecommunications and broadcasting networks tended to be national or public monopolies, but the last three decades have seen considerable economic liberalization, that has in turn led up to global controversies on flows of cultural content (Voon 2007). In addition, cultural content—films, television, music included—is now digital and its circulation over information networks expands the spheres of cultural controversies.

3.1 Battles over National Identity Preservation

Creative products such as movies, songs, or television productions provide a unique lens to examine the communicative basis of cultural identity politics.[8] They are readily understood as information technologies but being 'content' industries, they also facilitate cultural representations. Creative products are technologies of cultural representation, readily lending themselves both to interpretations from social actors and, in turn, responding to them. Even traditionally creative products such as heritage goods (museum, archives, cultural tourism) or fine and performing arts, linked in the past to ideas of patrimony or national identity, are now increasingly dependent on information technology networks. In a prior era, European music composers in the nineteenth century both responded to nationalism as well as shaped it. For instance, Sibelius' Finlandia composed around 1899–1900 gave expression to a rising tide of nationalism and substituted for the national anthem under Russian occupation. In our current era, it is becoming hard to find heritage 'goods' without a social media presence, and the forms of collective action they facilitate is hardly national anymore. For example, the 2013 Venice Biennale was widely reported as a twenty-first century information age exercise in curation that questioned all kinds of artistic and social boundaries (Wullschlager 2013).

The last two decades have featured fierce debates around these cultural products and information networks at the level of global governance and thus they approximate what STS scholars call *points of controversy* to unravel the deeper layers of the politics of technology (Pinch and Bijker 2012). These politics attend both to the affordances referred to above and the ways in which global actors alter the apparatus designs of technology through their use and advocacy.

Global governance organizations tend to feature nation-states prominently but a deeper look at two global organizations involved in flows of cultural products provides some appreciation for the deeper cultural identity battles. World Trade Organization has responded to global commercial organizations in facilitating networks that allow communication content to flow across borders. UNESCO has

[8] International relations scholarship concerned with the links between arts and entertainment and cultural identity includes Der Derian (1990), Goff (2007), and Sylvester (2009).

responded to nation-states and a few civil society actors that seek to protect 'national' borders from foreign cultural influences.

At a broader level though, both organizations are struggling to catch up with the affordances of new media. The WTO struggles to define rules for electronic commerce and other commercial flows over the Internet in an era when large parts of commerce do not obey the old-fashioned tariff rules: nation-state cannot enact tariffs or impose prices on cultural products that flow free in digital forms. Advocates of intellectual property have been the fiercest advocates for this regulation but they face fierce opposition (Sell 2013). Similarly, UNESCO processes argue for national processes to protect cultural identity when the latter may no longer be defined in national terms. The language of cultural diversity used in UNESCO literatures almost always speaks on behest of the nation-state.

A particularly interesting case of cultural identity debates at the international level is the UNESCO Convention on the Protection and Promotion of the Diversity of Cultural Expressions signed in 2005 with 148 votes for two against (United States and Israel). The Convention was the culmination of an effort to thwart efforts underway at the World Trade Organization (and, its predecessor, GATT) to liberalize trade in creative industry products that would allow cultural content (mostly films and television programs) to move across national borders with few restrictions (Singh 2011). In an earlier move, the WTO telecommunications accord of 1997, 95 % of the world's telecommunication 'conduits' over which this 'content' would flow had already been liberalized.

The WTO debates affected more than commercial concerns. The European Union invoked a "cultural exception" in arguing for the symbolic and representative importance of cultural content, but GATT and WTO, in its technical language, termed them 'audio-visual services'. The EU integration project had resulted in quotas on television programs since 1985 to encourage national representations on television (EU's Television Without Frontiers Directive) and TV programs from the United States were deemed a threat by EU negotiators both for the commercial viability of local television and also for the preservation of national cultural identities. The Uruguay Round of trade talks at the GATT (1986–1994) almost resulted in an impasse on this issue. In the final hours of the negotiations the EU, which bargains collectively on international trade issues, took an exemption on liberalizing trade in audio-visual products that was famously dubbed as a cultural exception. However, as subsequent moves within the WTO sought to liberalize global trade in creative products, European and other countries such as Canada moved toward creating another instrument that would supersede or bypass efforts at UNESCO, which culminated in the 2005 Convention.

Contestations are important for the social meanings that flow from and govern technologies. The negotiations around creative industries are hugely instructive. First, these negotiations foreshadow the way states endow meanings to their borders, which is important to preserving the legitimacy of the state (Goff 2007). At UNESCO, cultural identity became synonymous with national identity and for the Europeans centered on the use of national languages. In a prior era, Anderson (1983) spoke to the role of a communication technology such as the printing press in furthering this imaginary. In the UNESCO debates, nation-states turned to films, TV programs, and music to foster this imaginary, often in national languages. The important point is this: the efforts at UNESCO are often attempts by nation-states to preserve their legitimacy in a world of information networks where entertainment content easily crosses national frontiers. Similarly, the attempts at the WTO featured predominantly the prerogatives of global communication enterprises, which include Hollywood and Bollywood films, Korean and Latin American telenovelas, and the commercial prerogatives of the global music industry that includes four music firms that dominate sales (EMI, Sony, Universal, and Warner) or retailers such as Apple's iTunes.

Nation-states lead the debates on cultural identity at UNESCO and WTO. Even when civil-society is represented in these debates, as through the international non-governmental organization (INGO) network called "International Network for Cultural Diversity" at UNESCO, the final outcomes seldom reflect non-national prerogatives.[9] France, for example, is at the forefront of debates at UNESCO but constitutionally the nation-state is barred from admitting to anything but 'French' identity. Until recently, film or television subsidies could not be given for films other than in French language, while the state as a whole is silent on any form of sub-national identity. Therefore, cultural identities of France's minorities and religious groups are never officially recognized. These sub-national fissures are evident in adjoining states, Belgium and Spain, and also worldwide. Meanwhile, transnational identities—indigenous people's rights, for example, pose another challenge. The Internet is a counterpoint to these nationalisms: it is as alive on debates on global Muslim identities as on the Afro-indigenous identity movements in Colombia (Sandy 2010; Escobar 2001).

More generally, even a cursory look at the Facebook pages of one billion users would quickly reveal that their identities are neither limited to national languages nor always in conformity with due obedience to nation-states. If anything social media are often posed at loggerheads with the nation-state resulting in explicit bans on these global media in countries such as China, organizing protests against the state in the Arab and Persian countries, and leading to multiple identity formation in pluralist countries. Social media networks, rich in cultural content, have also been central to pushing and promoting the concerns of immigrant, ethnic, and sexual minorities almost all over the world. A film or a song about indigenous people's rights in Latin America sits uncomfortably across the kinds of national priorities articulated for the preservation of cultural diversity at UNESCO. As people imagine themselves as various genders, classes, and races in an age of social media, with its

[9] INCD was initially created and funded by the Canadian government. Its "Who We Are" website outlines the role for government in statements such as the following: "Governments have a right and responsibility to create policies which nurture domestic artists and develop the creative capacity and cultural industries of their societies, and not have them eroded by trade and investment agreements." http://www.incd.net/about.html accessed September 20, 2013.

rich cultural content, the stories they relate may or may not be about their national identities.[10]

Another way of examining these debates is to dissect further what nation-states choose to call 'cultural products' important for cultural diversity.

Table 1 provides a partial list of sectors at the WTO that are rich in cultural content. The European Union only focused on the Audio-Visual Content as the basis of cultural exception. Table 2 lists the countries that have liberalized their trade for various products that have cultural contents. Audio-visual has garnered the least number of commitments while tourism, the biggest industry in the world, has garnered 132 commitments. One might wonder, therefore, why a country might believe that its cultural identity is best preserved in protecting itself from a dozen foreign films but not millions of foreign tourists. With 79.5 million out of 983 million international tourist arrivals, France remained the top destination of international tourist arrivals in 2011 (UNWTO 2012). Tourism and associated industries such as travel are also highly dependent on information networks. France, however, remains at the forefront of debates on cultural diversity, most recently invoking the cultural exception for films and television for planned Transatlantic Partnership negotiations between the United State and the European Union launched in 2013.

3.2 Cultural Rights and the Challenge to Universality

The cultural diversity debates also resonate in those of cultural rights. The industrial revolution espoused the universal language of liberty, equality, and fraternity among all human beings, even though its actual realization had to wait another two centuries, post-colonialism, for all human beings to be included. Now individuals are imagining themselves along multiple dimensions and produce their own representations of identity through multimedia. Therefore, universal definitions of human rights are now supplemented with "less than universal" dimensions. Certainly the early advocacy for rights came from stigmatized, excluded, or marginalized groups: online spaces, in fact, allowed them to find each other and mobilize. In doing so they also redefined rights to mean cultural rights and not just universal rights (Rockhill and Gomez-Muller 2011).

Like security and property, human dignity is a representation. Thomas Paine's *Rights of Man* was such as aspiration, as are the current discourses over the Internet coming from groups that were denied rights or dignity. There is a plethora of communication scholarship, beginning with Said (1978), on how electronic images through film or television assign reason and rationality to white heterosexual males, while emotions and an inferior rank are assigned to women and other groups. Thus, a new language of cultural rights, arising from cultural identities (gender, ethnic,

[10] Please see Singh (2013b) on ways to think about cultural hybridity and identities in mediated forms.

Table 1 Partial list of creative sectors listed in WTO services sectoral classification list

Sectors and sub-sectors	Corresponding CPC
1. Business services	Section B
A. Professional services	
d. Architectural services	8671
F. Other business services	
a. Advertising services	871
p. Photographic services	875
r. Printing, publishing	88442
2. Communication services	
C. Telecommunication services	
D. Audiovisual services	
a. Motion picture and video tape production and distribution services	9611
b. Motion picture projection service	9612
c. Radio and television services	9613
d. Radio and television transmission services	7524
e. Sound recording	n.a.
9. Tourism and travel related services	
A. Hotels and restaurants (incl. catering)	641–643
B. Travel agencies and tour operators services	7471
C. Tourist guides services	7472
10. Recreational, cultural and sporting services	
A. Entertainment services (including theatre, live bands and circus services)	9619
B. News agency services	962
C. Libraries, archives, museums and other cultural services	963
D. Sporting and other recreational services	964

Note: This is a partial list. Technically, sectors such as telecommunications, listed without sub-categories here, are related to the creative sector. However, the sub-sectors listed above include some or large degree of aesthetic or imaginative content important for creative products
Source: World Trade Organization. Services Sectoral Classification List: Note by the Secretariat. MTN/GNS/W120. 10 July 1991

Table 2 Number of countries making market commitments in creative industry subsectors

Sub-sector	Number of countries
Professional services (including architectural services)	94
Other business services (including advertising, photography, printing, publishing)	90
Audio-visual	29
Tourism	131
Recreational, cultural and sporting	46

Source: World Trade Organization Services Database. Available at http://tsdb.wto.org/default.aspx Accessed March 15, 2009

religious, sexual, indigenous, to name a few), is now emerging alongside the notion of civic rights that were once cast in national terms. There is also considerable evidence that marginalized and stigmatized groups utilize the Internet more than their 'dominant' counterparts, and might even view national or global identities as top-down grand narratives. Therefore, national and cosmopolitan identities are now understood as forms of cultural identity alongside sexual, ethnic, racial, and gender identities often put forth in terms of cultural rights. Ernesto Laclau poses "incommensurability" between universal and particular cultural rights (Laclau 1995), while Seyla Benhabib (2004) resolves the question in favor of a process rooted in interactive contexts that foster deliberation and discourse ethics.[11]

4 Conclusion

Walt Whitman spoke of the "whispers passing under the sea" in referring to telegraph lines and asked in his *songs of Parting*:

> Are all nations communing? Is there but going to be one
> heart of the globe?
> Is humanity forming en-masse? For lo, tyrants tremble,
> Crowns grow dim

These lines from 1865, coincidentally the year International Telegraph Union (now ITU) was founded, could well describe the Arab Spring of 2011 or the coordinated protests worldwide against financial institutions later in October 2012. On the protests against financial institutions, one Chilean student leader remarked: "This is a world battle that transcends all frontiers." (quoted in Addley 2011).

Global collective action at large scales is now almost ubiquitous. Studies of global politics of technology therefore must understand this large-scale collective action and the ways in which existing institutions, including governments and international organizations, respond to these. However, not all politics is about such input–output models. This essay has argued that large-scale interactions fostered through information technologies are also changing the meanings of these inputs and outputs and the institutions that mediate them, although we are just beginning to observe and analyze these massive changes.

Large-scale interactions over information networks now challenge older forms of identity that legitimized the nation-state and forms of citizenship. The cultural politics in our midst are not settled though, and feature the overlapping of multiple meanings attached to actor identities and the issues that they contest. National identity, even when it dominates, now sits astride transnational and sub-national

[11] Benhabib shows how rights of immigrants have developed at the European Union level through successive deliberations even as their membership in the political community of nation-state remains incomplete.

identities. Similarly, a language of universal rights must now accommodate that of cultural rights from those who felt excluded from earlier universalisms.

Meta-power helps us understand the representational sites, and the cultural meanings of the information age. It attends to the communication and interactional processes among global actors that lead to meaning formation and reformation. But information technologies do not afford, in this sense, a growing clash of cultures. Rather, that the linear affixed meanings of the past are not replaced with another authoritative meaning might mean that multiplicity, at least for now, is the cultural outcome of large-scale information networks.

References

Addley, E. (2011, October 11). Occupy movement: From local action to a global howl of protest. *The Guardian*. http://www.guardian.co.uk/world/2011/oct/17/occupy-movement-global-protest. Accessed May 17, 2013.

Anderson, B. (1983). *Imagined communities*. London: Verso.

Barnett, M., & Duvall, R. (2005). *Power in global governance*. Cambridge: Cambridge University Press.

Baumgartner, T., Buckley, W., & Burns, T. R. (1975). Relational control: The human structuring of cooperation and conflict. *Journal of Conflict Resolution, 19*, 419–440.

Benhabib, S. (2004). *The right of others: Aliens, residents and citizens*. Cambridge: Cambridge University Press.

Bijker, W. E., Hughes, T. P., & Pinch, T. (2012). *The social construction of technological systems: New directions in the sociology and history of technology*. Cambridge, MA: The MIT Press.

Burns, T. R., & Hall, P. (Eds.). (2013). *The meta-power paradigm: Impacts and transformations of agents, institutions, and social systems*. New York, NY: Peter Lang.

Buzan, B., & Little, R. (2000). *International systems in world history: Remaking the study of international relations*. Oxford: Oxford University Press.

Castells, M. (1997). *The power of identity: The information age – economy, society and culture* (Vol. II). Oxford, UK: Blackwell Publishers.

Centola, D., Gonzalez-Avella, J. C., Eguiluz, V. M., & Miguel, M. S. (2007). Homophily, cultural-drift, and the co-evolution of cultural groups. *The Journal of Conflict Resolution, 51*(6), 905–929.

CheckFacebook. (2013). *Home*. http://www.checkfacebook.com. Accessed May 17, 2013.

Choucri, N. (2012). *Cyberpolitics in international relations*. Cambridge, MA: The MIT Press.

Cull, N. J. (2013). The long road to public diplomacy 2.0: The Internet in U.S. public diplomacy. *International Studies Review, 15*(1), 123–139.

Deibert, R. J., Palfrey, J. G., Rohozinski, R., & Zittrain, J. (Eds.). (2008). *Access denied: The practice and policy of global internet filtering*. Cambridge, MA: MIT Press.

Der Derian, J. (1990). The (s)pace of international relations: Simulation, surveillance, and speed. *International Studies Quarterly, 44*, 295–310.

Drezner, D. (2007). *All politics is global: Explaining international regulatory regimes*. Princeton, NJ: Princeton University Press.

Earl, J., & Kimport, K. (2011). *Digitally enabled social change: Activism in the internet age*. Cambridge, MA: The MIT Press.

Eriksson, J., & Giacomello, G. (2009). Who controls the internet: Behind the obstinacy or obsolescence of the state. *International Studies Review, 11*, 205–230.

Escobar, A. (2001). Culture sits in places: Reflections on globalization and subaltern strategies of localization. *Political Geography, 20*, 159–174.

Farrell, H. (2012). The consequences of the internet for politics. *Annual Review of Political Science, 15*, 35–52.
Fritsch, S. (2014). Conceptualizing the ambivalent role of technology in international relations: Between systemic change and continuity. In: Mayer M, Carpes M, Knoblich R (Eds.), *The global politics of science and technology* (Vol. 1, Concepts from international relations and other disciplines). Heidelberg: Springer.
Fung, A., Gilman, H. R., & Shkabatur, J. (2013). Six models for the internet + politics. *International Studies Review, 15*, 30–47.
Giddens, A. (1984). *The constitution of society: Outline of the theory of structuration*. Berkeley, CA: University of California Press.
Gil-Garcia, R. J. (2012). *Enacting electronic government success: An integrative study of government-wide websites, organizational capabilities, and institutions*. New York, NY: Springer.
Goff, P. M. (2007). *Limits to liberalization: Local culture in global marketplace*. Ithaca, NY: Cornell University Press.
Goldsmith, J., & Wu, T. (2006). *Who controls the internet? Illusions of a borderless world*. New York, NY: Oxford University Press.
Grewal, D. S. (2008). *Network power: The social dynamics of globalization*. New Haven, CT: Yale University Press.
Hughes, T. P. (2012). The evolution of large technical systems. In W. E. Bijker, T. P. Hughes, & T. Pinch (Eds.), *The social construction of technological system: New directions in the sociology and history of technology* (pp. 45–76). Cambridge, MA: The MIT Press.
Innis, H. (1950). *Empire and communications*. Oxford: Clarendon Press.
Karpf, D. (2012). *The MoveOn effect: The unexpected transformation of American political advocacy*. Oxford: Oxford University Press.
Keck, M. E., & Sikkink, K. (1998). *Activists beyond borders: Advocacy networks in international politics*. Ithaca, NY: Cornell University Press.
Kollman, K. (2007). Same-sex unions: The globalization of an idea. *International Studies Quarterly, 51*(2), 329–357.
Krasner, S. (1985). *Structural conflict: The third world against global liberalism*. Berkeley, CA: University of California Press.
Laclau, E. (1995). Universalism, particularism, and the question of identity. In J. Rajchman (Ed.), *The identity in question* (pp. 93–108). London: Routledge.
Livingston, S., & Walter-Drop, G. (Eds.). (2014). *Information and communication technology in areas of limited statehood*. Oxford: Oxford University Press.
McLuhan, M. (1964/1997). *Understanding media: The extensions of man*. Cambridge, MA: The MIT Press.
McPherson, M., Smith-Lovin, L., & Cook, J. M. (2001). Birds of a feather: Homophily in social networks. *Annual Review of Sociology, 27*, 415–444.
Nye, J. S., Jr. (2004). *Soft power: The means to success in world politics*. New York: Public Affairs.
Pinch, T., & Bijker, W. E. (2012). The social construction of facts and artifacts: Or how the sociology of science and the sociology of technology might benefit each other. In W. E. Bijker, T. P. Hughes, & T. Pinch (Eds.), *The social construction of technological systems: New directions in the sociology and history of technology* (pp. 11–44). Cambridge, MA: The MIT Press.
Rockhill, G., & Gomez-Muller, A. (2011). *Politics of culture and the spirit of critique*. New York, NY: Columbia University Press.
Rosenau, J. N., & Singh, J. P. (2002). *Information technologies and global politics: The changing scope of power and governance*. Albany, NY: State University of New York Press.
Said, E. (1978). *Orientalism*. New York: Vintage Books.
Sandy, K. Z. (2010). Muslims, identity and multimodal communication. *Contemporary Islam, 4*, 139–155.

Sell, S. K. (2013). Revenge of the "Nerds": Collective action against intellectual property maximalism in the global information age. *International Studies Review, 15*(1), 67–85.
Shirky, C. (2008). *Here comes everybody: The power of organizing without organizations.* New York, NY: The Penguin Press.
Singh, J. P. (2008). Agents of policy learning and change: US and EU perspectives on cultural policy. *Journal of Arts Management, Law, Society, 38*(2), 141–159.
Singh, J. P. (2011). *Globalized arts: The entertainment economy and cultural identity.* New York, NY: Columbia University Press.
Singh, J. P. (2013a). Information technologies, meta-power, and transformations in global politics. *International Studies Review, 15*, 5–29.
Singh, J. P. (2013b). Development remix: Representing poverty, culture, and agency in the developing world. *International Studies Perspectives.* doi:10.1111/insp.12023 doi:10.1111/insp.12023#Link to external resource: 10.1111/insp.12023.
Slaughter, A.-M. (2004). *A new world order.* Princeton, NJ: Princeton University Press.
Statistic Brain. (2013). *Twitter statistics.* http://www.statisticbrain.com/twitter-statistics/. Accessed May 17, 2013.
Sylvester, C. (2009). *Art/museums: International relations where we least expect it.* Boulder, CO: Paradigm Publishers.
Voon, T. (2007). *Cultural products and the World Trade Organization.* Cambridge: Cambridge University Press.
Wikipedia. (2013). *SinaWeibo.* http://en.wikipedia.org/wiki/Sina_Weibo. Accessed September 20, 2013.
World Tourism Organization. (2012). *UNWTO tourism highlights.* http://mkt.unwto.org/en/barometer. Accessed May 17, 2013.
Wullschlager, J. (2013). A Picasso for the Facebook age: At Venice Biennale, most national pavilions pale before the energetic vision of the compelling international exhibition. *Financial Times.* http://www.ft.com/intl/cms/s/2/6baa6c04-c881-11e2-acc6-00144feab7de.html#axzz2V45n0DMT. Accessed June 4, 2013.
YouTube. (2013). *Statistics.* http://www.youtube.com/yt/press/statistics.html. Accessed May 17, 2013.

How Much Is Enough? The Politics of Technology and Weaponless Nuclear Deterrence

Anne Harrington and Matthias Englert

Abstract This chapter contributes to the literature on nuclear proliferation by bringing theoretical resources from political philosophy and science and technology studies (STS) to bear on perennial questions for scholars of international relations (IR): "Why do states build nuclear weapons?" and "How many nuclear weapons are enough to sustain a credible nuclear deterrent?" We argue that states build nuclear weapons when they perceive the benefits of creating and maintaining a nuclear arsenal as outweighing the costs. If, therefore, a state is able to reap the benefits of having a nuclear weapon without actually creating one, it will have no reason to forego membership in the Nonproliferation Treaty. Furthermore, states with latent nuclear weapon programs are already actively engaging in a form of "weaponless" nuclear deterrence. Far from existing only in imagination, the "virtual" arsenals of the future are already observable in the nuclear security strategies of non-nuclear weapon states today. In making this argument, we draw on emergent theories of nuclear technopolitics that focus on the ambivalence of nuclear technology in constituting the political field of interactions.

Keywords Nonproliferation • Deterrence • International Relations • Uranium enrichment • Centrifuge • Technopolitics

This chapter contributes to the literature on nuclear proliferation by bringing theoretical resources from political philosophy and science and technology studies (STS) to bear on a perennial question for scholars of international relations (IR): "What are the determinants of proliferation?" or as Scott Sagan so eloquently put it,

A. Harrington (✉)
American Political Science Association, Washington DC, USA
e-mail: aih@uchicago.edu

M. Englert
Ianus Institute, Technical University Darmstadt, Darmstadt, Germany
e-mail: matthias.englert@dialectics.de

M. Mayer et al. (eds.), *The Global Politics of Science and Technology - Vol. 2*, Global Power Shift, DOI 10.1007/978-3-642-55010-2_17,

"Why do states build nuclear weapons?" (Sagan 1996). IR scholars believe that if they can discover what causes states to build nuclear weapons, they can make better recommendations about how to enhance nonproliferation policy. Their goal is to predict, or "forecast", nuclear proliferation behavior. To that end, the IR literature provides three competing answers to the question of why states build the Bomb: security, domestic politics, and prestige. Each of these motives is believed to account for at least one case of proliferation, but there is no overriding consensus that any one of them alone is sufficient to explain the variation across cases.

In contrast, the argument we develop in this paper proceeds from a different set of assumptions. Our premise is that there is no "causal logic" driving proliferation behavior. Furthermore, debating the relative contribution of one motive versus another will not ultimately improve nonproliferation policy beyond what is known today. In fact, the framing of the debate in terms of "proliferation"—as if nuclear weapons were self-reproducing—biases the field towards a search for the next bomb, anticipating it by watching for signs that indicate its imminent arrival, rather than engaging in a political analysis of what is both possible and desirable for policymakers.

Instead of searching for a concealed motive—"the hidden bomb" so to speak—in this paper we search on the surface for what is hidden in plain sight, namely that states build nuclear weapons when they perceive the benefits of creating and maintaining a nuclear arsenal as outweighing the costs. If, therefore, a state is able to reap the benefits of having a nuclear weapon without actually creating one, it will have no reason to proliferate. Contrary to the logic of proliferation, we start from the assumption that possessing nuclear weapons makes some states less, rather than more secure.

The appropriate question to answer from this point of view is not "What are the determinants of proliferation?" but rather, "How much is enough?" by which we mean, at what point does a state's nuclear program become a burden rather than a benefit? For some states, even a modest nuclear energy program is simply unobtainable without significant aid because they lack access to the necessary materials and expertise. For others, the answer is more ambiguous. So-called latent-nuclear states possess the scientific expertise and technology to build both robust nuclear energy and nuclear weapon programs, but prefer to stop developing those capabilities short of creating a nuclear warhead.

Our contention in this paper is that states with latent nuclear weapon programs are already actively engaging in a form of "weaponless" nuclear deterrence. Far from existing only in imagination, our contention is that the "virtual" arsenals of the future are already observable in the nuclear security strategies of non-nuclear weapon states today.

In making this argument, we draw on emergent theories of nuclear technopolitics that focus on the role of nuclear technology in constituting a field of political interactions. In contrast to Cold War-era technostrategic theories of nuclear deterrence and proliferation, we build a technopolitical explanation of nuclear politics. Technostrategic theories assign agency to technology. They reduce humans to rational actors whose job it is to discern and implement optimal strategies in response to technological incentives. Instead of positing a determinate role for technology, we contend that technology impacts human agency by altering the range of possible action. Technology does not dictate a correct strategy or answer, but it does constitute a range of available options.

In the case of nuclear weapons, the scientific discovery of nuclear fission brought into existence the technopolitical field of nuclear politics. However, nuclear technology does not determine actions within that field in a mono-causal way. Far from being determinate, nuclear technology is inherently ambivalent. It does not have a singular, politically neutral meaning. Rather, the fact that different groups can simultaneously read different meanings into the technology is its defining feature (Abraham 2010; Liebert et al. 1994). Thus, a country can be in compliance with its commitment as a non-nuclear weapon state under the Treaty on the Nonproliferation of Nuclear Weapons (NPT) and simultaneously be maintaining a weaponless nuclear deterrent.

This chapter has five sections, including the conclusion. We start out by introducing the nuclear nonproliferation regime, the primary feature of which is the boundary it creates between the managed systems of deterrence and abstinence. This is the battleground of nuclear technopolitics. Power struggles over where a state is entitled to locate itself vis-à-vis this boundary are the primary source of conflict. In the second section we describe the two technological pathways to the bomb, focusing in particular on the ultracentrifuges for uranium enrichment. Centrifuges are controversial because they convert natural uranium, which is widely available, into material appropriate to power a nuclear reactor or produce a nuclear weapon. The NPT does not specify what specific technologies a state is entitled to possess short of a nuclear explosive device, only that the technology must be intended for peaceful purposes. Therefore the only restriction on centrifuge enrichment is the difficulty of mastering the technology. Technical knowledge of these two pathways is the substance of subsidiary nonproliferation agreements detailing inspection and verification requirements. Familiarity with what is and is not technically possible is foundational knowledge necessary to understanding nuclear politics. In the third section we bring together the international political structure with the technology it is designed to regulate to show how the ambivalence of nuclear technology shapes political possibilities within the technopolitical field of nuclear politics. In the fourth section, we extend that same argument to the practice of weaponless nuclear deterrence. Finally, we draw out the political implications of our argument with respect to the contemporary conflict over Iran's nuclear program.

1 Nuclear Technopolitics and the Nonproliferation Regime

The field of nuclear politics is a system of governance in which a state's status and access to resources is linked to its ability to manipulate the ambivalence of nuclear technology in a battle for political ends. The most important aspect of understanding this field is unpacking how the nature of nuclear technology—what is known to be possible in the realm of scientific and technical knowledge about the ability to produce, detect, and manipulate matter—interacts with and conditions the political.

Unlike other international treaty regimes, in the realm of nuclear politics relationships between states are mediated through technology The ambivalence of nuclear technology allows for the immediate political purpose of nonproliferation agreements to remain ambiguous within the carefully choreographed boundaries of official diplomatic circles. Scientists always play an essential role in deciding what political leaders will (and will not) cover at international nuclear negotiations because the substance of these negotiations is technological in nature. Therefore, the boundaries of the negotiations, and the terms of the final agreement, take the form of permissive and restrictive technical barriers.

Even if uniformly applied, technical barriers impact states differentially given the technologies and materials available to each state. Most significantly, the NPT divides the world into two categories of states based on what technology they possessed as of January 1, 1967: nuclear weapon states and non-nuclear weapon states. In order to qualify as a nuclear weapon state, a country must have tested a nuclear explosive device prior to that date. Though this rule is applied equally to all states, it is de facto discriminatory because it impacts states differentially by fixing their status under the treaty to their technological achievements at a single point in time. Only five states are recognized nuclear weapon states (US, Russia, China, UK, France).

Rather than addressing the just or unjust nature of its particular political order, the nonproliferation regime bases its legitimacy on the common interest among states in combating the unique threat that nuclear war poses to the world. As Hans Morgenthau explained, nuclear war is no longer a first-order death in which individuals sacrifice their existence for the perpetuation of society, but rather a 'second death' that threatens to destroy meaning itself (Morgenthau 1961). Nuclear war is a cataclysmic event because it will not only end life within the war zone, but scientists predict that the fallout will black out the sun, lower the temperature of the earth and usher in a 'nuclear winter.' Previously, effects and unintended consequences of specific technologies were confined by geography and time. Today humanity holds survival in its own hands and has become responsible for its future.

As a counterbalance to this discriminatory framework, all states party to the treaty are obligated to work towards a world in which all states will become equal through nuclear disarmament. The legitimacy of the NPT is sustained by constantly reaffirming the primacy of the threat posed—not by nuclear weapon states, but by the technology itself—through a constant re-iteration of the desire for complete disarmament as the unifying mission that binds together all members (Harrington de Santana 2011).

Article VI of the Nonproliferation Treaty expresses this Cold War-era consensus about the desirability of disarmament but it was a consensus that postponed disarmament indefinitely. As codified, Article VI is not an agreement to disarm, but rather a meta-level commitment "to pursue negotiations in good faith." In other words, the commitment is to the pursuit itself, not actual disarmament.

Since the end of the Cold War and the indefinite extension of the NPT the commitment of the nuclear weapons states to the pursuit of nuclear disarmament, and the US commitment in particular, have come into question (Daase 2003). This has lead to multiple reaffirmations of the importance of this goal to the legitimacy of the NPT. The importance of this commitment to the pursuit of disarmament was reaffirmed at the 2010 NPT Review Conference in the final document, which emphasizes "that the strict observance of all the provisions of the Treaty remains central to achieving the shared objectives of the total elimination of nuclear weapons" (NPT 2010). US President Barack Obama has also reiterated the US commitment to nuclear disarmament. The Obama administration's 2010 Nuclear Posture Review spells out the perception that the existence of a common goal motivates compliance with the NPT, saying that "[b]y demonstrating that we take seriously our NPT obligation to pursue nuclear disarmament, we strengthen our ability to mobilize broad international support for the measures needed to reinforce the non-proliferation regime and secure nuclear materials worldwide" (United States Department of Defense 2010). Even for policymakers in the United States who do not perceive the vision of a nuclear-weapon-free world to be realistic in the short term, such as the Cold War-era American statesmen, Sam Nunn, George Shultz, William Perry and Henry Kissinger, the commitment to the vision is still essential because it ensures that the steps that the United States is advocating to limit access to nuclear materials will be "perceived as fair and urgent" (Nuclear Security Project 2010).

1.1 The Managed Systems of Deterrence and Abstinence

For the foreseeable future, the NPT divides the realm of nuclear politics into two interrelated technopolitical economies: An economy of deterrence, and an economy of nonproliferation, or what William Walker refers to as the managed systems of deterrence and abstinence (Walker 2000).

According to Walker, the managed system of deterrence consists, among other things of "a set of understandings and practices, expressed in the '*deterrence theories*' of Brodie, Schelling and others and enunciated in nuclear doctrine, of how military forces of various kinds should be deployed and managed to provide mutual vulnerability and restraint." Bernard Brodie was one of the primary authors of this system. In his contribution to the 1945 edited volume, *The Absolute Weapon*, Brodie formulated the two foundational principles of Cold War-era nuclear deterrence. The first principle is to ensure the capacity for retaliation in kind, the ability to absorb a nuclear attack and strike back. The second principle is that nuclear weapons, unlike other types of weapons, should be produced for the primary purpose of preventing their consumption through detonation in acts of war (Brodie 1945).

Back when Bernard Brodie first introduced the idea that the American military establishment should be reorganized for the purpose of preventing war his suggestion was considered crazy at best and dangerous at worst (Kaplan 1991). Today it is taken for granted within certain expert communities that deterrence "works." Deterrence is a doctrinal term. The American military under the Obama administration defines it as "The prevention of action by the existence of a credible threat of unacceptable counteraction and/or belief that the cost of action outweighs the perceived benefits" (United States Department of Defense 2014). Yet, in the 1950s, nuclear deterrence was (and still is for some) a controversial idea because it fundamentally altered the field of international politics. The existence of nuclear weapons foreclosed the possibility of total war as a mechanism for dispute resolution. Unlike the optimistic assessments of easy victory that inevitably precede the slow decent into a war of attrition, the pyrrhic victory of a nuclear war is easily foretold. Even for those who defend the concept of victory in a nuclear war, they cannot deny that both sides inevitably stand to lose more than they will gain. Therefore, countries that master the art and science of nuclear deterrence enter a new realm in which force is no longer the *ultima ratio* of international politics, yet at the same time there is no escape from the constant threat of nuclear war.

The managed system of abstinence is a function of the basic bargain of the NPT between nuclear weapon states and non-nuclear weapon states. Non-nuclear weapon states agree to abstain from developing nuclear weapons in return for access to nuclear energy technology and the promise that nuclear weapon states will work towards disarmament. The NPT forbids nuclear weapon states (NWS) to "assist" non-nuclear weapon states (NNWS) and NNWS to "manufacture" or "seek or receive any assistance" to "manufacture or otherwise acquire a nuclear weapon, or other nuclear explosive devices" (NPT 1968).

The NPT defines the boundary between these two systems as the manufacture of a "nuclear explosive device" because the ability to produce nuclear weapons was considered a necessary first step in creating a nuclear deterrent. Yet, even non-nuclear weapon states under Article V of the treaty are entitled to "potential benefits from any peaceful applications of nuclear explosions" and "the charge to such Parties for the explosive devices used will be as low as possible and exclude any charge for research and development" (NPT 1968).

Today the idea of a peaceful nuclear explosive device is considered risible, but the interpretation of all nuclear explosions as "weaponization" evolved slowly over time. In the 1960s when the NPT was under negotiation it was still believed that nuclear explosive devices might be used for purposes such as excavation, and ensuring access to these applications of the explosive potential was considered consistent with the Article IV guarantee that "Nothing in this Treaty shall be interpreted as affecting the inalienable right of all the Parties to the Treaty to develop research, production and use of nuclear energy for peaceful purposes" (NPT 1968).

1.2 The Battleground of Nuclear Technopolitics

Defending the boundary between the realm of deterrence and the realm of abstinence is the primary battleground of nuclear politics. What technologies non-nuclear weapons states are and are not permitted as part of the "inalienable right...to develop research, production and use of nuclear energy for peaceful purposes" is the space in which the struggle for political interpretation of nuclear technology plays out as countries joust over whether a certain technology will be coded as civilian or military for the purposes of political action. However, as we explain below, it is not possible to distinguish between a civilian and a military program simply on the basis of what technology a country possesses—although certain items and processes are more suspect than others. Establishing intent by answering the question of whether or not a program is civilian or military is an inherently political act, and the spread of the most sensitive elements of the fuel cycle is therefore considered by some to constitute a threat to the stability of the Nonproliferation Regime. For instance, Kofi Annan, then-Secretary-General of the United Nations, argued at the 2005 NPT Review Conference that sensitive technologies will destabilize the nonproliferation regime by undermining the function of the inspections carried out by the International Atomic Energy Agency (IAEA) (Zarate 2007 citing Annan 2005). The IAEA is the organization that verifies compliance with the NPT by upholding a regime of safeguards that comprise a series of technical measures to ensure that no fissile material known to the IAEA is diverted for military purposes. The measures range from cameras at certain positions, to the visit of inspectors to take probes and samples, count and weigh materials, tag and seal containers or conduct other visual inspections.

The allowed locations and methods of such inspections are limited under comprehensive safeguards agreements. Almost all countries party to the NPT have signed a safeguards agreement. Under these agreements the IAEA is charged with detecting in a timely manner the diversion of declared fissile materials. If a state has signed the next level of agreement, known as an additional protocol, it grants the IAEA complementary inspection authority to also detect possible undeclared activities (IAEA 2014).

All of these measures are designed to provide timely warning of noncompliance. However, as Albert Wohlstetter famously warned:

> If, in fact, technological transfers can bring a "nonnuclear weapons state" within weeks, days or even hours of the ability to use a nuclear explosive, [then] in the operational sense that "nonnuclear weapon state" will have nuclear weapons. The point is even more fundamental than the fact that effective safeguards [according to the IAEA] mean timely warning. A necessary condition for timely warning is that there be a substantial elapsed time. But if there is no substantial elapsed time before a government may use nuclear weapons, [then] in effect it has them (Zarate 2007 citing Wohlstetter et al. 1979).

This same line of reasoning is what led Mohammed el-Baradei, former IAEA director, to the conclusion that once non-nuclear weapon states master the nuclear fuel cycle, they become "virtual nuclear weapon states" (Baradei 2005). Virtual nuclear weapon states have mastered at least one of the technological processes for the production of fissile materials, a necessary step in the production of nuclear energy or nuclear weapons. They can produce Uranium 235, Plutonium 239, or both.

2 Pathways to the Bomb: The Technical Dimension of Nuclear Technopolitics

There are two technological pathways to the bomb: uranium enrichment and plutonium production. In this article we focus on enrichment. The argument about technological potentials and their effects on deterrence applies also to the plutonium path to the bomb. All nuclear reactors using uranium fuel produce plutonium as a bi-product of energy production and are therefore a proliferation concern. However, nuclear reactors large enough to be a proliferation concern are difficult to conceal. Additionally, extracting plutonium from highly radioactive fuel requires shutting down the reactor and accessing the nuclear fuel in the reactor core. Other than changing out the spent fuel and refueling the reactor, there is no reason to access the core. Therefore, timely warning is considered less of a concern with reactors than it is with uranium enrichment facilities.

Although natural uranium is plentiful in nature, to be weaponized it must be converted into a form that can be used to sustain a nuclear chain reaction. Natural uranium is made up almost entirely of two isotopes, one of which is the slightest bit heavier than the other. Only the lighter isotope, U-235, is useful for sustaining a chain reaction. Fortunately, at least from a nonproliferation perspective, natural uranium is 99.3 % U-238, so in order to be weapons usable it must be "enriched" to increase the percentage of desirable U-235 up to roughly 90 % (highly enriched uranium—HEU) by separating out the undesirable U-238.[1] Uranium enrichment facilities are still difficult to build, but a facility sufficient to produce several bombs worth of highly enriched uranium is relatively easy to hide.

Enrichment can be done by different physical or chemical methods. Today centrifuge enrichment is the most efficient and economical technology for uranium enrichment. In a centrifuge the lighter U-235 is separated from the heavier U-238. The natural uranium is first converted into the chemical molecule uraniumhexa-fluoride (UF6), in a separate conversion facility. Uraniumhexafluoride becomes solid at atmospheric pressures and temperatures below 56.5 °C (133.8 °F). The solid UF6 is then transported in heavy storage cylinders, each weighing several tons, to a centrifuge facility, where the solid is converted to gas form and fed into

[1] Instead of enriching the U-235 content, the U-238 can also be converted into plutonium, a completely different element, by irradiating U-238 with neutrons in a nuclear reactor.

underpressured pipes of a centrifuge facility. The gas is then pumped into the centrifuge—a stovepipe-like cylinder typically 10–20 cm in diameter and up to several meters long. Among the fastest spinning things on earth, centrifuges rotate more than 1,000 times per second. Their extreme centrifugal force presses the slightly heavier U-238 a little bit further toward the outside wall of the rotor than the lighter U-235. As a result, a very thin film of gas, now enriched with a marginal amount of additional U-235, forms on the wall. This enriched-gas film can be extracted and siphoned into another centrifuge for further enrichment. This process is repeated until the desired level (%) of enrichment is achieved.

The process of enrichment can be stopped at any point in the process making it possible to distinguish between different levels of enrichment for different purposes. The limit of 20 % enrichment is the accepted international norm that distinguishes military from civil use. Low-enriched uranium for civilian uses contains less than 20 % U-235, but once a state reaches that threshold, it has completed roughly two thirds of the work necessary to reach the minimum level for weapons-grade material: 90 % U-235.[2]

The percentage change achieved with only one centrifuge is quite low. To reach higher enrichment levels 10–20 stages (groups of centrifuges connected in parallel) have to be used in series, depending on the enrichment level of the final product. Additionally the gas stream entering and exiting a centrifuge is limited and many centrifuges have to be used in parallel in each stage. To maximize the efficiency, the enriched stream and the waste stream are both reused to squeeze out the maximum amount of U-235 and minimize the total gas flow forcing the serial and parallel arrangement of centrifuges into what is known as a cascade. Centrifuges are installed in rows in a cascade hall. The gas stream passes through steel pipes that connect one centrifuge to another. The shape of the cascade is created by the pattern of the pipes. A typical cascade of very simple first generation centrifuges would contain 164 centrifuges.

Centrifuges are a technological feat. The stress on the material due to the speed of the rotor wall challenges the metallurgical and fabricating skills of the manufacturer. High precision manufacturing is not only needed for the rotor, but also for the bearings and other rotating components. Any irregularities in the rotor will cause imbalances. Even a fingerprint on the rotor of a centrifuge would cause it to spin out of control and the centrifuge would mechanically explode.

With manufacturing equipment diffusing all over the world and with the help of a sophisticated supply network, the ability to manufacture centrifuges has come

[2] This can be visualized starting with a natural uranium mix (0.7 % enrichment) of 140 uranium-238 atoms (black balls) and 1 uranium-235 atom (red ball) in a salad bow, To enrich, black balls have to be sorted out. Taking out 70 black balls enriches to ~1.4 %, taking out another 46 black balls enriches to ~4 % which is roughly the enrichment used in commercial light water reactors. The mix contains 24 black balls and 1 red ball. We took out 116 black balls from a total of 141 balls, which is more than 2/3 of the work.

within the grasp of many countries. In addition to the five NPT-recognized nuclear weapon states, there are nine more countries with enrichment technology: Argentina, Brazil, Germany, India, Iran, Japan, the Netherlands, North Korea and Pakistan (International Panel on Fissile Materials 2013). Iran, for instance, bought centrifuge technology from Pakistan, which delivered the P1 design based on an early design stolen from the company URENCO in the 1970s. This design was known to be delicate to handle. Since the public disclosure of Iran's clandestine centrifuge program, Iran had to rely on reverse engineering, as it could no longer rely on advice from Pakistani experts. In hindsight, it might have been a better choice for Iran to develop an indigenous design right from scratch. However, beyond the centrifuge, an enrichment plant also requires a whole bundle of specialized equipment like pumps and high nickel alloy steel pipes to withstand the extremely corrosive UF6 environment in an enrichment facility. Iran partly bought this equipment from the international market but also manufactured it themselves and over the years became increasingly adept at handling the technology on its own.

To use a centrifuge facility for military purposes a country has three options (Hecker et al. 2012): It can use its declared facilities covertly; use a clandestine facility; or break-out and leave the NPT and use its facilities as it likes.

In order to covertly use a declared facility designed for low enrichment to enrich uranium to weapons-grade, a proliferator would either have to cycle UF6 through the same unaltered cascade several times to stepwise increase the enrichment (batch-recycling) or change the cascade shape and reconnect the gas tubes of the centrifuges to have more stages in series for a higher final enrichment. In a declared facility the only question is, when such a military use would be discovered. Another method of diversion is to use a declared facility to produce a stock of pre-enriched LEU that can be diverted to a secret small clandestine facility. As we showed above, most of the separative work is already done enriching to 20 %. A state could use the ambivalence of the technology to advance its latent capabilities for later use or just in case. (Arguably this is what Iran has done.)

Alternatively, clandestine facilities could be used secretly in parallel to a civil program or they could be fed by pre-enriched uranium stocks diverted from a civil program. It is almost impossible to remotely detect a clandestine uranium enrichment facility that uses centrifuge technology because their visible, thermal and radioisotopic signatures are very low. They are the same size as a larger supermarket, use no more electricity than a big box store, and do not emit many radioisotopes. As centrifuge facilities can be completely submerged underground, detection can only be based on intelligence information e.g. import of critical equipment or insider information etc. Plutonium, in contrast, must be generated by the neutron capture reaction of uranium-238. Therefore, producing plutonium requires an intense neutron source like a nuclear reactor to produce neutrons and irradiate uranium-238, the existence of which is much easier to detect, as it produces a lot of heat. Additionally the plutonium has to be separated from the uranium after irradiation, which releases detectable radioactive gases.

The third option is to withdraw from the NPT once a country decides to leave the NPT, it is legally entitled to do so as long it provides 3 months notice. At that point it could use its declared and possibly undeclared facilities as it wants to. However this option is likely to stir an international response by the UN Security Council and/or the General Assembly. The reaction of the international community would be very strong, up to and possibly including military action, so the incentive for doing so would have to be very high.

3 The Right to Nuclear Technology for Peaceful Purposes and the Centrifuge as a Hybrid Object: The Political Dimension of Nuclear Technopolitics

Non nuclear weapon states, including those without their own enrichment capabilities, often prefer to interpret their Article IV "inalienable right...to develop research, production and use of nuclear energy for peaceful purposes", to give them a *per se* right to sensitive technologies, including uranium enrichment facilities. Citing the Nonaligned Movement's (NAM) rationale, for instance, analyst Robert Zarate explains that, "Although Article IV never explicitly mentions enrichment, reprocessing and other nuclear fuel-making activities, some governments nevertheless interpret Article IV as implicitly recognizing the specific or per se right of signatories to any nuclear technological activities that can be conceivably labeled "peaceful"—short of actually inserting fissile material into a nuclear explosive device" (Zarate 2007). Of the 14 countries with uranium enrichment facilities as of 2013, 6 are NPT-recognized NNWS with enrichment capabilities: Argentina, Brazil, Germany, Iran, Japan, and the Netherlands (International Panel on Fissile Materials 2013).

However, technology for uranium enrichment is strictly controlled, and the inalienable right to its possession for peaceful purposes is hotly disputed (Ford 2005; Zarate 2007) because there are almost no technical hurdles to convert a declared centrifuge facility designed for civilian operation for military purposes. Zarate and others, including Christopher Ford, former US Special Representative for Nuclear Nonproliferation under President George W. Bush, contest the NAM's permissive interpretation of Article IV. As Special Representative, Ford argued that the development of sensitive technologies associated with the production of nuclear fuel goes beyond the minimum technology necessary to produce a peaceful program in compliance with the NPT. In essence, the NPT guarantees that NNWS will have access to nuclear fuel and therefore they should not produce it themselves. An independent nuclear fuel capability carries the implicit threat of an independent weapons program and therefore crosses the boundaries established between nuclear weapon states and non nuclear weapon states in Articles I, II, and III (Ford 2005).

Specifically, the ambivalence of centrifuge technology has two effects on the shape of the technopolitical field. First, it undermines the desire to draw a firm boundary between the managed system of deterrence and the managed system of abstinence—adjudicating a priori between "peaceful" and "military" programs enshrined in Article IV of the NPT. In other words, the technology itself enables *in potentia* two uses. The potential for a military use of nuclear energy technology does not have to be recalled immediately. Once a state has mastered all steps to produce centrifuges, build cascades, produce uraniumhexafluride and handle it, this state has the latent potential to use its assets for military purposes once it makes a decision to do so. Even though the use *in actus* can be monitored and verified as long as a state adheres to its obligations it is only possible to *de poste factum* distinguish military from civil use. Secondly, capabilities and potentials participate in the political sphere. Ambivalence means more than just "dual-use" or a technical option to be used at some point when a state decides to weaponize. The intrinsic potential that comes with the specifics of centrifuge technology opens a space of ambivalence and uncertainty that carries a certain agency that can be used politically. In Bruno Latour's words the centrifuge becomes a hybrid object. The centrifuge exceeds its material existence by entering the realm of politics. As we explain in the following section, this hybrid object can become a weaponless nuclear deterrent.

4 Nuclear Ambivalence and Weaponless Deterrence

The basic operational principles of Cold War-era of nuclear deterrence—the existence of a secure second strike-capability produced primarily for the purpose of preventing, not winning a war—are premised on the particular constellation of speed and destructive force that the combination of nuclear warheads loaded on intercontinental ballistic missiles embodies. When we think of nuclear deterrence, therefore, we usually think of the deterrent effect as stemming from the punishment that can be inflicted by the launch of a nuclear attack. It is the destruction of a nuclear explosion that generates "costs." However, achieving a deterrent effect does not actually require that a state impose those costs. It requires that a state maintain a securely-stored nuclear arsenal. In other words, as Albert Wohlstetter famously argued in his article "The Delicate Balance of Terror," if your nuclear arsenal is vulnerable, it will invite attack from an opponent who calculates that she can wipe out your capability by striking first. Therefore, a secure second-strike capability also requires an opponent to perceive that there are costs associated with overcoming the measures you have taken to protect your nuclear weapons. Fortifying your arsenal is a form of what Glenn Snyder identified as "deterrence by denial" in his 1959 classic, *Deterrence by Denial and Punishment*.

The US doctrinal definition quoted above—"The prevention of action by the existence of a credible threat of unacceptable counteraction and/or belief that the cost of action outweighs the perceived benefits"—captures these two different methods of operationalizing deterrence. The first, deterrence by punishment is the basic premise of mutual assured destruction: discouraging nuclear attack by creating a credible threat that aggression will be met with retaliation in kind. Likewise, deterrence by denial is also a strategy for discouraging military aggression, but rather than being based on a principle of reciprocity, denial is achieved through constructing superior defenses.

The defenses required for deterrence by denial do not necessarily have to be impenetrable, although it would be preferable if they were, just formidable enough that an enemy will assess the cost of mounting an attack as outweighing any possible benefit. Hardening underground missile silos, or evading detection by storing missiles on nuclear-powered submarines that can loiter at great depths are both examples of increasing the costs of attack by making the target more difficult to destroy. The more difficult a target is to destroy, the more likely a country is to take casualties and have to put its own assets at risk to interdict it. Constructing a nuclear missile defense shield, the ultimate form of deterrence by denial, has always been considered something of a pipedream. Even in the heyday of President Reagan's Star Wars missile defense shield, credible scientists argued that the goal remained out of reach. And although today perceptions of the effectiveness of missile defense are changing as systems become more robust, many technological problems remain. In traditional deterrence theory, deterrence by denial is a technical problem to be solved through demonstrated capabilities, but deterrence by denial, just like deterrence by punishment, is also a political battle for credibility.

This distinction between "deterrence by punishment" and "deterrence by denial" highlights the complex relationship between 'use-value' and 'threat-value' implicit in all definitions of deterrence (Harrington de Santana 2009). Punishment and denial are different kinds of use-value. They perform different operational functions because their material characteristics enable diverse sets of actions; use-value is defined by the universe of properties associated with the materiality of an object. It is an absolute form of value in the sense that it is measured relative to an accomplishment. So, for instance, a nuclear explosion can displace an impressive amount of earth or destroy a city. A solitary individual can realize the use-value of an object by benefiting from the actions and outcomes that technology enables regardless of anyone else's perceptions and behaviors. Threat-value, on the other hand, requires at least two actors and is what creates the deterrent effect. It is a relative form of value only indirectly related to the usefulness of the item. Threat-value is realized not through the application of force to change the behavior of another, although the use of force can certainly increase the credibility of a threat of more to come, but rather in terms of the ability to manipulate the behavior of another in order to achieve an end not achievable through force alone.

In other words, threat-value is never reducible to use-value because the distinction is not only one of means, but also of the relationship between means and ends. The means-end relationship expressed by use-value is realizable in the present (Schelling 1966). The means-ends relationship expressed by threat-value is always about manipulating the behavior of another now, in the present, by invoking the image of a future in which the use of offensive or defensive weapons against them make them worse off than they are now. Deterrence is the practice of manipulating the relationship between use-value and threat-value in order to create the perception that the cost of military aggression will outweigh the benefits. It is never about now—only about what comes next.

The fact that threat-value is always realized only through the potential for future use opens up the possibility of pushing that potential for use always further into the future. Cold War-era theories of nuclear deterrence are all based on the presumption that in order for deterrence to be effective, the speed with which nuclear war could occur would mean that the potential for future use always had to be immediately at hand in the form of nuclear weapons on hair-trigger alert. However, in theory, that potential for future punishment could be pushed ever further into the future, as long as there is a stable expectation that an attack on the potential can be deterred through denial. This is the essence of "weaponless" nuclear deterrence. Weaponless nuclear deterrence can be practiced with either a latent nuclear capability such as the one embodied in a centrifuge, or a virtual capability, which is embodied in the simulation technologies possessed by advanced nuclear weapon states, or some combination thereof (Cohen and Pilat 1998; Mazarr 1995; Schell 1982).

Cold War-era strategic nuclear deterrence relied almost entirely on the threat of punishment rather than denial. In fact, one of the most counterintuitive features of the Cold War was that while arsenals needed to be invulnerable to attack, leaving cities vulnerable to attack was considered a virtue. Mutual vulnerability contributed to strategic stability by imposing a measure of self-restraint on both the US and the Soviet Union and provided an intellectual justification for their mutual agreement to forego anti-ballistic missile (ABM) technology in the ABM treaty (Schelling 1966; Bohlen 2003).

Weaponless deterrence, in contrast, reverses the technological requirements for the use-value on which the deterrent threat relies. Instead of foreswearing defenses, a state forswears its weapons. Having adequate defenses, especially for a country's nuclear facilities is essential, but all that is required in terms of a weapons capability is a credible commitment to the potential for reconstitution.

5 Conclusion

Recognizing the fact that the technopolitical features of nuclear deterrence are already operating carries implications both for how to best negotiate the circumstances of current conflicts and for the larger issue of whether or not deterrence is possible in a world without nuclear weapons. The intention of the NPT is to prohibit non-nuclear weapon states from having not only the explosive device itself, but the deterrent effect that the device enables. However, capturing that effect does not necessarily require the construction of a weapon or explosive device. An effective nuclear deterrent, in the form of a latent nuclear capability defined as unfettered access to fissile materials through a domestic production capacity, can contribute to a strategy of "weaponless" nuclear deterrence. As long as the costs of taking military action to destroy that capability are perceived as outweighing the benefits, it is possible for a state to both reap the benefits of membership in the NPT as a non-nuclear weapon state, and participate in the economy of nuclear deterrence by securing diplomatic talks with the recognized nuclear weapon states.

In conclusion, to take a current example, one implication of this argument is that Iran is already drawing on the inherent ambivalence of nuclear technology to cross the boundary from the managed system of abstinence into the managed system of deterrence. Iran was unable to establish a direct diplomatic relationship with the US, despite its attempts to do so, until it was able to engage the US in negotiations about its nuclear program. Analysts who debate the dichotomous nature of Iran's nuclear program, presupposing that it must be either "peaceful" or "military" are blind to the way that Iran is manipulating the inherent ambivalence of the technology in order to gain the benefits of having a nuclear weapon capability without bearing the costs of losing membership in the NPT—a change of status that would have long-term implications for its regional security as it would give other countries a reason to build nuclear weapons as well. They are also blind to the fact that the ambivalence of the nuclear technology may also be holding together a diverse political coalition within Iran, allowing each faction to project onto the program the potential they would like to see in it. While it is important for the IAEA to offer its assessment of whether or not Iran has crossed the line between a peaceful program for the purposes of political action, the job of analysts is not necessarily to offer a political determination that participates as a political football in the policy process by lending legitimacy to the assessment of Iran's intentions as "peaceful" or "military." It is also to provide perspective on that political debate by building an accurate picture of the phenomenon under question and in the process, open up new possibilities for future political action.

References

Abraham, I. (2010). 'Who's next?' Nuclear ambivalence and the contradictions of non-proliferation policy. *Economic and Political Weekly, 45*(43), 18–20.

Annan, K. (2005, May 2). *Secretary-General's address to the nuclear non-proliferation treaty review conference*. New York, NY. http://www.un.org/apps/sg/sgstats.asp?nid=1427. Accessed January 7, 2014.

Baradei, M. E. (2005, May 6). U.N.'s ElBaradei warns of nuclear apocalypse. *Reuters*. http://today.reuters.co.uk/news/newsArticle. Accessed January 7, 2014.

Bohlen, A. (2003). The rise and fall of arms control. *Survival, 45*(3). http://faculty.maxwell.syr.edu/rdenever/PPA-730-27/Bohlen.pdf. Accessed January 6, 2014.

Brodie, B. (1945). *The absolute weapon: Atomic power and world order*. New York, NY: Harcourt, Brace and Company.

Cohen, A., & Pilat, J. (1998). Assessing Virtual Nuclear Arsenals. *Survival, 40*(1), 129–144.

Dasse, C. (2003). Der Anfang vom Ende des nuklearen Tabus: zur Legitimitätskrise der Weltnuklearordnung. *Zeitschrift für Internationale Beziehungen, 10*(1), 7–42.

Ford, C. A. (2005, May 18). NPT Article IV: Peaceful uses of nuclear energy. *New Paradigms Forum*. http://www.newparadigmsforum.com/NPFtestsite/?p=1518. Accessed January 6, 2014.

Harrington de Santana, A. (2009). Nuclear weapons as the currency of power: Deconstructing the fetishism of force. *The Nonproliferation Review, 16*(3), 325–345.
Harrington de Santana, A. (2011). The strategy of nonproliferation: Maintaining the credibility of an incredible pledge to disarm. *Millennium-Journal of International Studies, 40*(1), 3–19.
Hecker, S., Englert, M., & Miller, M. (2012). Nuclear non-proliferation. In D. S. Ginley & D. Cahen (Eds.), *Materials in energy* (pp. 162–177). Cambridge: Cambridge University Press.
IAEA. (2014). *Factsheets and FAQs: IAEA safeguards overview: Comprehensive safeguards agreements and additional protocols.* http://www.iaea.org/Publications/Factsheets/English/sg_overview.html. Accessed January 6, 2014.
International Panel on Fissile Materials. (2013). *Facilities: Uranium enrichment.* Last edited July 2013. http://fissilematerials.org/facilities/uranium_enrichment.html. Accessed January 6, 2014.
Kaplan, F. (1991). *The wizards of armageddon.* Stanford: Stanford University Press.
Liebert, W., Rilling, R., & Scheffran, J. (1994). *Die Janusköpfigkeit von Forschung und Technik - Zum Problem der zivil-militärischen Ambivalenz.* Marburg: BdWi.
Mazarr, M. J. (1995). Virtual nuclear arsenals. *Survival, 37*(3), 7–25.
Morgenthau, H. J. (1961). Death in the nuclear age. *Commentary, 32*(3), 231–234.
Nuclear Security Project. (2010). *About the project.* http://www.nuclearsecurityproject.org/site/c.mjJXJbMMIoE/b.3534665/k.5828/About_the_Project_Index.htm. Accessed August 25, 2010.
Sagan, S. D. (1996). Why do states build nuclear weapons? Three models in search of a bomb. *International Security, 21*(3), 54–86.
Schell, J. (1982). *The fate of the earth.* Stanford: Stanford University Press.
Schelling, T. (1966). *Arms and influence.* New Haven, CT: Yale University Press.
Treaty on the Nonproliferation of Nuclear Weapons (NPT). (1968). http://www.un.org/en/conf/npt/2005/npttreaty.html. Accessed January 6, 2014.
Treaty on the Nonproliferation of Nuclear Weapons (NPT). (2010). *2010 review conference of the parties to the treaty on the non-proliferation of nuclear weapons.* Final document, Vol. I. http://www.un.org/ga/search/view_doc.asp?symbol=NPT/CONF.2010/50%20(VOL.I). Accessed January 6, 2014.
United States Department of Defense. (2010, April). *Nuclear posture review report.* Washington, DC: U.S. Department of Defense.
United States Department of Defense. (2014). *Dictionary of military terms.* http://www.dtic.mil/doctrine/dod_dictionary/. Accessed January 6, 2014.
Walker, W. (2000). Nuclear order and disorder. *International Affairs, 76*(4), 703–724. http://www.jstor.org/stable/2626455.
Wohlstetter, A., Jones, G., & Wohlstetter, R. (1979). Why the rules have needed changing, Part I. *Towards a New Consensus on Nuclear Technology, 1*, 36–37.
Zarate, R. (2007). *The NPT, IAEA safeguards and peaceful nuclear energy: An "inalienable right," but precisely to what?* Unpublished manuscript. http://www.npolicy.org/article_file/The_NPT_IAEA_Safeguards_and_Peaceful_Nuclear_Energy.pdf. Accessed January 6, 2014.

<u>GPSR Compliance</u>

The European Union's (EU) General Product Safety Regulation (GPSR) is a set of rules that requires consumer products to be safe and our obligations to ensure this.

If you have any concerns about our products, you can contact us on ProductSafety@springernature.com

In case Publisher is established outside the EU, the EU authorized representative is:

Springer Nature Customer Service Center GmbH
Europaplatz 3
69115 Heidelberg, Germany

Batch number: 10739606

Printed by Printforce, the Netherlands